GLOBAL MARINE POLLUTION BIBLIOGRAPHY

Ocean Dumping of Municipal and Industrial Wastes

GLOBAL MARINE POLLUTION BIBLIOGRAPHY

Ocean Dumping of Municipal and Industrial Wastes

Michael A. Champ

The American University
Washington, D.C.

and

P. Kilho Park

National Oceanic and Atmospheric Administration
Rockville, Maryland

IFI/PLENUM • NEW YORK–WASHINGTON–LONDON

Library of Congress Cataloging in Publication Data

Champ, Michael A.
Global marine pollution bibliography.

Includes indexes.
1. Marine pollution—Bibliography. I. Park, P. Kilho (Paul Kilho), 1931–
II. Title.
Z6004.P6C49 1982 [GC1085] 016.3637394 82-18060
ISBN 0-306-65205-6

Cover illustration by Anne M. Lannak

A Division of Plenum Publishing Corporation
233 Spring Street, New York, N.Y. 10013

Printed in the United States of America

Wastes and the Ocean

To Professors M. A. Champ and P. K. Park

In the evolution of our civilization
Civitas (city) is formed where
Invention and industry blossom
Through cross-fertilization of minds.
There, some men and women are set aside
To produce science and philosophy,
To flower our art and literature.

In *civitas*, water is the medium of life,
For drinking, transportation, communication
And for the safe disposal of our wastes.
No waste discharge is total constipation
In which one succumbs amidst one's wastes.
We must continue our civil orderliness
Through which wastes are disposed safely.

The ocean, too, is used for waste disposal
As the land and the atmosphere are used
To bury, decompose, or disperse wastes.
Let us be anlytical and synthetical on this
To harmonize our civilization with the environ
So that our children see our wisdom,
Not inherit our wastes.

紅葉

Momiji
Terre Mariae
March 9, 1982

DEDICATION

This work is dedicated to two U.S. scientists who were among the first to contribute to research in the field of marine pollution in the United States:

Alfred C. Redfield
and
Bostwick H. Ketchum

Dr. Redfield was born on 15 November 1890 and was among the first to recognize that:

"... *Judgements* adverse to the use of the sea as a means of waste disposal are apt to be prejudiced by observations on the pollution of confined embayments where the exchange of water is greatly reduced.

... *Prejudice* also exists because of the very undesirable pollution of beaches which follows the discharge at sea of floating wastes such as oil or garbage. The capacity of offshore waters to receive and disperse soluble or suspended wastes without undesirable effects is very large and might properly be used more extensively (Redfield and Walford, 1951).

... *Each proposed operation should be examined critically*, to make sure that the waste is of a character to be safely treated as proposed, and that the *disposal site is properly chosen* in relation to the local hydrography, the disposition of fishing grounds, and the presence of other interests which might be adversely affected" (Redfield and Walford, 1951).

His most recent contribution, in his 90th year, was entitled "The Tides of the Waters of New England and New York," published by Woods Hole Oceanographic Institution (108 pages, 1980). It contributes to the understanding of the tidal processes which as we know affect the distribution of waste contaminants disposed in coastal waters. Dr. Ketchum, in his introduction to the book, wrote that "one of Redfield's important characteristics is his ability to focus on critical problems or mechanisms and ... explain complex problems in terms understandable by both novice and specialist ... so that the reasons behind the observed phenomena can be understood." Incidentally, Dr. Redfield was Dr. Ketchum's mentor at Harvard University.

Dr. Bostwick H. Ketchum was born on 21 January 1912 and has contributed more to the understanding of marine pollution and waste disposal than perhaps anyone in his time. He began writing papers on ocean dumping in 1948. His major early contributions were on physical processes as they affect the distribution and dispersion of wastes. He was among the first to use biological criteria for dumpsite selection, as evidenced by his writing "... the distribution of offshore fishing should be considered and every effort made to avoid disposal near fishing grounds. Before discharge operations are initiated a detailed study of the local circulation in the proposed disposal area by release of drift bottles is recommended" (Ketchum, WHOI Ref. 53-31, 1953).

His almost yearly contributions to the field of marine pollution from 1950 to 1980 are significant. His greatest contribution, in his 70th year, is perhaps that as senior editor he has been contributing to the development of the international ocean disposal symposium series, the first volume of which was published by Plenum Press in 1981. Six subsequent volumes from the second and third symposia are being published by Wiley-Interscience, New York.

PREFACE

This bibliography is an attempt to honor our fellow workers for their contributions to marine pollution as a scientific field. We also hope to increase the understanding and exchange of information from a global perspective. For the sake of younger workers, we hope that this exchange will prevent duplication of effort, the sad "reinvention of knowledge." An ulterior motive of ours is to create a living bibliography by not only having this document available as a printed text, but also as references with abstracts on word processor discs, so that researchers, teachers, and managers can update and expand this bibliography in their own fields of interest. We hope to be able to find a distributor for these discs in the near future.

To establish greater communication we had hoped to have an appendix with author's names and addresses; however, we found that we had less than half of the senior authors' current addresses. Therefore, for subsequent editions, we hope our fellow scientists will keep us informed of their work and current addresses, and will bring to our attention any errors or papers that we have inadvertently missed. Our aim is to produce another edition in three years.

ACKNOWLEDGMENTS

Many people have contributed to this effort. Extensive computer searches were conducted or made available to us by Robert Walter and Garry Adams (NOAA Library Services, Rockville, MD), William J. Rue, Jr. (Ecological Analysts, Sparks, MD), Wolfgang Gerwinski and Diether Schmidt (Deutsches Hydrographisches Institut, Hamburg, Federal Republic of Germany), and William Dunstan (Old Dominion University, Norfolk, VA). Extensive personal bibliographies were made available to us by H. Compaan (TNO, Delft, The Netherlands), Garry F. Mayer, Joel S. O'Connor, and Hal M. Stanford (U.S. National Oceanic and Atmospheric Administration, New York Bight Project), Jerry R. Schubel (State University of New York, Stony Brook, NY), Georg Koopmann, Niels-Peter Ruhl, and Fran Grabert-Carlson (Deutsches Hydrographisches Institut, Hamburg, Federal Republic of Germany), Gert Rauck (Bundesforschungsanstalt für Fischerei, Hamburg, Federal Republic of Germany), E. Rachor (Institut für Meeresforschung, Bremerhaven, Federal Republic of Germany), Gunter Weichart (Deutsches Hydrographisches Institut, Hamburg, Federal Republic of Germany), A.D. McIntyre, Derek Saward, Grant Topping, and A. Eleftheriou (Aberdeen Marine Biological Laboratory, Scotland), T.H. Pearson (Scottish Marine Biological Association, Oban, Argyll), N.J. Spijker, R.J. Dortland, and Mr. Van Vliet (North Sea Directorate, Rijkswaterstaat, The Hague, Netherlands), Peter C. Wood, Michael G. Norton, and Lindsay Murray (Ministry of Agriculture, Fisheries and Food, Burnham-on-Crouch, England), L.E. Ellis (U.K. Department of the Environment, London), Manfred K. Nauke (London Ocean Dumping Convention, London), Peter Hayward and Finn Bjerre (Oslo and Paris Commissions, London), Peter Anderson (U.S. Environmental Protection Agency, New York, NY), William C. Muir (U.S. Environmental Protection Agency, PA), Don W. Lear and Marria L. O'Malley (U.S. Environmental Protection Agency, Central Regional Laboratory, Annapolis, MD), John B. Pearce (U.S. National Marine Fisheries Service, Highlands, NJ), and Alan McIver (Canadian Environmental Protection Service, Darmouth, Nova Scotia).

Special thanks go to a few very special people - those librarians whose help made this effort a reality. Barbara A. Thompson (NOAA Interlibrary Loan), Janet McCappin (Burnham-on-Crouch), John Burn (Aberdeen), Daun Koffman (U.S. National Oceanic and Atmospheric Administration, New York Bight Project), Ida Lewis (U.S. National Marine Fisheries Service, Washington, DC), Donald Dennis (The American University), Meg Munro (U.S. Environmental Protection Agency, Central Regional Laboratory, Annapolis, MD), and Dorothy Booth (U.S. Army Corps of Engineers, Vicksburg, MS).

Sincere acknowledgments go to Martha C. Sager, John M. McKinley, and Richard Berendzen (The American University) and R.L. Swanson, Louis Butler, and T.P. O'Connor (U.S. National Oceanic and Atmospheric Administration) for their support in this study. We also wish to thank T.A. Wastler and William N. Hederman, Jr. (U.S. Environmental Protection Agency, Washington, DC) for their encouragement in this undertaking. We further wish to thank Professor Iver W. Duedall (State University of New York at Stony Brook) for permission to use three of his illustrations, one on the cover and Figures 2 and 3 in the introduction. Sincere thanks go to our word processing staff: Samar Manafzadeh, Nada Ghadry, Nancy Rossie, Zeb Neugebauer, and Nancy Edwards. We also want to thank Elna Crone for final editing. This work was supported in part by an IPA Agreement between the American University and the NOAA Ocean Dumping Program, Office of Marine Pollution Assessment, National Oceanic and Atmospheric Administration, U.D. Department of Commerce, Rockville, MD.

CONTENTS

INTRODUCTION

This bibliography was compiled as a result of growing global concern with the developing crisis in waste disposal, which, like a sleeping giant, poses an ever-present threat to our safety and well-being.

Three transformations have brought about this crisis:

(1) The shift of rural populations to metropolitan areas resulting from reduced labor requirements in agriculture and fisheries due to mechanization has greatly increased the volume of municipal waste.

(2) Advances in waste treatment processes have increased the recovery of solids and contaminants and reduced waste water effluent loading to receiving waters, but the subsequent offshore disposal of these materials has had a serious negative impact on water quality.

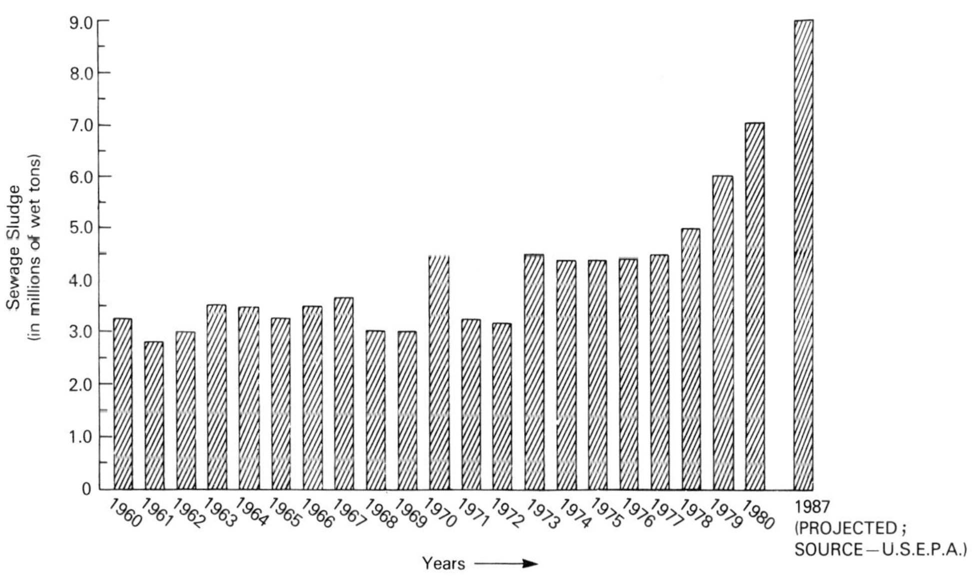

Figure 1. Quantities of sewage sludge ocean-dumped in the New York Bight over the past two decades (from Swanson et al., in press).

(3) The development of the industrial age, and in particular space age technology, has led to the need for tremendous quantities of raw materials for product development, accompanied by an even more rapid growth in the quantities of waste material produced.

For example, when we plot the production of sewage sludge for New York City wastes by year (Figure 1), the timing of the advances in waste treatment technology and the magnitude of the problem for coastal metropolitan areas is easily seen. Champ and Park (1981) and Duedall et al. (in press) have attempted to bring together global loading data for trend analysis. Figure 2 and 3 (from Duedall et al., in press) present the global ocean dumping permit tonnage for the years 1976-1979. The trend is clear: the quantities being ocean-dumped are steadily increasing.

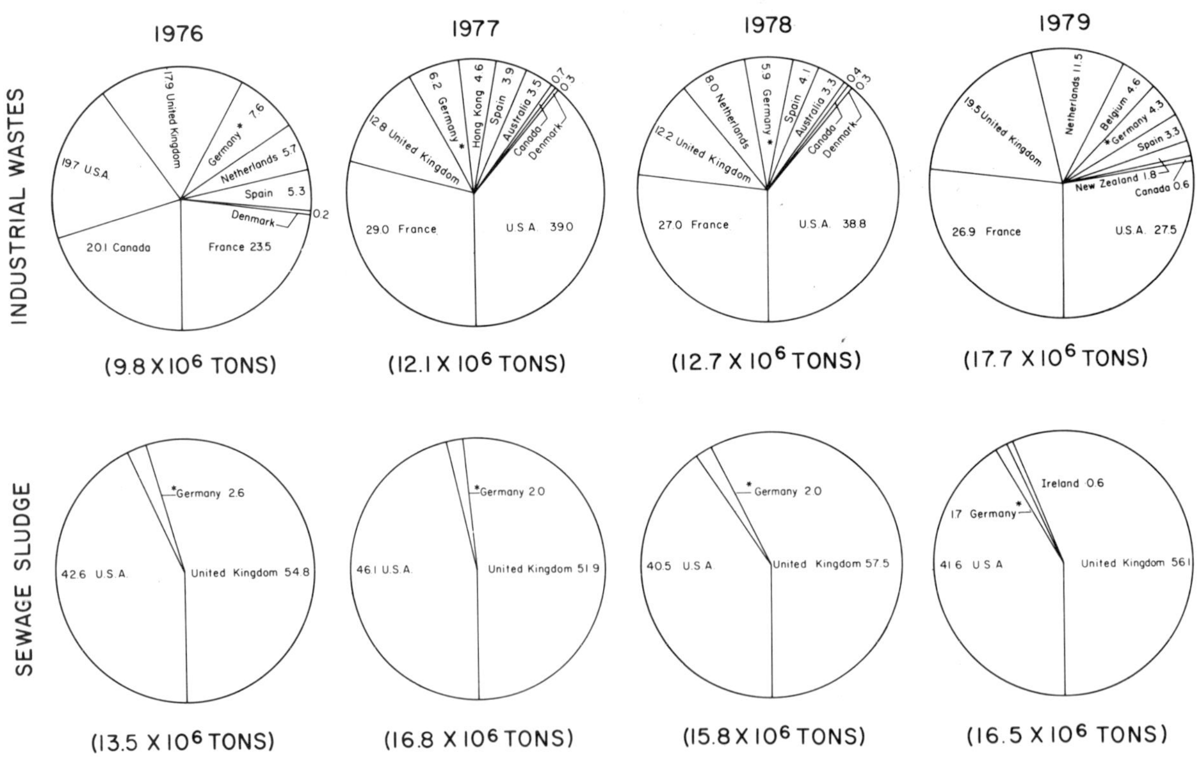

Figure 2. Quantities of municipal and industrial wastes (tons/year) on a global basis (as reported to the London Ocean Dumping Convention, for 1976-1979) (from Duedall et al., in press).

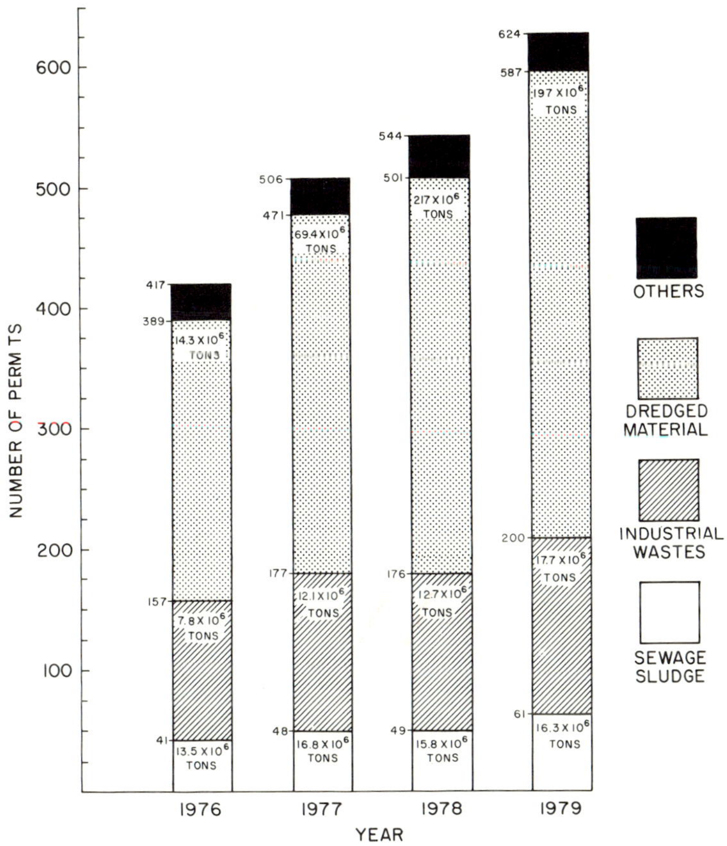

Figure 3. Number of ocean dumping permits on a global basis (as reported to the London Ocean Dumping Convention, for 1976-1979) (from Duedall et al., in press).

SCOPE

The initial approach of this bibliography was to focus intensively on municipal and industrial wastes that are ocean dumped at sea on a global basis. In 1981 we defined *ocean dumping* as *the transport of a waste material to sea for the sole purpose of disposal* (Champ and Park, 1981). However, we realized that a broader view of marine waste disposal would contribute to the field of marine pollution. Therefore, we amended our initial approach to include all papers on disposal from outfalls that were available to us. But we did not deliberately search out papers on outfalls as we did for ocean dumping. A global bibliography on marine outfall disposal would be a mammoth undertaking. We have also not attempted a deliberate search for references on ocean-dumped dredged materials on a global basis.

We have attempted to locate papers in the following categories:

Marine Pollution/Ocean Dumping

- General

Municipal Wastes

- Sewage sludge
- Outfalls/wastewater discharges
- Outfalls/sewage sludge discharges
- Solid wastes
- Garbage/baled refuse

Industrial Wastes

- General
- Acid wastes
- Alkaline/organic wastes
- Containerized wastes
- Coal wastes
- Cryolite (aluminum) sludges/red muds
- Dredged materials
- Drilling fluids/muds and cuttings
- Fish (seafood) wastes
- Munitions
- Ocean incineration
- Ocean mining wastes

- Petrochemical/petroleum
- Pharmaceutical
- Radioactive wastes
- Contaminant inputs/loads/sources

Legislation/Regulations

International Conventions

Geneva Conventions (The Law of the Sea Conventions)
London Dumping Convention (General-World)
Oslo Convention (Northeast Atlantic)
Helsinki Convention (Baltic Sea)
Barcelona Convention (Mediterranean Sea)
Paris Convention (The Convention for the Prevention of Marine Pollution from Land-Based Sources)

Ocean Dumping Criteria/Site Selection Studies

Waste Management Strategies

Biological Processes

- Bioavailability
- Bioaccumulation
- Phytoplankton/plants
- Zooplankton
- Invertebrates
- Fisheries/finfish/shellfish
- Field studies
- Bioassays/toxicity studies
- Microorganisms
- Assimilative capacity
- Field studies/monitoring/data reports

Chemical Processes

- Contaminant loading/sources
- Nutrients
- Contaminant exchange release (flux)
- Heavy metals
- Organochlorine compounds

Geological Processes

- Bathymetry
- Sediments (suspended and bottom)
- Particle dynamics and settling rates

Physical Processes

- Hydrography
- Dispersion
- Mixing

- Transport
- Currents/tides
- Sedimentation
- Salinity/temperature

Engineering Studies

- General engineering
- Design/construction
- Transportation/ships/barges
- Socioeconomic
- Modeling/predictive
- Remote sensing

Dumpsites (by countries)

Regions

- Canada
- New York (Bight/Apex)
- California
- 106-mile site
- UK - Dumpsites
- North Sea
- Baltic Sea
- Mediterranean/Black Sea

Bibliographies

Global Marine Pollution Bibliography

Ocean Dumping of Municipal and Industrial Wastes

1 - Aarseth, Inge. 1974. The planned red mud discharges in the Fensfjord, western Norway. Geological interpretation and tank experiments. ICES. C.M. 74/E:29.

2 - Abraham, G., and W. D. Eysink, et al. 1970. Full scale experiments on disposal of waste fluids into propeller stream of ships. Reprint FAO Conference. Rome.

3 - Achrem, T. J. 1973a. Coastal zone water quality monitoring in the New York Bight. Interstate Electronics Corp. Oceanics Div. Report 445-B1. 59p.

Abstract: A field case study of the New York Bight to determine the extent of and need for coastal zone water quality monitoring in the area is presented. Information is provided on organizations, their laboratory capabilities, data needs, and internal organization. Socioeconomic problems are discussed and recommendations are made for design and implementation of a coordinated coastal zone water quality monitoring network. Basic pollution problems in the area are complex and include sewer effluents, dredging, and industrial wastes. A selected bibliography and list of personnel concerned with water monitoring in the area are included.

4 - Achrem, T. J. 1973b. Ocean waste disposal in the New York Bight. Interstate Electronics Corporation. Oceanics Division. Report No. 4460C1559. 119p.

Abstract: Criteria are given for the control of ocean waste disposal in the New York Bight; an intensive field survey was conducted in the New York City metropolitan region during the spring of 1973. The topics surveyed include ocean dump site characteristics, their geographic location, type and volume of material dumped, method of disposal, description of disposal sites, monitoring procedures, local regulating agencies, estuarine economics, an extensive chronology of related major events, and alternatives and recommendations for ocean disposal in the New York Bight.

5 - Achrem, T. J., W. B. Merselis, and H. G. Stanley. 1975. Implementation of an initial network for an EPA ocean waste disposal information center in the New York Bight. Interstate Electronics Corporation. Environmental Engineering Division. Anaheim, CA. 18p.

6 - Ackerman, D. G., L. L. Scinto, R. J. Johnson, T. L. Sarro, and R. Scofield. 1981. Environmental assessment of the at-sea incineration of liquid silvex: Interim report. U.S. EPA-600/2-81-001a. Office of R&D. Research Triangle Park, NC. 205p.

7 - Adler, C. 1971. Ocean dumping. Pollution or fertilization. Water and Waste Engineering. April.

8 - Agg, A. R., and W. R. White. 1974. Devices for the pre-dilution of sewage at submerged outfalls. Proc. Instn. Civ. Engrs. No. 57. pp. 1-20.

9 - Agg, A. R. 1975. Criteria for marine waste disposal in Great Britain. In: E. A. Pearson and E. de Fraja Frangipane (editors), Marine Pollution and Marine Waste Disposal. (Supplement to Progress in Water Technology.) Proceedings of the 2nd International Congress. San Remo. 17-21 December, 1973. Pergamon Press. New York, NY. pp. 75-83.

10 - Ainsworth, G. 1974. Sludge disposal in the marine environment. In: Symposium on River Pollution Prevention, held at Music Hall & Assembly Rooms, Edinburgh (UK) on 13 March 1974. 12p.

11 - Aitsam, A. 1969. A stochastic model for marine waste disposal. In: S. H. Jenkins (editor), Advances in Water Pollution Research. Pergamon Press. New York, NY. pp. 837-842.

12 - Aitsam, A. M. 1975. Criteria for marine waste disposal in the USSR. In: E. A. Pearson, and E. de Fraja Frangipane (editors), Marine Pollution and Marine Waste Disposal. (Supplement to Progress in Water Technology.) Proceedings of the 2nd International Congress. San Remo. 17-21 December 1973. Pergamon Press. New York, NY. pp. 95-97.

13 - Akesson, B. 1970. *Ophryotrocha labronica* as test animal for the study of marine pollution. Helgolander wiss. Meeresunters. 20:293-303.

14 - Alexander, James E., and Elizabeth C. Alexander. (date unknown). The chemical oceanography of the New York Bight. 71p.

Abstract: Review of available information on chemical oceanography of the New York Bight. No extensive trace element data (except chlorophyll and urea) exists and until these data and other data needed for classical chemical parameters are collected, proper management of the area will not be possible. Suggested future studies are 1) processes of inorganic compulsion, 2) chemical fluxes through the region, and 3) the impact of life processes on the chemistry of these waters.

15 - Alexander, J. E., J. Foehrenbach, S. Fisher, and D. Sullivan. 1973. Mercury in striped bass and bluefish. New York Fish and Game Journal. 20(2):147-151.

Abstract: Bluefish and striped bass taken in the vicinity of Montauk Point on Long Island were analyzed for mercury content. In both species, there was a correlation between mercury and weight. Bluefish weighing less than 2.4 kilograms had concentrations of mercury below 0.5 milligrams per kilogram, while in those between 2.4 and 5.6 kilograms the concentration fluctuated above and below that level. All fish weighing more than 5.6 kilograms contained mercury in excess of 0.5 milligrams per kilogram. The corresponding weight categories for striped bass were: less than 3.2 kilogram; between 3.2 and 5.7 kilograms; and more than 5.7 kilograms, respectively.

16 - Alexander, J. E., R. Hollman, and T. White. 1974. Heavy metal concentration at the Apex of the New York Bight. N.Y. Ocean Sci. Lab. Montauk, NY. Rep. No. 4-35212.

17 - Alexander, J. E., and E. C. Alexander. 1977. Chemical Properties. MESA New York Bight atlas monograph 2. NOAA/MESA New York Bight Project and New York Sea Grant Institute. Albany, NY. 47p.

Abstract: ATLAS MONOGRAPH 2 reviews what is known about the chemical properties of New York Bight--we know relatively little. Temperature-salinity relationships show Bight waters to be two-layered in summer and vertically uniform in winter. In the slightly alkaline Bight waters, concentrations of bulk elements like calcium/, magnesium/, potassium/, and strontium/chlorinity ratios are lower than in the open ocean. Trace metal data are limited but are important in determining chemical input sources and in understanding how dangerous metals like mercury are concentrated in the food chain. Concentrations of micronutrients, nitrogen, phosphorus, and silica reflect responses to pollutions as well as to phytoplankton growth.

18 - Alexander, James E., Rudolph Hollman, and Theodore White. 1978. Heavy metal concentrations of the Apex of the New York Bight. NOAA TM ERL MESA-34. Boulder, CO. NTIS No. PB292615/AS. 38p.

Abstract: Particulate and total "soluble" cadmium, cobalt, copper, iron, and lead, total "soluble" mercury concentrations were determined for 146 seawater samples collected on November 8-9, 1973, in the Apex of the New York Bight. With few exceptions, particulate lead, cadmium, cobalt, and nickel concentrations were below the limits of detection for the analytical scheme used. The concentrations of particulate iron and copper and total "soluble" nickel varied as a function of tide and sample locations. Most of the iron was in the particulate form.

19 - Ali, Syed A., M. Grant Gross, and John R. L. Kishpaugh. 1975. Cluster analysis of marine sediments and waste deposits in New York Bight. Environ. Geol. 1(3):143-148.

Abstract: Cluster analysis of some chemical characteristics of marine sediments and associated waste deposits in the New York Bight reveals the existence of four cluster facies. Cluster facies I is the sediment containing dredged wastes from New York Harbor. Cluster facies II is the widespread sand and gravel deposit of the continental shelf. Cluster facies III is the finer-grained material in the Hudson Channel and the material accumulated in the sewage sludge disposal area. Cluster facies IV is not easily identified, it may correspond to some relict sedimentary feature in the area.

20 - Allied Chemical Corporation. 1972. A study of the dispersion effect of dumped waste muriatic acid on the ocean acid water background. Info. Tech. Report. No. 72-ET-7. Elizabeth Works, NJ.

21 - Allied Chemical Corporation. 1974. Outline of engineering report due 4/30/75 to USEPA. Exhibit A. Summary of 1974 analytical data - 12 calendar months by-product HCl disposed of in the ocean. Elizabeth Works, NJ.

22 - Allied Chemical Corporation. 1975. Report on the feasibility of using seawater as a neutralizing agent for waste hydrochloric acid before ocean dumping. Special Chemicals Division. Elizabeth Works, NJ.

23 - Anderson, A. R., and J. A. Mueller. 1976. Estimate of New York Bight future (2000) contaminant inputs. Environmental Engineering and Science Program. Manhattan College. Bronx, NY. 31p.

Abstract: The purpose of this study is to estimate future waste loads of selected water quality parameters from various sources and compare these with present load values and waste source distribution for New York Bight.

24 - Anderson, Andrew R., J. A. Mueller, and J. A. Hallden. 1979. The impact of marine coastal waters of the ocean discharge of municipal wastewater and its constituents. Case history of the New York Bight: Environmental Engineering Aspects. Hydroscience, Inc. Westwood, NJ. 115p.

Abstract: Dissolved oxygen is seriously depleted. High concentrations of coliform bacteria exist. These two problems must be relieved by improved treatment of wastewater. Largest amount of floating material is wood from derelict shore structures and has been serious enough to close beaches. Most of wastewater is discharged into the transect zone in waters that are utilized for recreation, shellfishing. Sludge is dumped 20 km at sea and is afforded greater dilutions. Sludge dumping is scheduled to be phased out in 1981.

25 - Anderson, E. S. 1968. The ecology of transferable drug-resistance factors in the Enterobacteriacae. Am. Rev. Microbiol. 22:131-180.

26 - Anderson, J. W. 1979. The transport of petroleum hydrocarbons from sediments to benthos. Pacific Northwest Laboratories. Sequin, WA. 22p.

Abstract: The examination of polycyclic aromatic hydrocarbons (PAH) and other slowly degradating pollutants. PAH in NY Bight marine environments, possible transfer routes from sediments to animals, and implications for the benthos and man.

27 - Anderson, J. W. (In Press). The transport of petroleum hydrocarbons from sediments to benthos and the potential effects. In: G. F. Mayer, A. Calabrese, F. A. Cross, D. C. Malins, D. W. Menzel, J. S. O'Connor, and F. J. Vernberg (editors), Ecological stress in the New York Bight: Science and Management. Estuarine Research Federation. Columbia, SC.

Abstract: Polynuclear aromatic hydrocarbons (PAH) enter the New York Bight region from multiple sources. Sewage sludge and bottom sediments collected from the Bight contain significant amounts (50-500 ppb) of various PAH components, while sediments from the area slated for dredging may contain very high levels (1-120 ppm). Whereas the surface microlayer in the Bight was relatively free of petroleum hydrocarbons, plankton contained 30 ppb of 2- and 3-ring aromatics. Fish and lobster flesh and mackerel liver contained relatively low levels of PAH. The detoxification organs of flounder and lobster contained the highest amount of PAH, and lobster digestive glands possessed contaminants in the full range of those detected (2- to 5-ring aromatics). Blue mussels exhibited contamination intermediate to the flesh and livers of fish and lobster. Possible routes of entry into marine species by PAH components are discussed. The implications of these data regarding tainting of food species and impacts on benthic biota from PAH and other pollutants are considered.

28 - Anderson, Kathleen J., and P. A. Read. 1974. Pollution studies in the Firth of Forth. Mar. Pollut. Bull. 5(8):121-125

29 - Anderson, P. W. 1978. Ocean dumping off New York. U.S. EPA Journal. 4(8):28-29.

30 - Anderson, Peter W., and Richard T. Dewling. 1981. Industrial ocean dumping in EPA Region II - regulatory aspects. In: B. H. Ketchum, D. R. Kester and P. K. Park (editors), Ocean Dumping of Industrial Wastes. Plenum Press. New York, NY. pp. 25-37.

Abstract: The Marine Protection, Research, and Sanctuaries Act of 1972 (PL 92-532) mandates that the Environmental Protection Agency "prevent or strictly regulate" the disposal, via vessel, of waste materials into the ocean.

The majority of ocean dumping, past and present, in the United States occurs at dump sites managed by EPA Region II. In 1977, for example, over 90 percent or 6.1 million metric tonnes of the U.S. ocean dumping occurred at sites managed by this Region.

EPA's implementation of this Act, beginning in April 1973, spurred industrial dumpers at sites managed by the Region to develop and construct land-based treatment facilities and/or implement other environmentally acceptable alternatives for handling their wastes. Of the roughly 150 industrial waste generators in 1973, as of April 1979 only 13 remain. Ten are under EPA orders to cease ocean dumping by the end of 1981. The other three, Allied Chemical Corp., NL Industries, Inc., and E. I. duPont de Nemours & Co. (Grasselli Plant), have demonstrated that their wastes are not harmful to the marine environment and that no feasible alternative exists at present which is more environmentally acceptable than ocean dumping. Thus, the level of ocean dumping activities in Region II represents a significant reduction from that prior to the passage of the Act.

31 - Andrulewicz, A., and J. E. Portmann. 1975. Determination of metal content of sediments in relation to dumping and discharge control. ICES. C.M. 75/E:22

32 - Anonymous. 1968a. Sludge disposal at sea. Civil Engineering. 38(8):62-63.

33 - Anonymous. 1968b. Radioactive waste disposal operation into the Atlantic. European Nuclear Energy Agency. Organization for Economic Co-operation and Development. 69p.

34 - Anonymous. 1969. A casual gas dump. Science News. Washington, D.C. 95(26):609-610.

35 - Anonymous. 1970a. Sewage disposal vessel is highly automated. Ocean Industry. 5(2):55-56.

36 - Anonymous. 1970b. Nerve gas saga. Mar. Pollut. Bull. 1(9):130-131.

37 - Anonymous. 1971a. Ocean pollution and marine waste disposal. Chemical Engineering. 78(3):60-67.

38 - Anonymous. 1971b. Controlled effluent disposal at sea. Effluent & Water Treatment Jour. 11(5):265-267.

39 - Anonymous. 1971c. Dumping of waste. British Institute of International and Comparative Law. London. Bulletin of Legal Developments No. 7:62.

40 - Anonymous. 1971d. First purpose built waste disposal ship for British owner launched in Rotterdam. Marine Engineers Review No. 33.

41 - Anonymous. 1971e. House approves ocean dumping bill, provides for research, sanctuaries. Environmental Reporter. Current Developments. 2(20):568-569.

42 - Anonymous. 1971f. World conference discusses marine pollution and its effects. Australian Fisheries. Canberra. 30(2):27-28.

43 - Anonymous. 1972a. Conference meets on dumping waste at sea. Nature. 240(5375):4.

44 - Anonymous. 1972b. Regulation of transportion for ocean dumping convention. Jour. of Maritime Law and Commerce. 3(2):423-426.

45 - Anonymous. 1972c. North sea. British Institute of International and Comparative Law. London. Bulletin of Legal Developments No. 3:31.

46 - Anonymous. 1972d. Tailing disposal today: Bigger safer dams for higher tonnages. World Mining. San Francisco, CA. 8(13):34-39.

47 - Anonymous. 1972e. Dumping of wastes. British Institute of International and Comparative Law. London. Bulletin of Legal Developments No. 12:126. (Ocean dumping, United Kingdom).

48 - Anonymous. 1973a. International guidelines proposed for disposal in the ocean. Solid Wastes Management Refuse Removal Jour. 16(6):23-24.

49 - Anonymous. 1973b. IMCO/FAO/UNESCO/WMO/WHO/IAEA/UN Joint group of experts on the scientific aspects of marine pollution (GESAMP): Report of the fifth session. GESAMP Report. Vol. 10. IMCO. London. 81p.

50 - Anonymous. 1973c. Law of the sea: Ocean dumping. Bulletin of Legal Developments. No. 17:9.

51 - Anonymous. 1973d. Ratification of proposed convention on the prevention of marine pollution by dumping of wastes and other matter. Government Reports Announcements. 73(6):107.

52 - Anonymous. 1973e. Marine pollution: Dumping. Bulletin of Legal Developments. No. 4:16.

53 - Anonymous. 1973f. Removal and disposal of digested sludge at lagoons 'A' and 'E', northeast water pollution control plan, City of Philadelphia, Pennsylvania. Government Reports Announcements. 73(3):112.

54 - Anonymous. 1974. German designed split-hull dump barges for carrying chemical wastes. Shipping World and Shipbuilder. No. 167. June.

55 - Anonymous. 1975a. Bibliography of literature on the pollution of the Baltic Sea area. 1970-1974. National contribution of Denmark. Copenhagen.

56 - Anonymous. 1975b. Marine pollution: Dumping of waste. Bulletin of Legal Developments. No. 21:221.

57 - Anonymous. 1975c. Bibliography of literature on the pollution of the Baltic Sea area. 1970-1974. National contribution of Finland. Unrev. version. [Mit Suppl. 1975]. Helsinki. 13p.

58 - Anonymous. 1975d. Bibliography of literature on the pollution of the Baltic Sea area. 1970-1974. National contribution of German Democratic Republic. Unrev. version 1975. [Suppl.] 41p.

59 - Anonymous. 1975e. How hot is ocean incineration? Environ. Sci. and Tech. 9(5):412-413.

60 - Anonymous. 1975f. Report of the IOC/GFCM/ICSEM International workshop on marine pollution in the Mediterranean. Monte Carlo, 9-14 Set. 1974. Workshop Rep. Intergovernm. Oceanogr. Commiss. No. 3. 6p.

61 - Anonymous. 1976a. Dumping sewage sludge in the ocean. Environ. Sci. and Tech. 10(6):530-531.

Abstract: The NOAA Ship George B. Kelez acts as the main platform for surveying and data acquisition of sewage sludge dumped at the designated New York Bight site off the coasts of New York and New Jersey. A short time after sewage sludge is dumped, very little evidence is found that the sludge was ever dumped there. Contaminating materials cannot necessarily be specified as having come from sewage sludge although the following adverse effects were noted in the New York Bight Apex: fin rot on certain fish; exoskeleton erosion and gill clogging on crabs and lobsters; DO often down to about 20% of saturation; and high concentrations (3%-5%) of TOC. Water pollution from the Hudson River, air pollution fallout, and waste disposal activities were also implicated as sources of pollution other than sewage sludge, NOAA does not recommend relocation of dumpsites to clean areas of the Middle Atlantic Shelf but rather suggests using the existing dumpsites until the 1981 ocean dumping phaseout occurs.

62 - Anonymous. 1976b. Ocean dumping phase-out in EPA Region 2. Jour. Water Poll. Control Fed. 48(10):2246-2248.

63 - Anonymous. 1976c. Bibliography of literature on the pollution of the Baltic Sea area. 1970-1974. National contribution of Federal Republic of Germany. Mit Suppl. 1.

64 - Anonymous. 1976d. Canada: Dumping at sea. Bulletin of Legal Developments. No. 2:15.

65 - Anonymous. 1976e. Dumping sewage sludge in the ocean. Environ. Sci. and Tech. 10(6):530-531.

66 - Anonymous. 1976f. Bibliography of literature on the Baltic Sea area. 1970-1974. National contribution of the USSR. [Mit Suppl. 1974]. Moscow.

67 - Anonymous. 1976g. Report of the IOC/FAO/UNEP International Workshop on Marine Pollution in the Caribbean and Adjacent Regions. Port-of-Spain, Trinidad, 13-17 Dec. Workshop Rep. Intergovernm. Oceanogr. Commiss. No. 11. 24p.

68 - Anonymous. 1976h. Koraalzee: Twin-screw, sea-going hydro dump barge. Holland Shipbuilding. 24(12):36-37.

69 - Anonymous. 1976i. Bibliography of literature on the pollution of the Baltic Sea area. 1970-1974. National contribution of Poland. [Mit Suppl. 1975.] Warsaw. No. 42. 5p.

70 - Anonymous. 1976j. Joint FAO(GFCM)/UNEP coordination project on pollution in the Mediteranean. Report No. 1. Circ. Gen. Fish. Coun. Medit. 4. 21p.

71 - Anonymous. 1977a. Ocean incineration anew. Envir. Sci. and Tech. 11:236-237.

72 - Anonymous. 1977b. Marine pollution: a multi-faceted, multi-agency detection, research and enforcement program. Sea Technol. 18(10):10-17.

73 - Anonymous. 1977c. Reports of the ACMRR/IABO Working Party on biological effects on pollutants, 1. Sess., Rome, Italy, 27-31 Oct. 1975. 2. Sess., Dubrovnik, Yugosl., 22-25. Nov. 1976. And report of the ACMRR/IABO expert consultation on bioassays with aquatic organisms. FAO Fish. Rep. No. 187. 28p.

74 - Anonymous. 1977d. Third Soviet-Swedish symposium on the pollution of the Baltic. Stockholm, Rosenon, September 15-21, 1975. Ambio Spec. Rep. 6. 294p.

75 - Anonymous. 1977e. Water pollution: A bibliography of reports. Aberdeen Marine Laboratory Library. Aberdeen, Scotland. 36p.

76 - Anonymous. 1978a. New research on waste dumping problem. Environ. Health. 86(2)45.

77 - Anonymous. 1978b. Report of the Int. Counc. for the Expl. of the Sea to the Oslo Commission, the Interim Helsinki Comm., and Inter. Paris Commission, 1977. Cooperative Research Report No. 76. Charlottenlund, Denmark. 37p.

78 - Anonymous. 1978c. Input of pollutants to the Oslo Commission area. Cooperative Research Report No. 77. Charlottenlund, Denmark. 57p.

79 - Anonymous. 1978d. Bibliography of literature on the pollution of the Baltic Sea area published in 1977. Deutshches Hydrographishches Institut. Hamburg, Federal Republic of Germany.

80 - Anonymous. 1978e. The effects of the ocean disposal of municipal waste. Southern California Coastal Water Research Project. El Segundo, CA. 27p.

81 - Anonymous. 1978f. Pollution in the marine environment. UNESCO, Paris, France. 35p.

82 - Anonymous. 1979. Marine, municipal and industrial waste water disposal. Progr. Water Technol. 291p.

83 - Amiel, A. J., and J. Navrot. 1978. Nearshore sediment pollution in Israel by trace metals derived from sewage effluent. Mar. Pollut. Bull. 9(1):10-14.

84 - Argaman, Y. et al. 1975. Use of jet pumps in marine waste disposal. Jour. of the Environ. Eng. Div. 101 EES pp. 703-711.

85 - Armstrong, Reed, S. 1979a. Bottom oxygen and stratification in 1976 and previous years. In: R. L. Swanson and C. J. Sinderman (editors), Oxygen Depletion and Associated Benthic Mortalities in New York Bight, 1976. NOAA Professional Paper No. 11. U.S. Dept. of Commerce. Rockville, MD. pp. 137-148.

Abstract: Based on historical oceanographic data, the D.O. content of bottom waters over the New Jersey-Cape May and Long Island continental shelves typically declines during spring and summer, reaching minimum values in August. The seasonal decline in D.O. closely parallels the development of density stratification, and the rate at which D.O. declines seems to correspond with the strength of stratification. Stratification tends to isolate bottom waters from vertical replenishment until September, at which time cooling of the surface and mixing begins to destroy the vertical structure. This results in increased replenishment of bottom D.O. Continued cooling and overturning through autumn and winter typically cause a steady increase in oxygen values to the annual maximum in March. The influence of stratification on rates of oxygen decline may be greater off New Jersey than in adjacent regions because of bathymetric differences. However, this assessment does not take into account any effects on D.O. concentrations that might result from unusual advective and biologic processes. Comparison with the normal cycle of bottom D.O. indicates that concentrations were already below normal by April in 1976 throughout the New York Bight.

In 1976, the early occurrence of increased river discharge accompanied by early warming led to the early development of stratification; above-normal stratification persisted through May. The early onset of stratification in 1976 would have contributed to the occurrence of abnormally low bottom D.O. in the New York Bight in three different ways;

1. If D.O. concentrations increased as per the normal trend into January 1976 (as indicated from observations made in December 1975), then with a 2-month earlier than normal onset of stratification, maximum concentrations for the year would have been in January at about 6.5 ml/l, which is 0.5. ml/l less than the usual March maximum of 7.0 ml/l. Given normal conditions through the rest of the year, the D.O. values in 1976 could have been somewhat below normal each month until autumn.

2. Normally, the season of strengthening stratification and declining D.O. lasts about 5 months (April-August). In 1976, this season was apparently lengthened as much as 2 months, because of the early onset of stratification. This means that 1976 had 7 months in which utilization of oxygen would have exceeded replenishment.

3. From March through May 1976, and probably beginning as early as February, stratification was stronger than normal and about typical for June through August. As a result, the decline in D.O. would have been greater than normal during as many as 4 months in 1976 and at typical rates during summer. Thus the cycle of bottom D.O. concentration would have proceeded at values below normal for as long as February through August.

Historical observations, stratification values, and bottom D.O. decline rates were used to develop a graphic model. The model was then used to estimate the cycle of bottom D.O. concentration for 1976. These estimates were compared with 1976 observations and indicated that stratification, with the early arrival of spring conditions, accounted for the below-normal D.O. values in April and May. The model-derived estimate further implied that D.O. would have become almost totally depleted over the New Jersey shelf in August. But D.O. measurements in the area during June indicate that concentrations fell much more rapidly than estimated, because of stratification alone. Similarily, model-derived estimates for Long Island shelf waters indicated that below-normal concentrations should have developed there in 1976, but not as such low values as off New Jersey. Also, in June, Long Island D.O. concentrations fell more rapidly than predicted from the model. Because stratification alone failed to explain the full magnitude of the oxygen depletion, there must be other contributory factors. Possibly the dramatic decline in bottom D.O. during June resulted from excessive demand from the presence of the large mass of *Ceratium* and, perhaps, from factors associated with the circulation.

86 - Armstrong, Reed. S. 1979b. Climatic conditions related to the fish kill and anoxia off New Jersey during the summer of 1976. In: Ocean Variability in the U.S. Fishery Conservation Zone. 1976. NOAA Technical Report NMFS Circular 427. Paper 17. pp. 289-300.

Abstract: A massive fish kill in the bottom waters over the middle continental shelf off New Jersey occurred during the summer of 1976. Observations in August 1976 indicated that the fish kill was probably related to the presence of exceptionally low oxygen concentrations in the bottom waters on the shelf.

87 - Arnesen, R. T., J. Knutzen and S. T. Kaliqvist. 1973. Undersökelse for Kronos Titan A/S, Fredrikstad. Jerninnhold i sedimenter og vekstforsök med alger. Unpublished report of NIVA. 22p.

88 - Arnesen, R. T. 1969. Vurdering av Glama som resipient for avlops-vann fra Titan Co. A/S, Undersökelser 1968-69. Unpublished report of the Norsk Inst. for Vann-forskning (NIVA). 23p.

89 - Arnold, Edgar L. and William F. Royce. 1950. Observations of the effect of acid-iron waste disposal at sea on animal populations. U.S. Dept. of Interior, Fish and Wildlife Service. Special Scientific Report, Fisheries No. 11. 12p.

90 - Arriaqe, L. 1976. Contamination en el Oceano Pacifico sur-oriental (Ecuador-Peru-Chile). Rev. Com. Perm. Pacifico Sur 5:3-62.

91 - Asplund, L. 1978. The inputs of substances to the Gulf of Bothnia from the coast of Sweden and from the atmosphere in 1976. Finn. Mar. Res. No. 244. pp. 158-164.

92 - Association Europeene Oceanique for Commission of the European Communities. 1977. Metallic effluents of industrial origin in the marine environment. Graham & Trotman. London. (ERJ 5331). 216p.

93 - Association of Metropolitan Sewage Agencies. 1981. The sludge management syndrome: Burn, barge, bury or bust. Conference Proceedings. Washington, DC. 300p.

Summary: Covers ocean, incineration and dumping, land utilization (agricultural uses and land reclamation), land fills, marketing and distribution, and major policy considerations.

94 - Atlas, E., J. Brooks, J. H. Trefry, T. Sauer, C. Schwab, B. Bernard, J. Schofield, C. Giam, and E. R. Meyer. 1980. Environmental aspects of ocean dumping in the western Gulf of Mexico. Jour. Water Poll. Control Fed. 52(2):329-350.

95 - Atlas, E., G. Martinez, and C. S. Giam. 1981. Chemical characterization of ocean dumped waste materials. In: B. H. Ketchum, D. R. Kester and P. K. Park (editors), Ocean Dumping of Industrial Wastes. Plenum Press. New York, NY. pp. 275-293.

Abstract: This paper describes methods for chemical characterization and organic "finger printing" techniques for complex ocean-dumped wastes. The methods employ wet chemical separations and gas chromatography combined with mass spectrometry and selective GC detectors. Application of these methods is demonstrated for the characterization of pharmaceutical waste and for tracking two types of chemical wastes in actual ocean dumping stations.

96 - Atwood, D. K., T. E. Whitledge, J. H. Sharp, A. Y. Cantillo, G. A. Berberian, J. M. Parker, P. G. Hanson, J. P. Thomas, and J. E. O'Reilly. 1979. Chemical factors. In: R. L. Swanson and C. J. Sindermann (editors). Oxygen depletion and associated benthic mortality in New York Bight, 1976. U.S. Dept. of Commerce. NOAA Professional Paper No. 11. Rockville, MD. pp. 79-123.

Abstract: Dissolved oxygen in bottom waters of the New York Bight shelf (especially in the Apex and off the New Jersey coast) was severely depleted in 1976 compared to other years for which there are data. Although D. O. depletion is an annual occurrence during the warm season (that is, when the density stratification is strong), it occurred earlier in 1976 and was more severe. In certain areas, D.O. values were zero or near zero.

Other than this severe oxygen depletion, no clear chemical differences were detected in the water column between 1976 and other years represented by the data base. Apparently, there was no exceptional nutrient input to stimulate productivity. In fact, there is some evidence that fewer nutrients may have been available at the shelf break in 1976. If this is true, it might have affected the types of organisms found in shelf waters; that is, it could have favored production of organisms such as *Ceratium tripos*. Also, there is no chemical evidence of exceptional inputs of organic carbon either as POC or DOC although there is strong biological evidence for a very dense plankton bloom (Ceratium tripos), which certainly contributed to the organic loading. The high DOC levels present in the Bight had the capacity to cause the depletion observed.

Clearly something was different in 1976, and perhaps this "something" was a natural occurrence. Since an adequate existing data base covers only a short time span (post-1970) and lacks many essential variables, we cannot ascertain what this was. It is significant that extraordinary DOC values exist in the Bight, especially in the Apex. Our limited data do not allow a comparison of 1976 DOC levels with those of previous years nor do they allow much insight into the chemical or physical nature or variations in space and time of the organic matter that is lumped into the category of DOC. More information on organic matter perhaps could provide insights into future events of a similar nature.

The observed chemical responses of Bight seawater to the oxygen depletion were as expected, based on our knowledge of other low oxygen or anoxic areas in the ocean. When oxygen was depleted, nitrate was reduced to nitrite and then ammonium, sulfate was reduced to sulfide, and the solubility of certain metals changed as the reducing environment developed and caused changes in their oxidation states.

97 - Aubert, M., and N. Desirotte. 1971. Etude en bassin hydraulique des rejets d'eau residuaire en mer. (Study in a hydraulic model of waste water rejects in the sea.) Revue Int. Oceanogr. Med. 22-23:5-40.

98 - Aubert, M. 1979. Toxicological studies in the field of oceanology. Progr. Water Technol. 4:33-39.

99 - Austin, Herbert M. 1975. An analysis of plankton from Deepwater Dumpsite 106. In: May 1974 Baseline Investigations of Deepwater Dumpsite 106. NOAA/ODP Dumpsite Evaluation Report 75-1. Rockville, MD. pp. 271-388.

100 - Austin, H. M. 1976. Distribution and abundance of ichthyoplankton in the New York Bight during the fall in 1971. New York Fish and Game Jour. 23(1):58-72.

Abstract: Sampling for phytoplankton, zooplankton, water chemistry, and physical parameters was conducted at a series of stations in the New York Bight in September and November of 1971. The relationships of the finfish eggs and larvae collected to the associated oceanographic conditions are discussed in terms of the sources of the parent stocks and the time and place of spawning. The bulk of the spawning that occurs in coastal New York waters is by migratory stocks, the principal species in the New York Bight being the menhaden, anchovy, weakfish, mackerel, and searobin.

101 - Ayers, John C., Bostwick H. Ketchum and Alfred C. Redfield. 1949. Hydrographic considerations relative to the location of sewer outfalls in Raritan Bay. Woods Hole Oceanogr. Inst. Ref. 49-23. Woods Hole, MA. 49p.

102 - Ayers, John C. 1951a. The "normal" flushing time and half-life of New York Harbor. In: John C. Ayers et al., The Hydrography of the New York Area. Status Report No. 1. (Contract N6 ONR 264, task 15). Cornell Univ. Ithaca, NY.

103 - Ayers, John C. 1951b. Tabulated temperature and salinity data of the New York area. In: John C. Ayers et al., The Hydrography of the New York Area. Interim Report No. 1. (Contract N6 DNR 264 task 15). Cornell Univ. Ithaca, NY.

104 - Ayers, John C. 1951c. The steady-state condition and the unbalanced condition in tidal estuaries, with special reference to the unbalanced condition of New York Harbor in March-April 1950. In: John C. Ayers et al., The Hydrography of the New York Area. Status Report No. 2. (Contract N6 ONR 264, task 15). Cornell Univ. Ithaca, NY.

105 - Ayers, John C. 1951d. An examination of the factor in correcting for the effect of bottom friction on the velocity of flow in tidal streams, and the manner in which it is affected by stage of tide. In: John C. Ayers, et al., The Hydrography of the New York Area. Status Report No. 3. (Contract N6 ONR 264, task 15). Cornell Univ. Ithaca, NY.

106 - Ayers, John C. 1951e. The average rate of fouling of surface objects and of submerged objects in the water adjacent to New York Harbor. In: John C. Ayers et al., The Hydrography of the New York Area. Status Report No. 6. (Contract N6 DNR 264, task 15). Cornell Univ., Ithaca, NY.

107 - Ayers, John C. 1951f. The transparency (by white Secchi disc) of the waters in and about New York Harbor. In: John C. Ayers et al., The Hydrography of the New York Area. Status Report No. 7. (Contract N6 DNR 264, task 15). Cornell Univ. Ithaca, NY.

108 - Ayers, John C. 1951g. A preliminary report upon the state of our knowledge of the waters in and around New York Harbor. In: John C. Ayers et al., The Hydrography of the New York Area. (Contract N6 DNR 264, task 15). Cornell Univ. Ithaca, NY.

109 - Ayers, John C., and Richard H. Backus. 1952. First annual report. In: John C. Ayers et al., The Hydrography of the New York Area. Progress Report No. 1. (Contract N6 DNR 264, task 15). Cornell Univ. Ithaca, NY.

110 - Ayers, John C. 1952a. Tabulation of reduced data of cruise Stirni I, July-September, 1951. In: John C. Ayers et al., The Hydrography of the New York Area. Data Report No. 1. (Contract N6 DNR 264, task 15). Cornell Univ. Ithaca, NY.

111 - Ayers, John C. 1952b. Tabulation of reduced data of cruise Stirni III, January-February 1952. In: John C. Ayers, et al., The Hydrography of the New York Area. Data Report No. 2. (Contract N6 DNR, task 15). Cornell Univ. Ithaca, NY.

112 - Ayres, P. A. 1977. The use of faecal bacteria as a tracer for sewage sludge disposal in the sea. Mar. Pollut. Bull. 8(12):283-285.

113 - Azarovitz, T., M. Silverman, V. Anderson, A. Thomas, and C. Aussicker. 1976a. Demersal finfish catches in the New York Bight by stations and species, R/V ATLANTIC TWIN, October 31-December 5, 1972. NOAA DR ERL MESA-11. Boulder, CO. 100p.

Abstract: Presents ADP summaries of finfish and associated data collected in the New York Bight on cruises conducted during this period. This report covers the catch on the inshore segment of the survey.

114 - Azarovitz, T., M. Silverman, V. Anderson, A. Thomas, and C. Aussicker. 1976b. Demersal finfish catches in the New York Bight by stations and species, R/V ATLANTIC TWIN, May 8-June 4, 1973. NOAA DR ERL MESA-12. Boulder, CO. 146p.

Abstract: Presents ADP summaries of finfish and associated data collected in the New York Bight on cruises conducted during this period. This report covers the catch on the inshore segment of the survey.

115 - Azarovitz, T., M. Silverman, V. Anderson, A. Thomas, and C. Aussicker. 1976c. Demersal finfish catches in the New York Bight by stations and species, R/V DELAWARE II and ATLANTIC TWIN, April 1-May 2, 1974. NOAA DR ERL MESA-14. Boulder, CO. 104p.

Abstract: Presents ADP summaries of finfish and associated data collected in the New York Bight on cruises conducted during this period. This report covers the catch on the inshore segment of the survey.

116 - Azarovitz, T., M. Silverman, V. Anderson, A. Thomas, and C. Aussicker. 1976d. Demersal finfish catches in the New York Bight by stations and species, R/V ALBATROSS IV and DELAWARE II, September 23-October 4, 1974. NOAA DR ERL MESA-15. Boulder, CO.

Abstract: Presents ADP summaries of finfish and associated data collected in the New York Bight on cruises conducted during this period. Covers the catch of both the inshore and offshore segments of this survey.

117 - Azarovitz, T., M. Silverman, V. Anderson, A. Thomas, and C. Aussicker. 1976e. Demersal finfish catches in the New York Bight by stations and species, R/V ALBATROSS IV and ATLANTIC TWIN, March 4-24, 1975. NOAA DR ERL MESA-16. Boulder, CO. 140p.

Abstract: Presents ADP summaries of finfish and associated data collected in the New York Bight on cruises conducted during this period. Covers the catch of both the inshore and offshore segments of the survey.

118 - Azarovitz, Thomas. R. et al. 1979. Effects on finfish and lobster. In: R. L. Swanson and C. J. Sinderman (editors), Oxygen Depletion and Associated Benthic Mortalities in New York Bight, 1976. NOAA Professional Paper No. 11. U.S. Dept. of Commerce. Rockville, MD. pp. 295-314.

Abstract: American lobster probably react to depressed oxygen levels like finfish do. McLeese (1956) found that oxygen uptake of small lobster is greater per unit weight than large lobster; lobsters are able to acclimate to low-oxygen concentrations; moulting lobsters are less resistant to low-oxygen than hard-shell lobster; and the minimum D.O. they can tolerate varies with temperature and salinity. As an example of the way the minimum D.O. varies, at 15°C with a salinity of 30°/oo the minimum D.O. is 0.54 ml/l; at 15°C with a salinity of 30°/oo the minimum is 0.19 ml/l.

During the 1976 RV Atlantic Twin cruise the bottom temperature in the anoxic area was 7°C to 12°C, while the corresponding salinities were about 30 to 33°/oo (Azarovitz, unpublished data). Applying McLeese's observations to these would seem to indicate that the mimimum D.O. level that lobster could survive would be between 0.26 ml/l and 0.43 ml/l. Much of the depressed D.O. water mass had oxygen levels below these. If lobsters resident to the areas affected did not leave, mortality would result.

Evidence is mixed on whether lobsters avoided or succumbed to anoxia. On June 27, 1976, divers observed that lobsters that were apparently residents of certain wrecks had climbed to the highest part of these wrecks. By July 4, 1976, these same lobsters were observed to be dead. In mid-July other lobsters were observed to be lethargic and lying outside their shelters, often with two or more close to each other. This observation is significant when we consider the aggressive behavior of lobsters. Late in September, lobsters again inhabited these same wrecks. The climbing behavior on the wrecks may indicate that lobsters can detect decreasing D.O. levels and areas of greater D.O. concentration.

Off southern New England and the middle Atlantic States the offshore lobster population has an inshore-offshore migration pattern (Uzmann et al. 1977). The on-shore migration is from March to August; the offshore migration is from October to December. Uzmann et al. (1977) estimated that at least 20 percent of the offshore population annually migrates. This migration significantly affects the inshore lobster fishery. It is reasonable to hypothesize that the anoxic water mass blocked the 1976 shoreward migration, because lobsters could probably detect the decreased oxygen level and are capable of traveling up to 9.3 km/d on a sustained basis (Uzmann et al. 1977).

The lobster landings at the southern end of New Jersey (Cape May County) did not seem to be affected and were up slightly in 1976. Since the southernmost end of New Jersey was not affected by the anoxic water conditions, this area may be loosely compared with the rest of the State. With this in mind, it seems evident that the decreased landings to the north can be attributed to the change in water conditions rather than to natural fluctuation in the lobster population.

119 - Babinchak, J. A., J. T. Graikoshi, S. Dudley, and M. F. Nitkowski. 1977. Distribution of faecal coliforms in bottom sediments from the New York Bight. Mar. Pollut. Bull. 8(7):150-153.

Abstract: The distribution of faecal coliform densities found in New York Bight sediments indicated that sewage sludge material has contaminated areas extending 11 km north and 37 km south from the disposal site. A high degree of confidence for enumerating faecal coliforms in marine sediments was demonstrated by standard water methodology since 82.3% of the faecal coliform group isolated were *Escherichia coli*. This study also indicated that marine sediments could be stored 4 days at 4°C without appreciable changes in the faecal coliform count, and that, in situ, faecal coliforms would persist longer when sediment temperatures were low.

120 - Backus, R. H. 1951. The distribution of dissolved oxygen in the New York Harbor area. In: John C. Ayers, et al., The Hydrography of the New York Area. Status Report No. 8. (Contract N6 DNR 264, task 15). Cornell Univ. Ithaca, NY.

121 - Baeyens, W., G. Decadt, and I. Elskens. 1979. A study of the distribution of mercury in the various compartments of the North Sea and Scheldt Estuary ecosystems. Oceanologica. Acta. 2(4):447-457.

122 - Baeyens, W., G. Decadt, and I. Elskens. 1980. Determination of dissolved, particulate and total mercury in the water column of the Southern Bight of the North Sea with adapted analytical procedure. ICES. C.M. 80/E:61.

123 - Bahn, G. S. 1978. Segregation of acid plume pixels from background water pixels. Signatures of background water and dispersed acid plumes, and implications for calculation of iron concentration in dense plumes. NASA Contract. Report No. 145372-N78-27487. 62p.

Abstract: Two files of data, obtained with the Bendix Modular Multiband Scanner, for an acid waste dump into ocean water, were analyzed intensively. Signatures were derived for background water at different levels of effective sunlight intensity, and for different iron concentrations in the dispersed plumes from the dump. The effect of increased sunlight intensity on the calculated iron concentration, calculated according to the ratio of Band 2 radiance to Band 4 radiance, was found to be relatively important at low iron concentrations and relatively unimportant at high values of iron concentration in dispersed plumes. It was concluded that the basic equation for iron concentration is not applicable to dense plumes, particularly because lower values are indicated at the very core of the plume, than in the surrounding sheath, whereas radiances increase consistently from background water to dispersed plume to inner sheath to innermost core. It was likewise concluded that in the dense plume the iron concentration would probably best be measured by the higher wave length variances. For dispersed plumes, a technique was devised to separate background water from plume pixels, and to adjust the plume pixel radiances for distortions caused either by effective increase in sunlight intensity or by an inexplicable edge effect along the right-hand side of the scene. It was found that dense plumes yield radiances in excess of the maximum distortions in the background data, so there is no difficulty in segregating the dense plume pixels when evaluating the average radiance levels of the background water. The techniques evolved in the course of this analysis should be generally applicable to similar sets of data.

124 - Balasha, E., H. Sperber and Z. Krischer. 1975. Marine disposal of Netanya industrial wastes. In: A. M. Wachs, (chairman), Proc. 5th Scientific Conf. Israel Ecological Society. Tel Aviv. 8-9 May, 1975. Haifa, The Israel Ecological Society. pp. G33-G43.

125 - Ball, J., T. D. Reynolds, and D. E. Beckett. 1974. Deep-sea ocean disposal of petrochemical plant wastes. Proceedings of 29th Industrial Wastes Conference. Purdue University. Lafayette, IN. pp. 223-229.

126 - Ball, J. and T. D. Reynolds. 1974. Ocean disposal of liquid petrochemical wastes. Proc. 10th Annual Meeting, Marine Technology Society. Washington, DC. pp. 97-106.

127 - Ball, J. and T. D. Reynolds. 1976. Dispersion of liquid waste from a moving barge. Jour. Water Poll. Control Fed. 48:2541-2548.

128 - Baltic Marine Environment Protection Commission-Helsinki Commission. 1981. Report of the Interim Commission (IC) to the Baltic Marine Environment Protection Commission. Baltic Sea Environment Proceedings No. 2. 36p.

129 - Bamber, R. N. 1980. Properties of fly ash as a marine sediment. Mar. Pollut. Bull. 11(11):323-326.

130 - Barber, R. T. and W. Kirby-Smith. 1973. The oceans as ultimate sinks for wastewaters and wastewater residue. In: F. E. McJunkin, and P. A. Vesilind (editors), Ultimate Disposal of Wastewaters and Their Residuals, National Symposium. Water Resources Research Institute of the University of North Carolina. Raleigh, NC. pp. 199-215.

131 - Bargman, R. D. 1975. Problems of meeting changing water quality requirements for large waste disposal systems. In: E. A. Pearson, and E. de Fraja Frangipane (editors), Marine Pollution and Marine Waste Disposal. (Supplement to Progress in Water Technology.) Proceedings of the 2nd International Congress. San Remo. 17-21 December, 1973. Pergamon Press. New York, NY. pp. 155-161.

132 - Barvenik, F. W., F. B. Hill, V. P. Aneja, and R. M. Felder. 1975. Hydrogen sulfide in bottom waste near a sewage sludge dumping site. Informal Report Brooknaven National Laboratory. Department of Applied Science. Upton, NY. 27p.

Abstract: An attempt was made to detect volatile sulfur compounds in bottom water at a sewage sludge dumping site in the Apex of the New York Bight. Water from two meters above bottom was pumped on board ship and then through a gas-liquid contactor. Volatile compounds transferred to the gas phase were passed to a gas chromatrograph sensitive to sulfur compounds. Significant peaks were found for hydrogen sulfide but peaks for other sulfur compounds were not clearly evident. Assuming an equilibrium distribution of H_2S between gas and liquid phases in the contactor, the concentration of total sulfide in solution in the bottom water was estimated to be 0.03-0.08 uM. The bottom water was found to be partially depleted of dissolved oxygen and contained a small population of sulfate reducers (Desulfovibrio desulfuricans). Additional experiments designed to permit further characterization of the marine environment in the vicinity of sewage sludge dumping sites insofar as sulfur compounds are concerned are suggested.

133 - Bascom, Willard. 1974. The disposal of waste in the ocean. Scientific American. 231(2):16-25.

Abstract: Contrary to some widely held views, the ocean is the plausible place for man to dispose of some of his wastes. If the process is thoughtfully controlled, it will do no damage to marine life.

134 - Bastian, D. F. 1974. Effects of open-water disposal of dredged material on bottom topography along Texas Gulf Coast. Miscellaneous Paper D-74-13. Hydraulics Laboratory. U.S. Army Engineer Waterways Experiment Station. Vicksburg, MS. NTIS No. A002659.

135 - Batey, G. 1979. Regulation and assessment of ocean dumping. Sea Technology 20(10):17-20.

136 - Batky, Robert J. 1975. Feasibility of using sewage sludge as fertilizer on a sandy beach. Masters Thesis. Adelphi University. Garden City, NY. 48p.

Abstract: There are three major benefits from sewage sludge fertilization:

1) Sludge provides nutrients to enhance fertilization.
2) More rapid growth traps more sediment, stabilizing beachfronts more quickly.
3) The project provides an alternative to disposal of an increasing volume of sewage sludge.

This study undertook to evaluate the practicality, and public health impact was difficult to judge because of the many inputs to Great South Bay. Although cheap, the operation is hard to perform without making recreation areas unpleasing aesthetically.

137 - Baumgartner, D. J., R. J. Callaway, and G. R. Ditsworth. 1968. Disposal of aluminum process wastes in the ocean. Working Paper No. 64, Pacific Northwest Water Lab, FWQA, Corvallis, OR. (Reprinted December 1969).

138 - Baumgartner, D. J. 1971. Disposal of liquid and particulate wastes to the ocean. Chemical Engineering Symposium Series 107. pp. 46-53.

139 - Baumgartner, D. J. 1972. A brief outline of a study of sewage sludge dumping in the New York Bight. U.S. Environmental Protection Agency. National Environmental Research Center. Pacific Northwest Environmental Research Laboratory. Corvallis, OR. PNERL Working Paper No. 5. 11p.

140 - Baxter, I. G., and J. H. Steele. 1973. Mortality of herring larvae in the Clyde Sea area. ICES. C.M. 73/E:29.

141 - Baxter, S. S. 1954. Sludge disposal in Philadelphia. Proc. of the Amer. Soc. Civil Eng. Sanitary Eng. Div. 85(SA6):127-141.

142 - Baxter, Samuel S. 1959. Sludge disposal in Philadelphia. Jour. of the Sanitary Engineering Division, Proceeedings of the Amer. Soc. of Civil Eng. No. SA6. Vol. 85. p. 127.

143 - Baxter, Samuel S., Carmen F. Guarino, Robert A. Erb, and Charles T. Davey. 1971. Philadelpha ocean sludge disposal experience and studies. Presented at the 44th Annual Conference of the Water Pollution Control Federation. 14p.

144 - Baylor E. R., and M. B. Baylor, Duncan C. Blanchard, Lawrence D. Syzdek, Curtis Apel. 1977. Virus tranfer from surf to wind. Science. 198(4317):575-580.

Abstract: Bubbles in the sea surf adsorb and carry viruses to the surface where they are propelled into the air on tiny jets of seawater when the bubble bursts. The ejected jets become tiny drops of aerosol. The bubble adsorption and virus concentration in the surf are analagous to industrial bubble levitation processes that concentrate metallic ores, enzymes, and finely divided organic crystals. Bubble levitation of viruses deliberately injected into the surf produced 200 times more virus per milliliter in the aerosol than were present in samples from the surf. Some aerosol drops created by the surf and carried by the wind fall out on the beach. The frequency of virus-bearing drops, that is, the number of plaques on seeded plates exposed on the beach, decreased exponentially with the distance downwind from the surf.

145 - Bayne, B. L. (Chairman). J. Anderson, D. Engel, E. S. Gilfillan, D. Hoss, R. Lloyd, and F. P. Thurberg. 1980. Physiological techniques for measuring the biological effects of pollution in the sea. In: International Council for the Exploration of the Sea. Rapports et Proces-Verbaux des Reunions. Vol. 179. Charlottenlund Slot, Denmark.

146 - Beardsley, R. C., W. C. Boicourt, and D. V. Hansen. 1976. Physical oceanography of the New York Bight. In: M. G. Gross (editor), Amer. Soc. Limnol. and Oceanogr. Spec. Symp. 2:20-34.

Abstract: Kinetic energy spectra from moored current meters in the Mid-Atlantic Bight reveal marked differences in current variability between the inner shelf and the outer shelf and slope regions. The nearshore subtidal current variability appears to be dominated by meteorological forcing. The amplitude of the semidiurnal and diurnal tidal peaks decreases in the offshore direction. Shallow water records show little or no inertial energy, while at the shelf break and over the slope, inertial motion contributes significantly to the current variance. A simple conceptual model is presented to explain how intense winter low pressure systems ("northeasters") drive strong alongshore currents which are coherent over much of the Bight. A map of "mean" currents measured in recent moored array experiments demonstrates subsurface water flow along the shore toward the southwest. The average currents generally increase in magnitude offshore and decrease with closeness to bottom. At most sites, the mean current veers toward shore with increasing depth. The alongshore volume transport measured at three transects across the Bight shows surprising uniformity, considering the possible sources for discrepancy. This transport (order $2.0 \times 10^5 m^3 s^{-1}$) of water within the 100-m isobath implies a mean residence time on the order of 3/4 year. Much of the shelf water observed flowing westward south of New England must originate in the Gulf of Maine-Georges Bank area.

147 - Beardsley, R. C., H. Mofjeld, M. Wimbush, C. N. Flagg, and J. A. Vermersch, Jr. 1977. Ocean tides and weather-induced bottom pressure fluctuations in the Middle Atlantic Bight. Jour. of Geophy. Res. 82(21):3175-3182.

Abstract: Five bottom pressure gages were deployed in the Middle-Atlantic Bight during the late winter of 1974. Analysis of the resulting pressure series and neighboring coastal tide gage series shows that tides are the dominant pressure signal in this section of the continental shelf. Most of the remaining pressure fluctuations appear to be forced by meteorological transients. During March 21, 1974, a developing cyclone moving up the coast excited a coherent group of sea level oscillations with characteristic periods of 5-7 hours, which are interpreted here as coastal-trapped edge waves. Spectra of the nontidal pressure series are red, however; most of the nontidal variability is caused by lower-frequency (subtidal) components. The subsurface pressure (SSP) fluctuations do appear coherent over the spatial extent of the array in the most energetic subtidal frequency bands, and estimates made of the relative horizontal SSP gradients indicate that cross-shelf gradient variations are significantly larger than alongshore gradient variations. Some consequences of these large weather-induced gradient fluctuations on the shelf circulation are discussed.

148 - Beardsley, R. C., W. Boicourt, L. C. Huff, and J. Scott. 1977. CMICE 76: A current meter intercomparison experiment conducted off Long Island in February-March, 1976. Woods Hole Oceanographic Institute Technical Report. Woods Hole, MA. 123p.

Abstract: A current meter intercomparison experiment (called CMICE 76) was conducted about 6 km off the southern coast of Long Island near 40° 47'N, 72° 30' W during February and March, 1976. A total of 20 current meters were deployed on 6 moorings set in a roughly linear array parallel to the local coastline and topography. The instruments included the Aandeaa RCM-4, the AMF VACM, the Brookhaven National Laboratory spar buoy system utilizing cylindrical and spherical Marsh-McBirney electromagnetic sensors, the EG&G 850 and CT-3, and the Chesapeake Bay Institute modified ENDECO 105. Local mean water depth was 27.8 m and current meters were clustered near 4 depth levels (3.5 m, 7.4 m, 15.7 m, and 25.0 m). Wave data was also obtained at the array site and 10 m wind and tidal data was obtained from nearby coastal stations. Intercomparisons of 1 hr vector average velocities measured with similar instruments deployed near the same depth level indicated sufficient horizontal homogeneity that most differences in the observed current data have been attributed to real differences in instrument and mooring characteristics and performance, and the effect of surface wave and wave-induced mooring motion on different measurement systems are presented.

149 - Beardsley, Robert C. and John Hart. 1978. A simple theoretical model for the flow of an estuary onto a continental shelf. Jour. of Geophy. Res. 83(C2):873-883.

Abstract: A simple theoretical model is developed to describe the steady flow of an estuary onto an adjacent continental shelf. A two-layer density stratification is assumed for the shelf water, and the fluid motion is driven by the positive (upper layer) and negative (lower layer) mass fluxes associated with a pair of point sources located at the mouth of the estuary. The dynamics are linear and include the effects of Coriolis acceleration, turbulent friction, and bottom topography. Analytic solutions for the one-layer single-source problem are found for two special depth profiles. A boundary layer solution is found for the power law depth profile h(y) = ay*, and a global solution is found for the log law profile $h(y) = H_o$ [1+ alpha $Ln(y/L_b)$] for small alpha << 1. Both solutions indicate that the far-field flow is asymmetrically concentrated toward the right-hand coast in the northern hemisphere, a consequence of the basic balance between topographic vortex stretching and bottom friction. This mechanism also applies when a constant along-shore current is present. In the two-layer case the flow in the upper layer generally tends to be concentrated toward the left-hand coast, since the upper fluid feels (1) the interface and not the bottom topography and (2) the interfacial drag exerted by the lower fluid toward the right-hand coast. A brief comparison is made between model predictions and observations for the Hudson and Chesapeake Bay estuaries.

150 - Bechtel Corporation. 1969. Bulk transport of waste slurries to inland and ocean disposal sites. (Summary). Bechtel Corporation. San Francisco, CA. 32p.

151 - Bechtel Corporation. 1969. Bulk transport of waste slurries to inland and ocean disposal sites - Volume I - The waste management concept - Inland and ocean disposal of selected wastes. Bechtel Corporation, San Francisco, CA. 156p.

152 - Bechtel Corporation. 1969. Bulk transport of waste slurries to inland and ocean disposal sites. Volume II. Criteria for waste management. Bechtel Corporation. San Francisco, CA. 166p.

153 - Bechtel Corporation. 1969. Bulk transport of waste slurries to inland and ocean disposal sites. Volume III. Technical aspects of pipelining of waste materials. Bechtel Corporation. San Francisco, CA. 208p.

154 - Bell, L. H. 1971. Effect of low pH on the survival and emergence of aquatic insects. Water Research. 5:313-319.

155 - Belton, Kathleen Elizabeth. 1979. Phytoplankton Distribution in New York Harbor. Ph.D. Dissertation. State Univ. of New York, Stony Brook, NY. 119p.

Abstract: Relative abundance of Phytoplankton distribution in New York Harbor was analyzed by intensive field sampling at 10 locations. Differential cell counts were used to determine the distribution succession, diversity and composition relative to the resident phytoplankton. Speciation was analyzed versus salinity and freshwater species were not significant to Harbor production. Phytoplankton densities were found to decrease progressively upon approaching the harbor proper. There is a winter peak of diatoms which is succeeded by chlorophytes and dinoflagellates in the summer.

156 - Benninger, L. K., D. M. Lewis and K. K Turekian. 1975. The use of natural Pb-210 as a heavy metal tracer in the river - estuarine system. In: Marine Chemistry in the Coastal Environment. American Chem. Soc. Sympsosium Series. No. 18:202-210.

157 - Benninger, L. K., R. C. Aller, E. P. Dion, and K. K. Turekian. 1977. Pb-210 distribution in sediments of Long Island Sound. EOS: Trans. Amer. Geophy. Union. 58(6):422.

Abstract: Application of excess Pb-210 Chronology to muddy sediments from Long Island Sound yields apparent sediment accumulation rates in the range 0.1-0.7 cm/y. Actual rates of sediment accumulation are probably much less variable, since long-term rates based on the total thickness of mud accumulated during post-glacial time lie between 0.05 and 0.10 cm/y. There is a direct correlation between the standing crop of excess of Pb-210 in each core and the Pb-210 apparent sediment accumulation rate, however. As this does not appear to be the result of real differences in sediment accumulation, it must be due to mixing processes. Two mixing regimes are distinguishable. The top few cm of the mud are everywhere rapidly reworked, tending to homogenize excess Pb-210 in surface sediment. Tidal currents tend generally to homogenize the suspended matter over the entire mud bottom of the Sound. High apparent sediment accumulation rates and large standing crops of excess Pb-210 are the consequences of efficient burial of excess Pb-210 below the surface zone of high-frequency disturbance. Deep dwelling organisms are the agents of this transfer. Metals and artificial radionuclides introduced into Long Island Sound can be expected to follow the route of Pb-210 on a long time scale even though the surface layers show the same concentrations.

158 - Bennett, John R., and Bruce A. Magnell. 1979. A dynamic analysis of currents near the New Jersey Coast. Jour. of Geophy. Res. 84(C3):1165-1175.

Abstract: A numerical model is used to analyze currents measured on the continental shelf near the shore of New Jersey. The model neglects longshore variations of current and all variations of density, but includes inertial accelerations and a nonlinear eddy viscosity. Local wind stress, sea level changes, and a constant longshore pressure gradient are the forcing terms. The model successfully reproduces most of the current variance; however, the predicted currents do not exhibit the dominant 4-hour response time of the observed currents, and the model sometimes misses energetic current events. These differences are ascribed to three-dimensional setup effects elsewhere in the New York Bight.

159 - Berg, G. 1975. Regional problems with sea outfall disposal of sewage on the coasts of the United States. Progr. Water Technol. Suppl. Paper No. 3. pp. 17-24.

160 - Berge, G., R. Ljoen and K. H. Palmork. (Unpublished manuscript). Containers with industrial waste into the North Sea. A problem to fisheries.

161 - Bergh, A. K., and R. S. Peoples. 1977. Distribution of polychlorinated biphenyls in a municipal wastewater treatment plant and environs. Sci. Total Environ. 8:197-204.

Abstract: Distribution of PCBs in sewage wastes at a municipal sewage treatment plant was studied, revealing most PCBs entering the plant become adsorbed onto the grit chamber solids and the sludge that is passed through from anaerobic digesters. When high concentrations of PCBs were present, significant amounts passed through with the effluents discharged from the plant. In secondary and tertiary plants, PCB concentrations were consistent with the limited solubilities of the PCBs. However, high concentrations were present in the sediments near the discharge and bioaccumulation in the fish does occur. Also described are quantitative data on PCBs in soils fertilized with PCB-contaminated sludge.

162 - Bergman, Edward F., and Tomas W. Pohl. 1975. A geography of the New York metropolitan region. Kendall/Hunt Publishing Company. Dubuque, IA. 205p.

Abstract: Using various working definitions for the New York Metropolitan Region, the book describes the development from 1626 to 1975. Includes population growth, ethnic and social patterns, transportation, economy; industry. Chapter 10 describes the demands on the environment, which includes air quality, water needs, sewage sludge disposal, and energy needs.

163 - Bernhard M. and A. Zattera. 1975. Major pollutants in the marine environment, metals; pesticides; PCB's hydrocarbons; detergents; radionuclides. In: E. G. Pearson and E. de Fraja Frangipane (editors), Marine Pollution and Marine Waste Disposal. (Supplement to Progress in Water Technology.) Proceedings of the 2nd International Congress, San Remo, 17-21 December, 1973. Pergamon Press. New York, NY. pp. 195-300.

164 - Berrien, Peter L. 1978. Eggs and larvae of Scomber scombrus and Scomber japonikus in continental shelf waters between Massachusetts and Florida. Fishery Bulletin: 76(1):95-115.

165 - Berrien, P. L., M. P. Fahay, A. W. Kendall, Jr., and W. G. Smith. 1978. Ichthyoplankton from the RV Dolphin Survey of continental shelf waters between Martha's Vineyard, Massachusetts and Cape lookout, North Carolina, 1965-1966. Sandy Hook Laboratory (NMFS/NOAA) Technical Series Report No. 15. Sandy Hook, NJ. 152p. 2 parts. pp. 1-67 and 68-152.

Abstract: Data are tabulated on ichthyoplankton and concomitant physical conditions collected during a survey of Atlantic continental shelf waters. Sampling information and laboratory procedures are described. Numbers and lengths of 87 species of larval fishes are tabulated by station; egg catches for 9 of these are included. Additionally, the presence of 79 categories of larvae, presently identified only to genus or a higher taxon, is noted by station.

166 - Betzer, P. R. A study of the sources, transport and reactions of suspended particles in waters of the New York Bight. NOAA TM ERL MESA-23. Boulder, CO. 39p. NTIS No. PB 282023/AS.

Abstract: Eighty-six samples of suspended matter collected on five cruises to the New York Bight have been subjected to chemical analyses. The results demonstrate that some of the major inputs of particulate matter to this system can be chemically typed and then traced. Seasonal changes in the distribution and composition of suspended materials are related to 1) river input, 2) diatom productivity, 3) sediment-water interactions, and 4) the dumping of anthropogenic materials.

167 - Bezanson, D. S., C. M. Moyse, and S. C. Byers. 1979. Research and related work on ocean dumping. An annotated bibliography. (Ocean Dumping Report 2). Department of Fisheries and Oceans, Ottawa, Ontario. 94p.

168 - Biggs, D. C., R. G. Rowland, H. B. O'Connors, Jr., C. D. Powers, and C. F. Wurster. A comparison of the effects of chlordane and PCB on the growth, photosynthesis, and cell size of estuarine phytoplankton. Environ. Pollut. 15(4):253-263.

Abstract: Repeated, daily doses of chlordane (a chlorinated insecticide, $C_{10}H_6Cl_8$) or Arochlor 1254 (a mixture of polychlorinated biphenyl compounds, PCB) at 5 or 10 parts per thousand million (ug/liter) reduced growth and

^{14}C-fixation per unit of chlorophyll a in mixed species, estuarine phytoplankton communities. Inhibition by chlordane was short-term (24 to 48 hours), and community cell-size distribution was not substantially altered. PCB, however, caused long-term effects, inhibiting the growth of phytoplankton larger than 8 um in equivalent spherical diameter (i.e., large centric diatoms) more strongly than that of smaller forms (pennate diatoms and microflagellates) and thus shifting community composition in favor of small-size algae.

169 - Birnie, Richard W., and Eric S. Posmentier. 1980. Identification of lateral spectral contrasts in the lower Hudson River Estuary using Landsat digital data. In: Proceedings of the Fourteenth International Symposium on Remote Sensing of Environ., 23-30 April 1980, San Jose, Costa Rica. Environ. Res. Instit. of Michigan, Ann Arbor, MI.

Abstract: An analysis of Landsat digital data has allowed us to detect and map lateral spectral contrasts in the Hudson River Estuary. These contrasts are believed to be related to turbidity in the Estuary with the western side being more turbid. It is clear that Landsat digital data can be mapped and used as an indicator of circulation and dispersion in the Hudson River Estuary.

170 - Bisagni, J. J. 1976. Passage of anticyclonic Gulf Stream eddies through Deepwater Dumpsite 106 during 1974 and 1975. NOAA/ODP Dumpsite Evaluation Report 76-1. U.S. Dept. of Commerce. Rockville, MD. 39p.

Abstract: The formational theory and characteristics of anticyclonic Gulf Stream eddies in the western North Atlantic Ocean are summarized. Their occurrence, trajectories, and structure are observed during XBT data and infrared satellite imagery to relate their passage through Deepwater Dumpsite 106 off the New York Bight and over the continental slope and rise. The frequency with which these eddies pass through the dumpsite, their residence times within the dumpsite, and the total time they occupy the dumpsite each year are computed to determine their potential effects on ocean dumping.

171 - Bisagni, James J. 1977a. Physical oceanography of Deepwater Dumpsite 106 February-March, 1976. In: Baseline Report of Environmental Conditions in Deepwater Dumpsite 106. NOAA/ODP Dumpsite Evaluation Report 77-1, Rockville, MD. Vol. 1. pp. 87-115.

Summary: The February 1976 characterization cruise to Deepwater Dumpsite 106 and vicinity provided physical oceanographic data from a warm core anticyclonic Gulf Stream eddy which had passed through the dumpsite in December 1975 and January 1976. The eddy was located with the aid of infrared satellite imagery and surveyed using standard physical oceanogrphic instruments. The warm, saline core of the eddy, located off the Virginia Capes in February was measured to be approximately 20 nautical miles in diameter. Depression of the 13°C isotherm below 200 meters and salinities greater than 35.6 ^{o}/oo were indicative of the eddy's core. The physical data together with the calculated geostrophic velocities of 0.5 knots or less indicated that the eddy was very weak at the time of the survey.

More intriguing than the eddy, however, were parcels of eddy-like water which appeared nearby and also within the confines of the dumpsite in the Slope Water region. These were identified at stations 14, 16, and 21 through analysis of temperature-salinity diagrams and comparison with temperature-salinity diagrams from the eddy. Ejection of water parcels from warm-core Gulf Stream eddies may represent an important large scale decay mechanism by which the eddies decrease in size as they move west and southwest, just beyond the edge of the continental shelf.

Future dumping at Deepwater Dumpsite 106, if continued, should be considered within the context of such large-scale physical processes and also the large physical variations which seem to be the rule rather than the exception within the Slope Water region of the western North Atlantic Ocean.

172 - Bisagni, James J. 1977b. Deepwater Dumpsite 106 bathymetry and bottom morphology. In: Baseline Report of Environmental Conditions in Deepwater Dumpsite 106. NOAA/ODP Dumpsite Evaluation Report 77-1. Rockville, MD. Vol. 1. pp. 1-8.

173 - Bisagni, James J., Steven W. Congdon, and Keith A. Hausknecht. 1977. A summary of the input of industrial waste chemicals at Deepwater Dumpsite 106 during 1974 and 1975. In: Baseline Report of Environmental Conditions in Deepwater Dumpsite 106. NOAA/ODP Dumpsite Evaluation Report 77-1. Rockville, MD. Vol. III. pp. 487-498.

174 - Bisagni, James J. 1981. Physical oceanographic studies at 106-Mile site, July 1977. In: Assessment Report on the Effects of Waste Dumping in 106-Mile Ocean Waste Disposal Site. NOAA/ODP Dumpsite Evaluation Report. 81-1. Rockville, MD. pp. 233-242.

175 - Bisagni, James J., and Dana R. Kester. 1981. Physical variability at an east coast United States offshore dumpsite. In: B. H. Ketchum, D. R. Kester and P. K. Park (editors), Ocean Dumping of Industrial Wastes. Plenum Press. New York, NY. pp. 89-107.

Abstract: Three hydrographic stations conducted during a 1977 cruise to Deep Water Dumpsite 106 inadvertently sampled an established anticyclonic Gulf Stream ring both prior to and after it interacted with the Gulf Stream. Continuous STD data and discrete dissolved oxygen measurements in the ring were analyzed using temperature-salinity (T-S) and temperature-oxygen (T-O_2) diagrams. These data showed good correlation with T-S and T-O_2 diagrams from Gulf Stream, Sargasso, and Slope Waters obtained during a cyclonic Gulf Stream ring study. Apparently, a new entrainment of Gulf Stream Water around the ring occurred within the sampling period and was manifested by increased salinity and decreased levels of dissolved oxygen. Combined satellite surveillance and at-sea measurement of temperature, salinity, and oxygen provide an accurate method of describing this highly dynamic and variable region in and around Deep Water Dumpsite 106.

176 - Bisagni, James J. 1981. Lagrangian measures of near-surface waters at the 106-Mile dumpsite. NOAA TM OMPA-11. Boulder, CO. 23p.

Conclusions: Based on these few Lagrangian studies, a slow southwesterly drift of the near-surface water from the 106 Mile Dumpsite seems apparent. This slow, perhaps wind-driven, current was measured to be less than 0.5 knots (25 cm/sec) and is somewhat consistent with Bumpus' (1972) drift bottle data for the shelf region. The drift of waters originally located in the dumpsite to a position located off Cape Hatteras thus requires about one month's time.

Onshore flow of subsurface Slope Water of close to 1.0 knot from the dumpsite was observed during a period when offshore Ekman flow of surface Shelf Water caused a stratified condition to occur in late winter.

Both types of wind-driven currents appear, however, to be perturbed aperiodically by the swifter geostrophic currents (2.0 knots) associated with anticyclonic Gulf Stream (warm-core) rings. Observations from such an episode showed strong northeasterly drift. Waters entrained in such currents could be swiftly advected away from the 106 Mile Dumpsite.

The current regime at the 106 Mile Dumpsite is extremely variable with speeds spanning at least one order of magnitude and widely different current directions. This current variability is attributable to the location of the site itself (i.e., its proximity to strong oceanic density fronts, the movements of which are not well understood) and the seasonal changes in both the near-surface waters and the associated wind regimes.

177 - Biscaye, Pierre E., and Curtis R. Olsen. 1976. Suspended particulate concentrations and compositions in the New York Bight. In: M. G. Gross (editor), Amer. Soc. Limnol. and Oceanogr. Spec. Symp. 2:124-137.

Abstract: Suspended particulate concentrations in the New York Bight during 1973-1975 decrease seaward across the shelf, decreasing more rapidly in surface than in bottom waters. Resuspension of fine-grained sediments causes local high concentrations of suspended particles in near-bottom waters. Horizontal displacement of these high concentration plumes from their sources suggests short residence times for suspended particles. Vertical mixing of resuspended particles is limited by the thermocline. Along the upper continental slope over a depth range of approximately 1,000 m, there is a minimum in near-bottom suspended particulate concentrations of trace metals of anthropogenic origin associated with organic particles, Fe-Ti (oxide) coatings and discrete Ti (oxide) particles are potential tracers of particle dispersion paths and transport processes. Different types of organic particles exhibit different interparticle and trace element associations and appear to have different geographic distributions. Some of this variability may be seasonal. Surface-water-suspended matter has a higher proportion of biogenic (inorganic skeletal as well as organic) particles than near-bottom suspended matter which is dominantly nonbiogenic (principally aluminosilicate). Skeletal debris is primarily siliceous in shelf waters, becoming more carbonate-rich seaward of the shelf break. Aluminosilicate-suspended particles in shelf waters are predominately K-rich whereas Mg-Ca-K-Fe-rich aluminoscilicates dominate beyond the shelf break.

178 - Bischoff, J. L., and R. J. Rosenbauer. 1978. Chemical changes in seawater caused by resuspension of deep sea sediments from DOMES sites A, B, and C. (metals). Mar. Min. 1(4):283-304.

179 - Bishai, H. M. 1960. The effect of hydrogen ion concentration on the survival and distribution of larval and young fish. Sonderdruck aus "Zeitschrift fur wissenschaftliche Zoolgie". Bd 164, Heft 1/2.

180 - Bishop, J. M. 1975. An analytical sea current model for coastal regions with application to the New York Bight. Dept. of Transportation. U.S. Coast Guard Oceanographic Unit Technical Report 75-2. 27p.

Abstract: Seasonal coastal currents on a continental shelf are modeled for use in Search and Resource planning. The model considers a balance of Coriolis, pressure gradient, and frictional forces. Input parameters are the climatological wind and density fields. Comparisons of results to currents depicted on climatological atlases for the New York Bight indicates the validity of the approach. In this light, one might extend this approach to other geographical regions where anologous oceanographic conditions preveil.

181 - Bishop, W. P. 1975. Seabed disposal: Where to look. Sandia Labs. Nuclear Science Abstracts. 31(2):417.

182 - Blom, B. E., T. F. Jenkins, D. C. Leggett, and R. P. Murrmann. 1976. Effect of sediment organic matter on migration of various chemical constituents during disposal of dredged materials. DMRP contract D-76-7. U.S. Army Cold Regions Research and Engineering Laboratory. Hanover, NH. 161p.

Abstract: The disposal of both freshwater and saltwater sediments in open waters has been experimentally investigated in the laboratory. The emphasis of the experimental work was directed towards identifying the role that sediment organic carbon has in controlling various water-quality parameters at disposal sites.

Long-term studies examined the possibility of transport of material from sediments into an overlying water column as well as monitoring the changes in the aqueous phase which initially contained large amounts of suspended matter. For the latter case, metal and nutrient concentrations were initially high, but decreased significantly with time. Exceptions to this behavior were noted.

Specific components were identified which migrated from the sediment into the water column. They include ammonium-nitrogen, orthophosphate, cadmium, and manganese, the latter only is seawater media. Organic carbon and inorganic nitrogen underwent transformations with the systems investigated.

Both sediment organic carbon and soluble organic matter were generally found to have no demonstrable effect on water quality. Exceptions to this general conclusion were found in two sediments containing significant amounts of petroleum hydrocarbons.

183 - Blum, R. 1968. Considerations techniques sur la realisation d'emissaires d'eaux residuaires en mer. Rev. Int. Ocenogr. Med. 9:123-126.

184 - Boehm, P. D., D. L. Fiest, and A. Elskus. 1979. Organic chemical measurements-dredged material tracking experiment. ERCO Final Report - Contract No. NA-79-RAA-03401. Submitted to NOAA MESA-New York Bight Project. Cambridge, MA. 63p.

Abstract: A description of a field sampling and analytical program to examine time-dependent changes in the organic chemical composition of the water column. The aims of the program were to design a sampling rapid on-board filtration system to obtain suspended particulate material; to determine the levels of polynuclear aromatic hydrocarbons and PCB in dredge spoil, pre-dump and post-dump water samples; and to use these organic tracers to determine the residual time of the dredge spoil plume in the water column.

185 - Boesch, Donald F. (In Press). Consequnces of alternations of benthic community structure and function in the New York Bight region. In: G. F. Mayer, A. Calabrese, F. A. Cross, D. C. Malins, D. W. Menzel, J. S. O'Connor, and F. J. Vernberg (editors), Ecological Stress in the New York Bight: Science and Management. Estuarine Research Federation. Columbia, SC.

Abstract: Structure and function are divorced in most studies of the ecology of benthos; furthermore, the interaction of the benthos with the remainder of the ecosystem is seldom considered. Data on the structure and function of benthos in the New York Bight region are drawn together, and inferences are made about the role of benthos in these ecosystems and the consequences of human alterations of benthic communities.

Alterations to the macrobenthic communities in the Hudson-Raritan estuary, the Christiaensen Basin, and upper Hudson Shelf Valley have resulted in reductions in populations of species that are predominant food items for demersal fishes and invertebrates. The Christiaensen Basin supports a dense and productive benthic community, but little of this productivity appears to be transfered to high trophic levels. Energy flow through the seabed has been increased due to waste additions, but this may result in localized depletion of oxygen in botton waters during the summer when water masses are stratified. Nutrients regenerated from the seabed do not appear to be influential in supporting the high primary productivity in the Bight Apex, and the seabed may actually be an important nutrient sink. Benthic biological processes affecting toxicants concentrated in sediments may determine their availability to benthic and demersal organisms, but are probably not responsible for significant reintroduction of these toxicants into the water column, except as they affect resuspension of bottom sediments.

186 - Bogost, M. S. 1973. Hawaii's experiment with ocean disposal of baled wastes. Mar. Technol. Soc. Jour. 7:34-37.

187 - Bokn, T., J. Knutzen, and B. Rygg. 1974. Influence of freshwater, industrial waste and domestic sewage on bottom fauna and benthic algae in the Hvaler Archipelago (S.E. Norway). Manuscript of a paper presented at the "Symposium on the Influence of Fresh-Water Outflow on Biological Processes in Fjords and Coastal Waters", Geilo (Norway).

188 - Bokuniewicz, H. J., J. Gerbert, and R. B. Gordon. 1976. Sediment mass balance of a large estuary, Long Island Sound. Estuar. and Coastal Mar. Sci. 4:523-536.

Abstract: Acoustic reflection profiles and bottom sampling are used to measure the volume of sediments accumulated in Long Island Sound. There is present 1.0×10^{10} m^3 of sediment of which 5.3×10^9 m^3 is marine mud and 4.9×10^8 m^3 is probably of pre-marine, lacustrine origin. The balance consists of reworked sand derived from glacial drift.

189 - Bokuniewicz, H., M. Dowling, J. Gebert, R. Gordon, P. Kaminsky, C. Pilbeam, and C. Tuttle. 1977. Aquatic disposal field investigations, Eaton's Neck disposal site, Long Island Sound. Appendix A: Investigation of the hydraulic regime and the physical characteristics of bottom sedimentation. Dredged Material Research Program, Tech. Rept. D-77-6. Environmental Effects Laboratory. U.S. Army Engineer Waterways Experiment Station. Vicksburg, MS. 108p.

Abstract: The major goal of the Eatons Neck disposal site field investigation was to evaluate the effects of aquatic disposal of dredged material on organisms and water quality, including the significance of physical, chemical, and biological factors that influence the rate of disposal site recolonization by benthic animals. A comprehensive research program was planned and conducted at Eatons Neck in order to evaluate cause-and-effect relationships associated with the impacts of open-water disposal. This volume of the study presents the investigation of the hydraulic regime and the physical characteristics of bottom sedimentation. Acoustic-reflection profiles and mechanical analysis of core and grab samples of the bottom were used to define the sediment-type distribution of the area. Results were confirmed by penetrometer tests and bottom and profile photographs. Since the study was made to determine the possible movement of dredged material placed on the bottom, the currents were studied using both averaging and instantaneous recording current meters. The study concluded that there was no physical evidence of significant dispersion of dredged material from the Eatons Neck disposal site; no previously deposited material was detected outside the designated disposal area. Appendix A details navigation procedures and permanent buoy construction used in the study. Appendix B are the 16 reflection-profile photos. Sediment analysis data are presented in Appendix C. Appendix D shows the profile camera photos and Appendix E shows the bottom photos.

190 - Bokuniewicz, H. J., and C. T. Fray. 1979. The volume of sand and gravel resources in the lower bay of New York Harbor. Special Report 32. Marine Sciences Research Center. State Univ. of New York, Stony Brook, NY. 34p.

Abstract: The shallow (<100 ft) stratigraphy of the Lower Bay floor was investigated in order to estimate the volume of sand and gravel deposits under the Lower Bay. Four types of information were studied. These were: (1) core and boring logs, (2) seismic reflection profiles, (3) the surficial sediment distribution on the floor of the Lower Bay, (4) the stratigraphy in Long Island, Staten Island, northern New Jersey, and the New York Bight. In general, marine sands overlie glacial outwash sands which, in turn, overlie unconsolidated Cretaceous sediments. Along the margins of the Bay, sands are known to rest on fine-grained deposits at depth, the composition of layers underlying the surficial sand deposits in the central and eastern Bay is unknown. Surficial mud deposits are confined primarily to Raritan and Sandy Hook bays. They may be as much as 150 feet thick. Sand deposits were identified that have a total volume of 3,429 million cubic yards.

191 - Bokuniewicz, H. J., and R. B. Gordon. 1979. Containment of particulate wastes at open-water disposal sites. In: H. G. Palmer & M. G. Gross (editors), Ocean Dumping and Marine Pollution. Dowden, Hutchinson and Ross, Inc. Stroudsburg, PA. pp. 109-129.

Abstract: The degree of retention of particulate silt-clay size waste at open-water disposal sites depends primarily on the sedimentary regime at the site and the height of the waste deposit. In estuarine, lacustrine and coastal localities particulate waste released at the water surface is found to deposit on the bottom within a radius of about 200 m from its impact point. This radius is insensitive to water depth and the volume of material released. After deposition the waste deposit self-consolidates and expels interstitial water. Bio-aggregation decreases the erosion resistance of the deposited material; bioturbation keeps a surficial layer of the waste exposed to the ambient water. As the deposit is made higher, it becomes more susceptible to disturbance by waves and currents and may itself alter the local hydraulic flow field. A necessary condition for the retention of particulate waste at a disposal site is that it be below the wave-affected zone. In marine localities the tidal stream will disperse waste placed in a deposit rising above natural topographic features of the bottom unless its surface is armored by erosion-resistant material or the natural sedimentation rate exceeds a critical value that depends on pile height.

These factors are evaluated for dredged material disposal sites in Long Island Sound. Silt-clay is found to be dispersed at the Cornfield Point site, where the natural sedimentation rate is zero, but retained at the New Haven site, where natural sedimentation is 0.8 kgm/(m^2yr). At the latter site 1.2 x 10^6 m^3 of dredged material was placed in a conical mound 8 m high. Self-consolidation occurred over several weeks and caused an outward advection velocity of interstitical waters as great as 5 x 10^{-5} cm/sec. The top of the waste pile penetrates the wave-affected zone but, because it is armored with sand and clods of cohesive silt, it is not dispersed. The lower part of the waste pile is unarmored. Tidal resuspension of the neighboring muddy bottom and the waste deposit occurs very tidal cycle and is increased by as much as deposit occurs every tidal cycle and is increased by as much as a factor of ten during severe storms. A layer of wastes less than 1 cm thick is intermixed with sediments from the neighboring sea floor within a few weeks after deposition is completed; subsequent entrance of wastes into the water column is to biological mixing in the top 10 cm of the waste deposit. The results show that permanent waste deposits may be constructed in coastal localities with favorable natural sedimentary regimes provided close control over the deposition process is attained.

192 - Bopp, R. F., H. J. Simpson, and C. R. Olsen. 1977. PCB's and Cs-137 in sediments of the Hudson Estuary. EOS: Amer. Geophy. Union Trans. 58(6):407.

Abstract: During the past two decades, significant quantities of both PCB's and Cs-137 have been introduced into the Hudson River. The two major sources of Cs-137 are global fallout from nuclear weapons testing and low level release from a nuclear reactor situated approximately 70 km upstream from the southern tip of Manhattan. The single largest source of PCB's in the Hudson River is the discharge from two capacitor plants located approximately 60 km upstream from Albany, N.Y.

Cs-137 is adsorbed to fine-grained particles and has been used to trace recent sedimentation in the Hudson Estuary. PCB's are also associated with fine-grained sediments and we have found excellent corelation between the amounts of PCB's and Cs-137 in sediment samples. The correlation suggests that the relatively easy Cs-137 measurement can be used to screen large numbers of smples to determine which ones would be appropriate for the much more difficult PCB analysis.

The preliminary results of our measurements indicate that recent sediments of the Hudson Estuary are contaminated with PCB's to levels of several ppm on a dry weight basis. This represents a severe contamination problem which has already resulted in the closing of the Hudson River to commercial fishing. Areas most affected include New York Harbor and marginal areas where much of the recent sedimentation takes place. The continuous dredging of the river channel and shoaling areas in the harbor has resulted in a significant transport of PCB's to the continental shelf where the dredge spoils are dumped.

193 - Bopp, R. F., H. J. Simpson, and C. R. Olsen. 1978. PCB's and the sedimentary record of the Hudson River. Lamont-Doherty Geological Observatory. Palisades, NY. 12p.

Abstract: Polychlorinated biphenyls (PCB's) are among the most important anthropogenic pollutants of the Hudson River system. The association of PCB's with recent, fine-grained sediments permits the use of ^{137}Cs distributions in the sediments as an indicator of PCB accumulation patterns. Sediments of the Hudson constitute a major, widely distributed source of PCB's to the water column. PCB transport in the river occurs both as an absorbed phase on suspended matter and associated with the water phase. PCB contamination of the Hudson has had a direct impact on the shelf because of significant transport associated with Hudson River water and with sediments dredged from New York Harbor.

194 - Bopp, R. F., H. S. Simpson, C. R. Olsen, and N. Kostyk. 1979. PCB's in Hudson River sediments. Lamont-Doherty Geological Observatory. Palisades, NY. 10p.

Abstract: The Hudson River has received a major point source discharge of polychlorinated biphenyls (PCB's) between the early 1950's and 1976 from the effluent of two General Electric capacitor and transformer manufacturing plants located approximately 60 km upstream of the southernmost dam on the river at Troy, N.Y. Contaminated water and suspended matter passing over the dam has been the single dominant source of PCB's entering the tidal Hudson for the past two decades.

This report presents a summary of the PCB data for the more than 200 sediment samples from the Hudson that we have analyzed over the past two years.

195 - Bostrom, R. C. and M. A. Sherif. 1970. Disposal of waste material in tectonic sinks. Nature. London. 228:154-156.

196 - Bostrom, R. C. 1977. Marine disposal of waste material: potential and limitations. In: N. T. Monney (editor), Ocean Resources Utilization. Winter annual meeting. American Society of Mechanical Engineers, New York, NY. December 5-10, 1976. Vol. 3. 23p.

197 - Botton, Mark L. 1979. Effects of sewage sludge on the benthic invertebrate community of the inshore New York Bight. Estuar. and Coastal Mar. Sci. 8:169-180.

Abstract: Benthic invertebrate community structure including species diversity, biomass, and trophic structure was examined at a sewage sludge disposal site and nearby 'control' site in the inner New York Bight off New York City. Although both sites are characterized by soft bottoms with high organic carbon, and support similar deposit feeding assemblages, species diversity is lower at the sludge site, reflecting decreased species richness and evenness. Likewise, total biomass as well as the proportional distribution of biomass among trophic groups, differed between sites. A previously suggested indicator species, *Capitella capitata*, is present in greater numbers at the control station than the sludge station. It is therefore unsuitable as an indicator species of pollution tolerance in the portion of the New York Bight examined in this study. A small polychaete belonging to the family Ampharetidae, *Amage auricula*, is by far the most abundant animal at the sludge disposal site, but its potential use as an indicator species is complicated by the fact that it exhibits a great tendency for aggregation. Ampeliscid amphipods appear sensitive to the presence of sludge and may be useful as indicators of contamination. Most of the common animals at both stations were aggregated, and it is recommended that future sampling in the area should be by the use of a larger number of small grabs such as the Shipek, rather than a smaller number of large grabs, as this will increase the probability that patchy fauna are encountered.

198 - Bourcier, M. and Zibrowius, H. 1973. Les "boues rouges" deversees dans le canyon de la Cassidaigne (region de Marseille). Obserations en soucoupe plongeante SP 350 (juin 1971) et resultats de dragages. Tethys 4:811-842.

199 - Bovee, E. C. 1973. A preliminary report on the amoebae of the Atlantic coast from Virginia to Massachusetts, including the New York Bight. Contract Report. U.S. Dept. Commerce. Natl. Mar. Fish. Serv. Sandy Hook, NJ. 19p.

200 - Bovee, E. C. 1979a. A preliminary report on amebas found in marine coastal waters of Virginia, New Jersey, and Massachusetts. Jour. Protozool. 26 (Pt. 1), 26A.

201 - Bovee, E. C. 1979b. A preliminary report on the amebas found in marine coastal waters of Virginia, New Jersey, and Massachusetts. Jour. of Protozoology. Vol. 26, No. 3, Pt. 1, p. 26A, #73.

Abstract: A survey of amebas found in collections from Chincoteague and Assateague Bays in Virginia, Sandy Hook Bay in New Jersey and Buzzard's Bay, Eel Pond, Salt Pond at Woods Hole, Massachusetts, produced a total of 17 genera including 26 already known species and 26 amebas that are unknown to the literature and are probably new spp. These included: *Pelomyxa* sp; *Saccamoeba sphaerarum*, *S. gumia*, 4 other *Saccamoeba* spp; 2 *Vahlkampfia* spp.; *Mayorella conipes*; *M. corlissi*; *M. smalli*, 5 other *Mayorella* spp.; *Oscillosignum* sp.; *Vexillifera telmathalassa* *V. otto*, *V. browni*, *V. pagei*, 2 other *Vexillifera* spp.; *Acanthamoeba gigantea*, *Acanthamoeba* sp.; *Striolatus tardus*; *Triaenamoeba jachowskii*; *Flabellula hoguei*, *Flabellula* sp; *Vannella sensilis*, *V. mira*, *Vannella* sp.; *Lingulamoeba leei*; *Platyamoeba langac*, *P. murchelanoi*, *P. weinsteini*.; *Clydonella rosenfieldi*, *C. wardi*, *Clysonella* sp.; *Unda maris*, *U. schaefferi*, *Unda* sp.; 3 unidentified spp. of *Striamoeba*; *Mastigamoeba schultzei*, an unidentified sp. of *Mastigamoeba*. All the *Mayorella* spp. (except *M. conipes*), the *Vexillifera* spp., the *Oscillosignum* sp., the *Acanthamoeba* spp., *Triaenamoeba jachowskii* and the *Mastigamoeba* spp. were found only from Chincoteague or Assateague Bays, suggesting adaptation to warmer waters. The other genera and most species of them are distributed at all three locales.

202 - Bowker, David E. and William G. Witte. 1977. The use of LANDSAT for monitoring water parameters in the coastal zone. Proceedings of the AIAA 1977 Joint Conf. on Satellite Applications of Marine Operations. November 15-18. New Orleans, LA.

Abstract: Landsats 1 and 2 have been successful in detecting and quantifying suspended sediment and several other important parameters in the coastal zone, including chlorophyll, particles, alpha (light transmission), tidal conditions, acid and sewage dumps, and in some instances oil spills. When chlorophyll a is present in detectable quantities, however, it is shown to interefer with the measurement of sediment. The Landsat banding problem impairs the instrument resolution and places a requirement on the sampling program to collect surface data from a sufficiently large area. A sampling method which satisfies this condition is demonstrated.

203 - Bowman, M. J. 1975a. Pollution prediction model of Long Island Sound. Proceedings of the Specialty Conference on "Civil Engineering in the Oceans/III." June 9-12, 1975. Amercian Society of Civil Engineers, University of Delaware, Newark, DE. pp. 1084-1103.

Abstract: A one-dimensional, steady state, numercial mass balance model suitable for the study of sewage effluent dispersion in Long Island sound is presented. The predicted values of contaminant concentrations and transport for both summer and winter conditions reflect the combined effects of dispersion by tidal mixing and gravitational convection, advective transport, net addition across the lateral boundaries, and first-order biochemical decay.

High longitudinal (east-west) concentration gradients of inorganic nutrients exist in western Long Island Sound and are shown to be primarily due to a large influx of sewage effluent from the East River, which connects the Sound to New York Harbor. Winter nutrient transport is primarily eastward throughout the Sound, but seasonal reversals in direction occur in the eastern regions, due to large increases in the biochemical decay rates during the summer. Nutrient input from agricultural sources via runoff is shown to be negligible compared to that transported by sewage effluents.

204 - Bowman, M. J. 1975b. Scales of variability of the non-tidal Characteristics of the East River, New York. Trans. Amer. Geophys. Union, Abstr. Publ. EOS. 56(6):382.

Abstract: The East River is a cooscillating tidal strait approximately 25 km long that connects New York Harbor to Long Island Sound. The tide dominated by semidiurnal components, with a mean tidal flux of approximately 2,800 m^3/sec. Historical evidence also exists of a highly variable nontidal transport. Fourteen estimates of this net flux made at different times over the lst 103 yr are reviewed. Values range from +22% to -40% of the mean tidal flux (a positive flux is considered toward the Harbor). A 5-d experiment of transport measurements was made by NOS in Oct. 1959 with 12 current meters strung in a linear array across the river. The nontidal flux computed from these current meter records shows a significant correlation with the concurrent nontidal hydraulic head computed from hourly NOS tide gage data gathered at the Battery and Willets Point. Hourly computations of the nontidal hydraulic head were also made from the NOS tide gage data for 1 yr beginning July 1, 1958. The resulting time series has the essential characteristics of band-limited random noise. The SD of the fluctuations is approximately ±5 cm, which is approximately 5% of the mean tidal hydraulic head. Frequent reversals in the sign of the nontidal head are observed. Successive zero crossings occur over periods ranging from 2 to 40 d. An investigation of the monthly mean hydraulic head spectrum over the 1941-59 epoch was also made. The spectrum has significant peaks at 1 yr^{-1} and several other lower frequencies.

205 - Bowman, M. J. and L. D. Wunderlich. 1976. Distribution of hydrographic properties in the New York Bight Apex. In: M. G. Gross (editor), Amer. Soc. Limnol. and Oceanogr. Spec. Symp. 2:58-68.

Abstract: Seasonal hydrographic cycles are interpreted through a series of isometric block diagrams and planimetric projections illustrating the distributions of temperature, salinity, and density in the Bight Apex. Seasonal cycles are typical of those found in coastal seas near river mouths in drowned river valleys in temperature latitudes. The prevailing southwest coastal drift, the right-angle bend of the coastline, seasonal and short term wind patterns, and the presence of the Hudson Shelf Valley, all influence the nearshore circulation.

206 - Bowman, M. J. 1976a. Tidal locks across the East River: an engineering solution to the rehabilitation of western Long Island Sound. Estuarine Processes VI. Academic Press. New York, NY. pp. 28-43.

Abstract: Water quality in western Long Island Sound and New York Harbor is seriously degraded. A major source of pollutants is sewage released into the East River, a cooscillating tidal strait connecting the Sound to the Harbor. Very rapid and significant improvements in water quality could be attained by constructing ship locks across the Upper East River to increase the circulation of the sea through the Harbor and Sound. During ebb tide these locks would be opened, allowing an unhindered flow of Sound water into the Harbor. After six hours the locks would be closed at slack water, blocking the following flood tide from re-entering the Sound. The net result would be a strong pulsating unidirectional flow (approximately 2500 m^3 sec^{-1}) of relatively clean central Long Island Sound water, pumped by the semidiurnal tides through New York Harbor, and the Lower Bay, out into the New York Bight.

Simple models indicate that the concentration of conservative contaminants in the western Sound and the Harbor would drop by approximately 88% and approximately 45%, respectively, from present levels, within a month of operation. The accompanying decreases in inorganic nutrient concentrations are calculated and tabulated for both winter and summer conditions.

The major physical effects of blocking Hudson River water from entering the Sound through the East River would be to change the essential estuarine characteristics of western Long Island Sound to those of a coastal embayment, and increasing the salinity of the western Sound and New York Harbor both by approximately 4°/oo.

207 - Bowman, M. J. 1976b. The tides of the East River, New York. Jour. of Geophy. Res. 81:1609-1616.

Abstract: The East River, New York, is a cooscillating tidal strait connecting New York Harbor to western Long Island Sound. Linear wave theory is applied to the river in an attempt to interpret the tidal characteristics. The theory accurately describes the observed longitudinal variation in amplitude of the tidal currents, but the simple model used does not adequately predict the amplitudes of the tidal heights or the overtides. The frictional resistance to flow through the river is characterized by an effective value of De Chezy's coefficient of approximately 26 m^{12} s^{-1}. Tidal current records gathered over a 5-day period from a linear array placed across the Upper East River have been analyzed to investigate the harmonic content of the tidal transport. The semidiurnal frequency and its third harmonic (M_6) are dominant.

208 - Bowman, M. J. 1976c. The hydrodynamic characteristics of the East River tidal strait, New York. Memoires Societe des Sciences de Liege. 6 serie. 10:165-174.

Abstract: The East River, New York, is a co-oscillating tidal strait, connecting New York Harbor to Western Long Island Sound.

Linear wave theory is applied to the River in an attempt to interpret the semidirunal tidal characteristics. The theory accurately describes the longitudinal variation in amplitude of the horizontal tides but does not adequately predict the amplitudes of the vertical tides.

The frictional resistance to flow through the River is characterized by an effective value of De Chezy's coefficient, approximately 26 $m^{1/2}$ sec^{-1}.

209 - Bowman, M. J., and L. D. Wunderlich. 1977. Hydrographic properties. MESA New York Bight Atlas Monograph 1. MESA New York Bight Project and New York Sea Grant Intitute. Albany, NY. 78p.

Abstract: Cycles of temperature, salinity, and density characteristic of the Middle Atlantic Bight, which includes New York Bight, are determined by seasonal patterns of insolation, river runoff, evaporation minus precipitation, winds, ocean currents, and shelf/slope exchange processes. Shelf temperatures have a large annual nearshore range around 25°C; the water column is unstratified in winter but is dominated by a strong thermocline in summer. A cool pool of bottom shelf water on the middle and outer continental shelf remains during the summer as a remnant of the previous winter and persists throughout the entire Middle Atlantic Bight.

The salinity of shelf water varies from year to year; these variations are often larger than the seasonal cycle. The principal salinity gradients are vertical and cross-shelf, salinities are lowest near shore. Salinities increase with depth on the shelf, but the vertical gradient reverses over the continental slope. Salinities are highest in winter and lowest in late summer. Total freshwater volume on the shelf equals an average of 18 months' discharge of river water.

The density field in winter is principally determined at all depths by salinity; patterns of surface density throughout the year are similar to those of surface salinity. The summer thermocline converts the winter's unstratified density field into one with strong vertical stratification.

The principal winter hydrographic feature of the outer shelf is a strong temperature/salinity front separating shelf water from slope water. During summer the thermocline eliminates any surface manifestation of the temperature front, but the salinity front persists throughout the year. A temperature/salinity inversion underlying the fronts is formed of shallow slope water and inclines shoreward, intersecting the shelf break at about 150m.

Shelf/slope exchange processes and slope frontogenesis are dominated by small-scale dynamics, vary greatly with time and space, and are difficult to measure. Gulf Stream meanders and eddies occasionally pass through the outer Bight, creating major disruptions in the hydrographic properties of the slope waters.

210 - Bowman, Malcolm, J. 1977. Nutrient distributions and transport in Long Island Sound. Estuar. Coastal Mar. Sci. 5:531-548.

211 - Bowman, M. J. 1978. Spreading and mixing of the Hudson River effluent into the New York Bight. In: J. C. Nihoul (editor), Hydrodynamics of Estuaries and Fjords. Elsevier. Amsterdam. pp. 373-386.

Abstract: Results are presented form three Hudson River plume sampling cruises made in the New York Bight, in August 1976. The data show that the set and shape of the spreading effluent vary widely over time periods approximately 6 days, and are clearly influenced by local wind stress. Application of Takano's model of a steady state plume spreading into a stagant ocean suggests a horizontal eddy viscosity 10^8 cm^2 sec^{-1}, and a strong anticyclonic deflection of the plume. This value is considered to be an overestimate, since interfacial shear stress is neglected in the model. More careful measurements and calculations are needed to separate out the effects of horizontal and vertical viscosities, Coriolis force, advection by a prevailing coastal current and local wind stress, on plume dynamics.

212 - Boyce Thompson Institute. 1977. An Atlas of the Biological Resources of the Hudson Estuary. Yonkers, NY. 104p.

Abstract: This atlas presents information on the abundance, frequency of occurrence, seasonal variation, and distribution of important species of plants and animals found in the Hudson estuary over a 120-km reach from New York Harbor to Poughkeepsie.

213 - Boyd, M. B. et al. 1972. Disposal of dredge spoil-problem identification and assessment and research program development. Technical Report H-72-8. U.S. Army Engineer Waterways Experiment Station, Vicksburg, MS. NTIS No. AD-757599.

214 - Brannon James M., Russell H. Plumb, Jr., and Issac Smith. 1978. Long term release of contaminants from dredged material. U.S. Army Corps of Engineers. DMRP T.R. D-78-49. Vicksburg, MS. 65p.

Abstract: Man's activities in the Nation's coastal zones and waterways have served to contaminate sediments in these areas, generating concern that dredging and disposal may exert adverse effects on water quality and aquatic organisms. The long-term effects of dredged material on water quality at the disposal site remain an area of particular concern.

To study the magnitude and predictability of long-term releases from dredged material, an 8-month aerobic leaching study was conducted in the laboratory. A large number of sediments (32) representing broad geographical and pollutional variation were used to ensure wide applicability of study results.

Under the aerobic chemical conditions likely to prevail at aquatic disposal sites, total organic carbon, orthophosphate-P, and zinc exhibited the most consistent net releases to the water column. Very little net mass release of mercury (Hg), lead (Pb), cadmium (Cd), or arsenic (As) into the water column was observed regardless of the composition of the sediments. Worst case evaluation of the potential effects of contaminant releases, conducted by comparing results of this study with the most stringent water quality criteria available, indicated that sediments used in this study would not be expected to cause significant long-term water quality problems.

The Elutriate Test procedure (modified for aeration) and interstitial water analyses demonstrated considerable utility as predictors of potential long-term net mass releases of contaminants from sediments. The Elutriate Test was related to long-term net mass release of copper (Cu), total organic carbon, phosphorus (P), As, Cd, Pb, and Hg after 4 months. Interstitial water analyses were related to net mass release of zinc (Zn), As, Hg, and ammonium-nitrogen (NH_4-N) after 4 months of leaching, and Hg, total organic carbon, and NH_4-N after 8 months of leaching. The ability of the Elutriate Test and interstitial water analyses to predict release decreased slightly as time after initial mixing increased, apparently due to the increased significance of the aquatic chemistry of each constituent.

Results from limited sediments incubated under both agitated and quiescent conditions indicated that mechanical agitation did not appreciably enhance long-term net mass release of most constituents. However, one sediment suspension from Oakland Inner Harbor, California, when incubated under aerobic, agitated conditions, exhibited a marked drop in pH from near 8.0 to 3.6. The acidic condition was accompanied by a high net release of trace metals. The same sediment, incubated under aerobic, quiescent conditions did not lower the overlying water pH and did not release significant amounts of trace metals. If disposed of in an upland and drained disposal site, such a sediment could oxidize, become acidic, and pose a potentially severe environmental problem.

Manganese (Mn) and NH_4-N exerted no long-term effects on water quality. Large amounts of Mn and NH_4-N were released during the Elutriate Test and presumably when the leaching columns were prepared. However, at the end of a 4-month incubation both constients were usually present in lower concentrations than in the initial disposal site waters. These results indicate that Mn and NH_4-N were being actively removed from the water column.

Consideration of net mass release during the Elutriate Test procedure and other short-term extractants can provide useful information on the probable direction (i.e., sediment to water, to sediment) of long-term net mass release when concern exists over possible contaminant releases at aquatic disposal sites. However, because the relationships are based on laboratory observations and have not been field verified, caution in their use is recommended.

215 - Bray, R. N. 1979. Dreding: A handbook for engineers. Publ. by Edward Arnold. London, England. ISBN: 0-713-3412-7.

216 - Breittmayer, J., G., Flatau, M. Barelli, F. Laumond, and D. Puel. 1975. Incineration en mer de residus chlores liquides. [Burning of liquid chlorinated residues in the open sea.] Revue Int. Oceanogr. Med. 39-40:213-224.

217 - Bressette, W. E. 1978. Effect of sun elevation upon remote sensing of ocean color over an acid waste dump site. Annual Remote Sensing of Earth Resources Conference, 7th. University of Tennessee. Tullahoma, TN. 27-29, March. 16p.

Abstract: Photographic flights were made over an ocean acid waste dump site while dumping was in progress. The flights resulted in wide angle, broadland, spectral radiance film exposure data between the wavelengths of 500 to 900 nanometers for sun elevation angles ranging from 26 to 42 degrees. It is shown from densitometer data that the spectral signature of acid waste discharged into ocean water can be observed photographically, the influence of sun elevation upon remotely sensed apparent color can be normalized by using a single spectral band rationing technique, and photographic quantification and mapping of acid waste through its suspended iron precipitate appear possible.

218 - Bridges, W. A. 1967. Ocean disposal of wastes. In: Background papers prepared for the National Conference on Pollution and Our Environment. Oct. 31-Nov. 4, 1966. Montreal. Sponsored by the Canadian Council of Research Ministers. Montreal. Background Paper B17-1-4.

219 - Briggs, Philip T. 1973. Comparison of the fish fauna over sand-filled and natural bottoms of Great South Bay, New York. In: Proceedings of a workshop on egg, larval, and juvenile stages of fish in Atlantic Coast Estuaries. NOAA Middle Atlantic Coastal Fisheries Center. Technical Publication #1. pp. 220-246.

Abstract: We captured a total of 297,795 fish (weighing 303,477 grams) of 29 species in 62 seine hauls made over natural bottoms in Great South Bay between mid-May and mid-October in 1967 (exclusive of the month of June). Atlantic silverside, fourspine stickleback, and mummichog were principal species encountered over this environment. A total of 320,665 fish (weighing 486,201 grams) of 24 species were captured with equal effort during the same time period over adjacent sand-filled bottoms. Atlantic silverside and striped killifish dominated the catches made over sand-filled bottoms. Several species found in large numbers in one environment were either absent entirely or taken in considerably lower quantities in the other. In addition, we observed different size distributions for the same species in each of the two bottom types.

220 - Briggs, Philip T. 1977. Status of tautog populations at artificial reefs in New York waters and effect of fishing. N.Y. Fish and Game Jour. 24(2):154-167.

Abstract: Population estimates indicate that as many as 18,898 tautog were attracted to the Kismet artificial reef in Great South Bay. There, the fish averaged only 216 millimeters long, while at the Fire Island artificial reef in the Atlantic Ocean they averaged 280 millimeters. Tag returns indicate that a migration of tautog from Kismet reef began in October each year and that these fish overwintered in the inshore waters of the northwestern corner of the New York Bight. Data for mortality, growth rate, and age at sexual maturity indicate that the population studied is not achieving its potential productivity. Assuming this population to be representative of that in New York waters generally, in order that females spawn at least once, it would be necessary to raise the minimum size of tautog that may be taken to 9 inches (228 millimeters) and apply this size limit to both the sport and commercial fisheries.

221 - Briggs, Philip T. 1978. Black bass in New York waters. N.Y. Fish and Game Jour. 25(1):45-58.

Abstract: Black sea bass were found to be slow growing, reach an average total length of only 378 millimeters at annulus IV. The ratio of males to females did not differ significantly from 1:1. Larger fish were mostly males. Migration southward from New York waters appears to begin in October. Local fish winter at least as far south as off the Delaware coast. Northward migration back to New York waters apparently begins in late April and early May. To assure that black sea bass reach sexual maturity and spawn at least once, it would be necessary to raise to at least 9 1/2 inches the minimum size for fish that may be bought or sold and apply it to sport-caught fish as well as those traded in commerce.

222 - Brooks, James M., Denis A. Wiesenburg, Guy Bodennec and Theodor C. Sauer, Jr. (In Press). Volatile pharmaceutical organics at the Puerto Rico dumpsite. In: I. W. Duedall, B. H. Ketchum, P. K. Park and D. R. Kester (editors), Wastes in the Ocean, Vol. I: Industrial and sewage wastes in the ocean. John Wiley and Sons, Inc., New York, NY.

Abstract: Waste materials from pharmaceutical production were discharged into the ocean at a dumpsite which is located approximately 74 km north of Arecibo, Puerto Rico, overlying 6000 to 8000 m of water. Volatile organic material constitutes the major non-aqueous fraction of the approximately 3.0×10^8 kg y^{-1} (wet wt) discharged into the surface waters at this site. Three samples of composite waste discharged at different times indicate significant changes in the waste composition over a two-year period (1978-1980). Waste samples obtained in February 1978 consisted mainly of dichlomethane, butanol, toluene, and dimethylaniline. These compounds were measured in the discharged waste plume and were also found in the dumpsite area three days after dumping. Dimethylaniline was observed in background measurements at the dumpsite at a concentration of approximately 100 ng liter^{-1}. Background concentrations of dichloroethane, benzene, chloroform, toluene, ethylbenzene, and zylene at the dumpsite area were between 5 and 30 ng liter^{-1}. These concentrations are six orders of magnitude less than the 1 to 10 mg liter^{-1} observed in the plume sampled 30 min after the waste was dumped.

Composite waste samples taken in October 1979 indicated significant compositional differences from the February 1978 samples. The majority of the volatiles in the 1979 waste consisted of ketones and alkyl substituted one-ring aromatic compounds (toluene, xylenes, C_3-benzenes, and C_4-benzenes) with no significant amounts of alcohols or substituted anilines being observed. The volatile hydrocarbon "fingerprint" in the waste plume and in the surface waters north of Puerto Rico and in the 1979 waste consisted mainly of these alkyl benzenes. A volatile organic "fingerprint" from the discharged waste was observed over a 23,000 km^2 area north of Puerto Rico at concentrations ranging from 0.1 to 8 μg liter^{-1}. The observed "fingerprint" suggested that the impact of organic volatiles from previous dumps still remained in the area. Samples taken from depths down to 200 m indicated that the dumpsite wastes had penetrated the thermocline.

223 - Brooks, N. H. 1956. Predictions of sedimentation and dilution of digested sludge in Santa Monica Bay. Report to Hyperion Engineers. April.

224 - Brower, Jr. William A. 1977. Climatic study of New York Bight. In: Baseline Report of Environmental Conditions in Deepwater Dumpsite 106. Vol. 1. NOAA/ODP Dumpsite Evaluation Report 77-1. Rockville, MD. pp. 117-218.

225 - Brown, J. C. 1977. Sludge disposal - A specialised operation. Water. July. pp. 2-9.

226 - Brown, Mary F., Dana R. Kester and Joyce M. Dowd. (In Press). Automated iron measurements after acid-iron disposal. In: I. W. Duedall, B. H. Ketchum, P. K. Park and D. R. Kester (editors), Wastes in the Ocean, Vol. I: Industrial and Sewage Wastes in the Ocean. John Wiley and Sons, Inc. New York, NY.

Abstract: An automated analytical procedure to measure iron in seawater based on a colorimetric method using the ferrozine reagent was developed. The method, applicable to iron concentrations in the range 0.02 to 10 umole kg^{-1}, was used to track an acid-iron waste plume that was formed after an acidic industrial waste highly concentrated in iron was dumped into the ocean at the deepwater dumpsite off the continental shelf of New Jersey. This automated system was used in conjunction with a submersible shipboard pumping system to obtain continuous real-time iron measurements in the waste plume. Real-time iron measurements were used to track an acid-iron waste plume for over 50 h after a dump; a tenfold dilution of iron of the waste plume was observed during this period. Changes in the horizontal distribution of iron in two waste plumes discharged into the ocean within 2 km of each other were monitored for a period of 5 h after the dump. The vertical dispersion of the waste in the ocean observed immediately after the dump and after the waste plume had mixed with seawater for over 50 h suggests that the acid-iron waste does not penetrate the seasonal thermocline.

227 - Brown, R. P., D. D. Smith, and C. G. Gunnerson. 1969. Status of oceanic disposal of solid wastes--a preliminary review. Conf. on Man in His Environment. Anaheim, CA. April 20-24.

228 - Brown, R. P., and D. D. Smith. 1969. Marine disposal of solid wastes--an interim summary. Mar. Poll. Bull. 18:12-16.

229 - Brown, R. P., and E. H. Shenton. 1971. Evaluating waste disposal at sea - the critical role of information management. (Paper presented at and published in the Proceedings of the Ann. Mtg. Marine Technology Society. 16-18 August. Washington, DC. pp. 353-363.

230 - Brown, Robert P. 1971. Ocean disposal of barge-delivered liquid and solid wastes from U.S. coast cities. U.S. Environmental Protection Agency. Solid Waste Management Office. Cincinnati, OH. Publ. No. SW-19C. 120p.

231 - Brungs, W. A. and D. I. Mount. 1978. Introduction to a discussion of the use of aquatic toxicity tests for evaluation of the effects of toxic substances. ASTM Spec. Techn. Publ. 657. Washington, DC. pp. 15-32.

232 - Buelow, Ralph W., B. H. Pringle, and J. L. Verber. 1968. Preliminary investigation of waste disposal in the New York Bight. U.S. Health, Education and Welfare. P.H.S. Bur. Disease Prevention and Environ. Control. 33p.

233 - Buelow, R. W. 1968. Ocean disposal of waste material. Transactions, National Symposium on Ocean Science and Engineering of the Atlantic Shelf. Marine Technology Society. Philadelphia, PA. pp. 311.

234 - Bumpus, Dean F. 1965. Residual drift along the bottom on the continental shelf in the Middle Atlantic Bight area. Limnol. Oceanogr. Vol. X (suppl.):R50-R53.

235 - Bumpus, Dean F., and Louis M. Lauzier. 1965. Surface circulation on the continental shelf of eastern North America between Newfoundland and Florida. Serial Atlas of the Marine Environment. Folio 7. American Geographical Soc. New York, NY. 4p. 8 pls.

236 - Bumpus, D. F., R. E. Lynde, and D. M. Shaw. 1972. Physical oceanography. In: Coastal and Offshore Environmental Inventory Cape Hatteras to Nantucket Shoals. Marine Publication Series No. 2. University of Rhode Island. Kingston, RI. pp. 1-72.

237 - Bumpus, Dean F. 1973. A description of the circulation on the continental shelf on the East Coast of the United States. In: B. A. Warren (editor), Progress in Oceanography. Vol 6. Pergamon Press, Ltd. Oxford. pp. 111-157.

Abstract: The circulation on the continental shelf of the east coast of the United States is discussed, including the historical development of concepts, and the surface and bottom circulation based on drift-bottle data and sea-bed drifter data. Suggestions for future research to further our understanding of the circulation problem are offered.

238 - Burd, R. S. 1968. A study of sludge handling and disposal. U.S. Dept. of Interior. Fed. Water Poll. Control Auth. Publ. WP-20-4.

239 - Burdick, G. E. 1954. An analysis of the factors, including pollution, having possible influence on the abundance of shad in the Hudson River. NY. Fish Game Jour. 1(2):188-205.

240 - Burnett, W. C. and O. A. Schaeffer. 1980. Effect of ocean dumping on $^{13}C/^{12}C$ ratios in marine sediments from the New York Bight. Estuar. and Coast. Mar. Sci. 11(6):605-611.

241 - Burns, B. 1978. Effect of a deep ocean mining test. ICES. C. M. 78/E:44.

242 - Bulliaux, R. 1968. Le devesement en mer des eaux residuaires. (Emptying sewage into the sea.) Bull. Inst. Nat. Sante Rech. Med. 23. pp. 623-638.

243 - Byers, S. C., and R. O. Brinkhurst. 1979. Report on ocean dumping R and D, Pacific Region Fisheries and Environment Canada 1977-1978. Insitute of Ocean Sciences. Sidney, British Columbia. (Pacific Marine Science Report 79-5). 76p.

244 - Byram, K. V. 1971. The barged ocean disposal of wastes: A review of current practice and methods of evaluation. U.S. Environmental Protection Agency. Pacific Northwest Water Laboratory. Corvallis, OR. 122p.

245 - Cabelli, V. J., A. P. Dufour, M. A. Levin, and P. W. Habermann. 1976. The impact of pollution on marine bathing beaches: An epidemiological study. In: M. G. Gross (editor), Amer. Soc. Limnol. and Oceanogr. Spec. Symp. 2:424-432.

Abstract: As part of a national program to develop health effects criteria for marine recreational waters, the U.S. Environmental Protection Agency conducted a prospective epidemiological-microbilogical study at bathing beaches in the vicinity of New York City, specifically at 20th St. on Coney Island and 67th St. and Riis Park at the Rockaways. The most consistent findings over the first 2 years of this study were that, for most of the water quality indicators examined, the mean densities at the Coney Island beach were appreciably and significantly higher than those at the Rockaways, and that the rate of gastrointestinal (GI) symptoms was significantly higher among swimmers relative to nonswimmers at the Coney Island beach but not at the Rockaways. When the data from two summers at both beaches (four points) were examined, good agreement was obtained between the mean *Escherichia coli* and enterococcus densities and the differential (swimmers minus nonswimmers) rate of GI symptoms. This preliminary finding addresses the objective of the study: relating illness as measured by symptomatology to some indicator of water quality.

246 - Cabelli, V. J. (In Press). Environmental health. Section II. MESA Atlas Mongraph 32. MESA New York Bight Project and New York Sea Grant Institute. Albany, NY.

247 - Calabrese, Anthony, Edith Gould, and Frederick P. Thurberg. (In Press). Heavy metal effects in marine animals of The New York Bight: some laboratory observations. In: G. F. Mayer, A. Calabrese, F. A. Cross, D. C. Malins, D. W. Menzel, J. S. O'Connor, and F. J. Vernberg (editors), Ecological Stress in the New York Bight: Science and Management Estuarine Research Federation. Columbia, SC.

Abstract: This report summarizes the results of eight years of experimental exposure studies on the effects of heavy metals in marine animals. The studies were conducted at the National Marine Fisheries Service Laboratory at Milford, Connecticut, and focused on organisms from the New York Bight. Of the metals tested, mercury and silver were most toxic to early-life stages bivalve embryos and larvae, juvenile crustaceans, as determined by bioassay methods. Cadmium and mercury were most toxic to adult teleosts and crustaceans, as determined by physiological and biochemical measurements. Tissue uptake was far greater for mecury than for cadium. Lead was more toxic than cadmium for early-life forms, but far less toxic than either mercury or silver. Chronic exposure to 6 ppb cadmium or 6 ppb mercury induced metabolic disturbances in the lobster *Homarus americanus* with somewhat more significant effects seen in cadmium-exposed animals despite greater accumulation of mercury in body tissues. Analogous results were seen during chronic exposures of winter flounder Pseudopleuronectes americanus to 10 ppb of either cadmium or mercury; similar exposures of flounder to five times as much lead produced less metabolic disturbance. Low salinity enhanced or altered some effects and, in combination with suboptimal temperatures, increased metal toxicity to early forms. Chronic exposure of adult animals to sublethal concentrations of toxic metals drains energy reserves and may reduce the capacity to adapt to and survive natural environmental stresses.

248 - Calamari, D. 1977. Selected bibliography on studies and research relevant to pollution in the Mediterranean. FAO Fish Techn. Pap. No. 165. 93p.

249 - California State Water Pollution Control Board. 1960. Summary of marine waste disposal research program in California. California State Water Pollution Control Board. Pub. No. 22. 77p.

250 - California State Water Quality Control Board. 1964. An investigation of the effects of discharged wastes on kelp. The Resources Agency of California. California State Water Quality Control Board. Pub. No. 26. 124p.

251 - Callahan, J. D. and J. B. Palmer. 1978. A multivariate procedure for comparing control areas to possible impact areas, graphically displaying the results. In: K. L. Dickson, J. Jr. Cairns, and R. J. Livingston (editors), Biological Data in Water Pollution Assessment: Quantitative and Statistical Analyses. ASTM STP 652. American Society for Testing and Materials. Washington, DC. pp. 78-91.

252 - Callahan, M. A. et al. 1979. Water related environmental fate of 129 priority pollutants. Volume 1: Introduction and technical background metals and inorganics, pesticides and PCBs. 459p. U.S. EPA - 440/4-79-029a. Vol. II Halogenated aliphatic hydrocarbons, halogenated ethers, monocyclic aromatics, phthalate esters, polycylic hydrocarbons, nitrosamines, and misc. compounds. EPA-440/4-79-0296) 669p. U.S.EPA Office of Water Planning and Standards. Washington, DC.

253 - Callaway, R. J., A. M. Teeter, D. W. Browne, and G. R. Ditsworth. 1976. Preliminary analysis of the dispersion of sewage sludge discharge from vessels to New York Bight waters. In: M. G. Gross (editor), Amer. Soc. Limnol. and Oceanogr. Spec. Symp. 2:199-211.

Abstract: New York City sewage treatment plant wastes discharged to the New York Bight Apex average 2.6% solids with an average solids density of 1.50 g cm^{-3}. Bulk waste density is about 1,000 g cm^{-3}, whereas the density of surface seawater in the sludge dumping area ranges from 1.019-1.025. Solids concentration of the wastes in the sludge vessels ranged from 5 to 50 g $liter^{-1}$. Correlation of extinction coefficient from a 10-cm light-path beam-transmissometer with total suspended matter (TSM) allowed continuous profiling of TSM. STD and beam transmittance profiles were made either by towing the instrument through a sludge path or wake or by making vertical profiles. Dilution from a near-instantaneous release was on the order of 1,000 within 10 min of release. Dilution in the wake of a release ranged from 500-1000.

The time for TSM to reach background or equilibrium values (0.5-2mg liter^{-1}) depends on initial concentration. Equilibrium time was approached exponentially for well mixed conditions in about 5.5 h. Pycnocline formation in the upper 8 m caused a similar approach to equilibrium time; below that depth TSM increased slightly with time.

Settling velocities for the larger flocculated sludge particles averaged about 0.5 cm s^{-1}. Values of 0.01-0.3 cm s^{-1} were obtained from plots of the center of mass of the waste field. The remainder of the dispersing sludge field had velocities of 10^{-3} cm s^{-1} and less.

TSM from New York Harbor can reach the permit area, but oceanographic conditions in the Apex usually prevent this.

Relocating the permit area to other deeper areas would cause the affected bottom area to increase in proportion to the increased depth, but concentrations of settled-out material would be inversely proportional, if the oceanographic environment was similar.

254 - Cambray, R. S., D. F. Jefferies, and G. Topping. 1976. The atmospheric input of trace elements to the North Sea. ICES. C.M. 76/E:17.

255 - Camp Dresser and McKee - Environmental Engineers. 1976. Phase 2 Report of technical investigation of alternatives for New York-New Jersey Metropolitan area sewage sludge disposal management program. Boston, MA. 319p.

256 - Capuzzo, J. M. 1980. The effects of pollutants on marine zooplankton at deepwater dumpsite 106; toxic effects and bioaccumulation of Dupont Edgemoor waste. ICES. C.M. 1980/E:58. 9p.

257 - Capuzzo, J. M., and B. A. Lancaster. 1981. The effects of pollutants on marine zooplankton at Deepwater Dumpsite 106- Preliminary Findings. In: B. H. Ketchum, D. R. Kester, and P. K. Park (editors), Ocean Dumping of Industrial Wastes. Plenum Press. New York, NY. pp. 411-419.

Abstract: The toxicity of DuPont's Grasselli waste to the marine copepods *Pseudocalanus* sp. and *Centropages typicus* has been evaluated in 96 hour bioassays. Test organisms were exposed to concentrations of waste ranging from 1 to 1000 ppm and the resulting effects on survival and feeding rates were monitored. Compelte mortality of *Pseudocalanus* was observed at an exposure concentration of 1000 ppm; no mortality was detected among control organisms or those exposed to lesser concentrations. Significant reductions in filtration activity, however, were detected with exposure to 10 and 100 ppm. Significant mortality of *Centropages* was detected at 100 ppm and exposure to higher concentrations resulted in 100% mortality. Feeding rates of Centropages as measured by both consumption of Artemia nauplii and fecal pellet production rates were significantly lower than control values with exposures to 50 and 100 ppm. The findings of other investigators dealing with the effects of dumped wastes on marine zooplankton are discussed.

258 - Capuzzo, Judith M. 1981. Predicting pollution effects in the marine environment. Oceanus. 24(1):25-33.

Abstract: Discussion on the assessment of the impact of pollutants through an understanding of the reponses of both individual organisms and whole populations as well as the adaptive nature of such responses. In addition to bioassays, responses of pollutants should be determined at levels of biological organization, such as: biochemical and cellular reponses. Organismal responses, including the integration of physiological and behavioral changes, simulated community, poplulation dynamics, community dynamics and structure, etc. with discussion of each.

259 - Caracciolo, J., and F. W. Steimle, Jr. 1977. An atlas of the distribution and abundance of dominant benthic invertebrates in the New York Bight Apex with a review of their life histories. U.S. Department of Commerce. NOAA NMFS Northeast Fisheries Center. Sandy Hook Laboratory. Sandy, Hook, NJ. 106p.

Abstract: Distributional, quantitative and life history summaries are given for 57 dominant species of benthic invertebrates collected in the New York Bight Apex during five sampling cruises during 1973 and 1974. These species showed affinities to major community types that have been previously identified in the Middle Atlantic Bight and some showed varying degrees of tolerance or intolerance to restricted areas in the Apex where the dumping of New York Harbor dredge spoils and New York Metropolitan area sewage sludge occur. Capitella capitata, a species often associated with extreme pollution stress, dominated the sewage sludge dumpsite.

260 - Caracciolo, J., J. Pearce, M. Halsey, and L. Rogers. 1978. Distribution and abundance of benthic organisms in the New York Bight. First and Second monitoring cruises, November 1975 and March 1976. NOAA DR ERL MESA-40. Boulder, CO. 48p.

261 - Carleton, H. R., and F. F. Y. Wang. (In Press). Testing and evaluation of stabilized coal wastes for ocean disposal. In: I. W. Duedall, B. H. Ketchum, P. K. Park, and D. R. Kester (editors), Wastes in the Ocean, Vol. I: Industrial and Sewage Wastes in the Ocean. John Wiley and Sons, Inc. New York, NY.

Abstract: A variety of compositions of fly ash and filtercake scrubber sludge stabilized with lime were evaluated in the laboratory and at an offshore site over a period of two years. The primary purpose of this work was to establish the suitability of stabilized coal wastes for ocean disposal through examination of the physical properties of this material over an extended period, and to determine those properties that adequately reflect physical integrity.

Groups of test blocks were prepared from fly ash and scrubber sludge obtained from two eastern coal-burning power plants. Blocks were divided into groups for placement at an Atlantic Ocean test site (at 25 m depth) and for laboratory tests in seawater tanks. The apparent elastic modulus of these materials was evaluated using a time-of-flight ultrasonic intervalometer operating at 162 kHz developed for this application. All blocks were ultimately tested for compressive strength in the laboratory to obtain correlation between compression strength and elastic modulus. Results showed that the ultrasonic method was a sensitive indicator of internal physical changes. The accuracy of the method was sufficient to differentiate block-to-block variations in density and to reflect changes in water absorption and cementation.

262 - Carls, E. G. 1978. Recreation. MESA New York Bight atlas monograph 19. MESA New York Bight Project and New York Sea Grant Institute. Albany, NY. 32p.

Abstract: Marine recreation in the New York Bight region is a varied and extensive field for study. Although its shoreline resources are limited, the Bight region is increasingly in demand as a coastal recreational area. Federal, state, and local authorities currently provide significant park and open space opportunities for recreation and related services. Private and commercial enterprises provide important services and facilities as well (e.g., marinas and charter boat rentals) and may become increasingly important in meeting future demand. Recreational activities include swimming, fishing, and boating. Extensive information is available on coastal recreation in the Bight region but not in a form for integration of sources or overall regional synthesis. Future study should center on problems associated with shoreline access, environmental quality, planning and management procedures, and economic analysis. The future of coastal recreation in New York Bight is tenuous. Accurate prediction of recreation potential is made impossible by a set of ongoing and interacting changes, e.g., population size and distribution, water quality, and demand for shoreline uses. The future of recreation is tied very closely to and will depend on the dynamics of the region's broader social, economic, and environmental conditions.

263 - Carlson, R. O., and M. V. Brown. 1955. Seismic refraction profiles in the submerged Atlantic coastal plain near Ambrose Lightship. Bull. Geol. Soc. Amer. 66(8):969-976.

264 - Carmody, D. J., 1972. The distribution of five heavy metals in the sediments of the New York Bight. PhD Dissertation. Columbia University. New York City, NY. 121p.

Abstract: This study provides new evidence about the landward movement of wastes toward Long Island and movement eastward in the Hudson Submarine Valley. Unfortunately, measuring seasonal changes in the metal concentrations was found to be impractical.

Since it has been assumed until recently that the wastes remained stationary on the bottom, these findings may be of importance in helping to formulate environmental policy decisions.

265 - Carmody, D. J., J. B. Pierce, and W. E. Yasso. 1973. Trace metal in sediments of New York Bight. Mar. Pollut. Bull. 4(10):132-135.

Abstract: Concentrations of chromium, copper, lead, nickel, and zinc in superficial sediments in New York Bight are ten to a hundred times greater near waste disposal areas than in uncontaminated sediments. Some wastes are transported by currents north-east towards Long Island and south-east along the Hudson Submarine Valley.

266 - Carnes, R. A. and D. A. Oberacker. 1976. Pesticide incineration, U.S. EPA news of environmental research in Cincinnati. Cincinnati Municipal Environmental Research Lab. Cincinnati, OH.

267 - Carpenter, J. H. 1973. Determining ultimate capacity of the coastal zone for wastewater and wastewater residuals. In: F. E. McJunkin and P. A. Vesilind, (editors), Ultimate Disposal of Wastewaters and Their Residuals. National Symposium. Raleigh, North Carolina, Water Resources Research Institute of the University of North Carolina. April 26-27. pp. 216-225.

268 - Carritt, D. E., et al. 1959. Radioactive Waste Disposal into Atlantic and Gulf Coastal Waters. NAS-NRC Report No. 655. Washington, DC.

269 - Carter, H. H. 1975. Prediction of far-field exclusion areas and effects. Progr. Water Technol. Suppl. Paper No. 36. pp. 363-370.

270 - Carter, L. 1964. Effects of acid and alkaline effluents on fish in seawater. Effluent and Water Treatment Jour. October. pp. 484-486.

271 - Carter, L. 1973a. An industrial approach to waste disposal in estuaries. Jour. Envir. Planning and Pollution Control. 1(3):5-11.

272 - Carter, L. 1973b. Disposal to the sea is sewage treatment. Effluent and Water Treatment Jour. Oct. pp. 647-649.

273 - Carter, L. 1976a. Relation of treatment and ecological effects - disposal of wastes to tidal waters. Pure and Appl. Chem. 45(3/4):205-206.

274 - Carter, L. 1976b. Marine pollution and sea disposal of wastes - the disposal of wastes to tidal waters. Chemistry and Industry. 19:825-829.

275 - Casey, John C., and John M. Hoenig. 1977. Apex predators in Deepwater Dumpsite 106. In: Baseline Report of Environmental Conditions in Deepwater Dumpsite 106. NOAA/ODP Dumpsite Evaluation Report 77-1. Vol. II. Rockville, MD. pp. 309-376.

276 - Cassidy, J. E. 1975. Coastal contamination from sewage and industrial effluent discharges. Oceanology International 75. London. pp. 296-301.

277 - Catanzaro, E. J. 1976. Mass spectrometri-isotope dilution determinations of copper and lead in Hudson River water. In: J. O. Nriagu (editor), Environmental Biogeochemistry. Volume 2 - Metals Transfer and Ecological Mass Balances. Ann Arbor Science Publishers, Inc. Ann Arbor, MI. pp. 513-517.

Abstract: The ultimate goal of this research is to characterize the distribution and migration of copper and lead in the various phases in natural waters. This preliminary report describes a precise and accurate analytical technique for the measurement of microquantities of copper and lead in water samples, and presents some preliminary results on sampling and distribution.

278 - Chamberlin, J. L. 1977. Monitoring effects of Gulf Stream meanders and warm core eddies on the continental shelf and slope. International Commission of Northwest Altantic Fisheries. pp. 145-153.

279 - Chamberlin, J. L., and H. Diamond. 1978. Passage of Gulf Stream warm core eddy recorded by the vertical temperature array on a NOAA buoy. Gulf-Stream. 3(12):6-7.

280 - Champ, Michael, A. (editor). 1974. OPERATION SAMS: Sludge acid monitoring survey. CERES Publication No. 1. The American University. Washington, DC. 169p.

281 - Champ, Michael A. 1974. Concentrations of trace elements on the continental shelf, a review of previous research. Proceedings of the Estuarine Research Federation Outer Continental Shelf Conference and Workshop on Marine Environmental Implications of Offshore Oil and Gas Development in the Baltimore Canyon Region of the Mid-Atlantic Coast. pp. 171-183. University of Maryland. College Park, MD.

282 - Champ, Michael A. 1975. The environmental effects of dumping waste into ocean waters and the Great Lakes. House of Representatives Committee on Science and Technology. Pub. in Proceedings No. 55. Ninety-Fourth Congress. pp. 3-24.

283 - Champ, Michael A. 1975. Current status of nutrient loading in the nation's estuaries. Estuarine Pollution Control and Assessment Conference. EPA No. P4-01-03874. EPA Report to Congress. Vol. 1. 237-257.

284 - Champ, Michael A. 1976. Ocean dumping of industrial and municipal wastes in the Mid Atlantic Bight. Oversight Joint Hearings on the Marine Protection, Research, and Sanctuaries Act of 1972. U.S. House of Representatives Committee on Merchant Marine and Fisheries. Proceedings Serial No. 94-25. Nintey-Fourth Congress. pp. 42-69.

285 - Champ, Michael A., and P. Kilho Park. 1981. Ocean dumping of sewage sludge: A global review. Sea Technology. 22(2):18-24.

286 - Champ, Michael A., Thomas P. O'Connor, and P. Kilho Park. 1981a. Ocean dumping of seafood wastes as a waste management alternative. In: W. S. Otwell (editor), Seafood Waste Management in the 1980's. Conference Proceedings. Sea Grant. University of Florida. Gainesville, FL. Report No. 40. pp. 103-115.

287 - Champ, Michael A., Thomas P. O'Connor, and P. Kilho Park. 1981b. Ocean dumping of municipal and industrial wastes in the United States: An analysis of environmental effects. Paper presented at Oceans 81. Abstract pub. Vol. II. pp. 720.

Abstract: The volume of ocean-dumped municipal wastes (sewage sludge) has increased 32%, while that of industrial wastes has declined 43% in the United States since the enactment of P.L. 92-532, the Marine Protection Research and Sanctuaries Act of 1972 (for the period of 1973 to 1980). In 1980, the total tonnage of industrial and municipal wastes ocean-dumped was over 10 million metric tons. The environmental effects of the disposal of these wastes is determined by the nature of the waste (quality and quantity) combined with the dynamics of natural physical, chemical, and biological processes. The most pronounced environmental effects have occurred at waste accumulating and low dispersing dumpsites. At sewage sludge dumpsites, benthic community structure and function have been altered by the build-up of organic matter in bottom sediments. The same sediments exhibit enhanced concentration of: 1) several metals (Cr, Ni, Cu, Zn, and Pb) and organic compounds (PCBs), 2) coliform and fecal coliform bacteria, 3) bacteria resistant to antibiotics and metals, and 4) pathogenic protozoans of the genus Acanthamoeba.

Dumping of industrial wastes, inorganic (acid-iron wastes) and organic (process and pharmaceutical wastes) leaves amounts of contaminants in the water column which depend on the rate and frequency of dumping and the flushing characteristics of dumpsites. Phytoplankton response (laboratory bioassays with cultered organisms) varies with species, and growth inhibition occurs usually at concentrations greater than those existing in the ocean subsequent to dumping. Field studies indicate that there may be a short term, small scale alteration of phytoplankton communities. The response of zooplankton is manifested in decrease of fecundity which, again, is of small spatial scale and, since subsequent generations are unaffected, are not of long term consequence.

288 - Champ, Michael A., Thomas P. O'Connor, and P. Kilho Park. 1981c. Ocean dumping of seafood wastes in the United States. Mar. Pollut. Bull. 12(7):241-244.

Abstract: In the United States fish wastes can be ocean-dumped without an Ocean Dumping Permit if the wastes contain no additives (flocculating or preserving agents, etc.). Fish wastes are currently ocean-dumped off American Samora, Louisiana and Puerto Rico. Fish wastes are characteristically high in protein, fats, dissolved and suspended organic materials. Water quality parameters affected by these wastes are biochemical oxygen demand, total dissolved solids, chemical oxygen demand, oil and grease, pH, and turbidity. Other key factors include organic and nutrient enrichment, the attractant of undesirable predator species (i.e. sharks) and the natural oxygen regeneration processes.

289 - Champ, Michael A., Orterio Villa, and Robert C. Bubeck. 1981d. Historical overview of the freshwater inflow and sewage treatment plant discharges to the Potomac River Estuary with resultant nutrient and water quality trends. In: R. Cross and D. Williams (editors), Proceedings of the National Symposium on Freshwater Inflow to Estuaries. U.S. Fish and Wildlife Service, Office of Biological Services. FWS/OBS-81/04. Vol. 2. pp. 350-373.

290 - Chan, A. T., E. Ozturgut, R. Jacobson, and G. C. Anderson. 1980. Environmental investigation of the effects of deep-sea mining on the phytoplankton in the tropical Eastern Pacific Ocean, with special attention on primary productivity. Univ. of Washington, Seattle, WA. Dept. of Ocean. Spec. Rept. No. 92. 48p.

291 - Charnell, Robert L., and George A. Maul. 1973. An oceanographic observation of New York Bight from ERTS-1. NOAA TR ERL 262-AOML 9. Boulder, CO. 8p.

Abstract: The Earth Resources Technology Satellite (ERTS-1) made a transit over New York Bight on August 16, 1972. Imagery from this transit shows several oceanographic features that demonstrate the usefulness of remote sensing over a large area for the synoptic observation of changes in water quality in the coastal zone. Both the extent and turbulent character of the Hudson River plume are discernible in the image. Residue from a dump of waste acid is visible over a 5-mile area in the Apex of the Bight. Little dispersion of this residue has occurred, suggesting that this feature will be a persistent signature in images from future satellite transits.

292 - Charnell, R. L., D. V. Hansen, and R. I. Wickland. 1973. Surface and bottom water movement in New York Bight. Final Report on the Effects of Waste Disposal in the New York Bight, for the Coastal Eng. Res. Ctr., U.S. Army Corps. of Eng., by National Marine Fisheries Service. Sandy Hook Laboratory. Highland, NJ.

Abstract: A sampling grid of 23 stations was established in the New York Bight between 73° 30'W and 74° 00'W, north of 40° 20'N. Periodically at these stations, measurements were made to determine values for temperature and salinity at depth intervals of four meters, and for dissolved oxygen near the bottom. Useable records of current speed and direction were obtained at three fixed stations, each recording near the bottom and approximately 40 feet above the bottom. Estimates of surface and bottom particulate transport were made with the use of seabed and surface drifters which were released at 21 of the fixed grid stations on each of the regular cruises.

293 - Charnell, R. L., J. R. Apel, W. Manning, and R. H. Qualset. 1974. Utility of ERTS-1 for coastal ocean observation: The New York Bight example. Mar. Techn. Soc. Jour. 8(3):42-47.

Abstract: An Earth Resources Technology Satellite (ERTS-1) image taken over New York Bight on 16 August 1972 was enhanced with the view of evaluating the sensor package for oceanic observation. Although ocean information is substantially less dynamic than land-related information, several important ocean features were detected. This examination leads to the conclusion that satellites such as ERTS can be used to study movement and structure of particulate plumes from river outflow, to monitor deposition and dispersion of waste dumps, to observe some water mass boundaries, and to detect the presence and measure the characteristics of internal gravity waves.

294 - Charnell, R. L., and D. V. Hansen. 1974. Summary and analysis of physical oceanography data colleted in New York Bight Apex during 1969-1970. NOAA MESA TM No. 74-3. Boulder, CO. 44p. NTIS No. PB 247906/AS.

Abstract: This report presents an analysis of physical oceanopraphy data collected on a monthly basis in the Apex of the New York Bight during 1969 and early 1970. Data include temperature and salinity values, recovery information on surface and seabed drifters, and current meter observations. Hudson River discharge and wind data from Ambrose light station are also included. The data show Apex water to be stratified three-fourths of the year caused by high river runoff and insolation. During winter, heat loss and wind mixing destroy and impede reformation of stratification. There is a strong northward flow of water in the lower layers along the axis of the Hudson shelf channel; some of this bottom water flows parallel to Long Island. Eventually, this eastward flow turns and joins the southwest flow of shelf water, suggesting that an anticyclonic circulation exists in the Apex most of the year. Surface flow exhibits high seasonality in response to surface winds, with northward flow during spring and summer and southeast movement during fall and winter. Surface flow from Raritan Bay flows south along the New Jersey coast most of the year.

295 - Charnell, R. L., and D. A. Mayer. 1975. Water movement within the Apex of the New York Bight during summer and fall of 1973. NOAA TM ERL MESA-3. Boulder, CO. 29p. NTIS No. PB 247901/AS.

Abstract: An investigation of the water movement within the New York Bight Apex was made during the summer and fall of 1973. Observations from this study produced a fairly consistent picture of a general clockwise circulation for the mean and low frequency (periods greater than 40 hr) current motions. Additionally, the effects of stratification on the vertical structure of the flow showed the difference between the summer and later fall flow regimes. There also appeared to be, for low frequencies, excellent spatial coherence between stations in the Apex area and a fairly strong relation between meteorological forcing and current motions; currents tended to lag wind by up to 18 hr.

296 - Charnell, R. L. (editor). 1975a. Assessment of offshore dumping in the New York Bight; technical background: physical oceanography, geological oceanography, chemical oceanography. NOAA TR ERL 332-MESA 3. Boulder, CO. 83 p. NTIS No. COM 75-11386/AS. Supercedes NOAA TM ERL MESA-1.

297 - Charnell, Robert L. (editor). 1975b. Assessment of offshore dumping in the New York Bight, technical background: physical oceanography, geological oceanography, chemical oceanography. NOAA TM ERL 332-MESA-3. Boulder, CO. 83p.

Abstract: Physical Oceanography: Analysis of historic and MESA data shows two distinct circulation regimes. 1) Near the harbor mouth and along the New Jersey coast, New York Harbor discharge flows southward parallel to the New Jersey Coast; at depth, there is a return flow of external water into the estuary. In spite of its importance in the oceanographic and ecological systems of the region, this superposed flow system is recognized in measurements as only a slight imbalance of much stronger ebb and flood tidal currents. 2) Outside the region of strongest influence from river discharge, a persistent clockwise circulation or eddy appears to exist. In the eddy's most inshore portion, flow is toward the north and east, counter to the general flow over the continental shelf adjacent to this part of the coast. Details of its horizontal extent, its vertical structure, and its persistence are imperfectly known at present.

Geological Oceanography: Fine-grained waste dumped in New York Bight is entrained in a clockwise circulation pattern and is dispersed to the north. A significant portion is deposited in the low area (Christiaensen Basin) immediately northwest of the dumpsites. The fraction of finer dredge spoils and the bulk of sewage sludge is widely dispersed through the Bight Apex in highly diluted form. There are no "pure" sludge beds; there has been no detectable aggradation of the sewage sludge site since 1936. However, the Christiaensen Basin, a natural zone of mud deposition, appears to be significantly contaminated with sewage sludge. Near Long Island beaches, there are scattered, thin, small patches of mud, physically indistinguishable from natural mud patches on similar, less populous coast. Microscopic examination of samples indicates that less than 3 percent of the resolvable particles in these patches are of artificial origin, the lowest contamination level detected by microscopic technique in the Bight Apex. No evidence for a front of sewage sludge approaching the Long Island shore was observed.

Chemical Oceanography: Data from water sampling show that nutrient (nitrates, nitrites, silicates, and phosphates) distributions are dominated by the lower New York Bay outflow, with dumped sewage sludge contributing very small amounts. Bottom grab samples were analyzed for total organic carbon and total carbohydrates. The carbohydrate/TOC ratio indicates that the whole Bight contains some sewage-derived materials with the greatest concentration in the Hudson Shelf Valley, in the Christianensen Basin, and north of the geographical sewage sludge dumpsite. This is consistent with the distribution expected if sewage sludge were disseminated throughout the Apex by the hypothesized current pattern (gyre). Although carbohydrate/TOC ratios in sediments throughout the area close to Long Island suggest the presence of sewage-derived material, low TOC values found in all but isolated pockets demonstrate that contaminant material comprises only a small fraction of the sediments.

298 - Charnell, R. L., M. E. Darnell, G. A. Berberian, B. L. Kolitz, and J. B. Hazelworth. 1976. New York Bight Project Water Column Characterization Cruises 1 and 2 of the NOAA Ship Researcher, 4-15 March, 5-14 May 1976. NOAA DR ERL MESA-18. 225p. Boulder, CO. NTIS No. PB 271485/AS.

299 - Chasan, Daniel Jack. 1981. The Water Link, A History of Puget Sound as a Resource. Puget Sound Books. University of Washington Press. Seattle, WA. 179p.

Preface: I started this book with the idea of explaining how we got here from there, how the early uses of and attitudes toward Puget Sound evolved into the uses and attitudes of today. When I decided to cover events that had occurred as recently as 1977, I realized that I would have no way of judging their ultimate importance and that most of the people who had participated in them would be alive, well, and fully capable of challenging my interpretations. It is inconvenient to write "history" about recent events. I believe, though, that it's well worth the inconvenience. The history of Puget Sound didn't end with the gold rush, the demise of the sailing ship, or the invention of the chainsaw, and of all the reasons for studying history, learning how the present evolved from the past isn't the least valuable.

300 - Chase, R. R. P. 1979. The coastal longshore pressure gradient: Temporal variations and driving mechanisms. Jour. of Geophy. Res. 84 (08):4898-4904.

Abstract: Along the northern coastal boundary of the Mid-Atlantic Bight, synthetic subsurface pressure records derived from 1975 tide gage and airway weather observations are examined for fluctuations having monthly time scales. Additional calculations are performed to determine the factors controlling variance in these pressure data. Similar analyses are applied to the longshore gradient in synthetic subsurface pressure. Results of the study indicate that synthetic subsurface pressures are controlled by three mechanisms, two each at either end of the coastal segment. Fluctuations in Gulf Stream position have a significant effect on monthly mean pressures, which are uniform along the coast. At the western end of the coastal segment, east-west winds are also important, while changes in water density contribute significantly to subsurface pressure variance at the eastern end. Coastal steric differences are found to be a minor contributor to monthly fluctuations of longshore pressure gradient, while the coastal mean east-west winds effect a linear response in the gradient. Using this relationship and upon removal of geodetic leveling errors, the annual mean longshore sea surface slope is calculated to be 1.6×10^{-7}, forming a westward pressure gradient along the northern boundary of the Mid-Atlantic Bight.

301 - Chen, C. W. 1970. Effects of San Diego's wastewater discharge on the ocean environment. Jour. Water Poll. Control Fed. 42:1458-1467.

302 - Chiang, E. 1974. The implication of sediment dynamics and heavy metals dispersion to ocean disposal at the "NB" buoy, south of Fire Island, N.Y. Masters Thesis. Adelphi University. Garden City, NJ. 65p.

Abstract: Baseline studies of a proposed dumping area at the "NB" buoy, located 10 nautical miles south of central Fire Island, N.Y., were undertaken to evaluate the environmental impact of dumped material on the surrounding area. Bottom photographs taken with diver-operated and bottom bounce cameras show distinct ripple fields separated by areas of smooth sands. Physical properties of the substrate were determined through granulometric analysis of 48 bottom grab samples. A current meter situated 1/2 mile south of the buoy provided current information.

Median grain size of the substrate ranged from 1.12 0 units (.480 mm) to 2.68 0 units (.155 mm), or granular medium sands to fine sands. These sands experience current velocities (weather or wave induced) ranging from 0 to 65 cm/sec.

Fifteen "NB" samples were paired with 15 samples from the periphery of waste disposal sites of the New York Bight Apex for comparison of heavy metal content (Mn, Fe, Cu and Zn). Although the physical properties of each paired sample was similar, Fe, Cu and Zn were found to differ significantly in concentration between the two areas.

Results of the analysis for the physical properties of the sediments, coupled with current meter data at the "NB" buoy show that the "NB" site is characterized by a highly mobile and therefore transient substrate. Should disposal occur, trace metal results indicate that the contamination of the bottom sediments cannot be avoided, but that contamination of the sediments may be minimized by disposal in a relatively high energy environment. Of greater concern however, would be the solids and suspended matter which would undergo a net direction of transport to the northwest. These materials would pose a serious potential hazard to Long Island's shorelines.

303 - Chiang, E., and A. E. Cok. 1974. Correspondence of sediment dynamics and heavy metal dispersion south of Fire Island and in the New York Bight Apex. Abstracts with Programs. GSA. 6(1):11.

304 - Chipperfield, P. N. J. 1962. Some thoughts on the discharge of industrial effluents to tidal waters. Chemistry and Industry. pp. 1660-1666.

305 - Churchill, J. H., B. H-G Pade, and K. R. Peal. 1981. Water velocity measurement from nearsurface to 110 m depth at Deepwater Dumpsite #106, using acoustically tracked drogues and conventional current meters. Technical Report WHOI-81-11. Woods Hole, MA. 29p.

Abstract: A system has been developed recently at W.H.O.I. for tracking nearsurface drogues equipped with sonobuoys using an acoustic navigation system. Surface and submerged drogues of mean depths ranging from 0.15 m to 4.88 m were tracked in the vicinity of deepwater dumpsite #106. A least squares linear regression technique was used to determine drogue velocities over 2 hour periods. Water velocities at depths from 8 - 110 m were measured using a ship-deployed current meter coupled with acoustic shear in the surface mixed layer. There were two regions of strong shear at greater depths, one associated with the main thermocline and the other presumably associated with a halocline.

306 - Chytalo, Karen Nadia. 1979. PCBs in dredged and benthic organisms in Long Island Sound. Master's Thesis. Marine Sciences Research Center. State Univ. of New York. Stony Brook, NY.

Abstract: The occurrence and distribution of polychlorinated biphenyls (PCBs) were examined in dredged materials at Long Island Sound disposal sites and in naturally accumulating sediments, and the levels of PCBs were assessed in deposit-feeding organisms. Three historical disposal sites (Eatons Neck, Bridgeport, and New London) were selected for study. For each of the three disposal sites, a control site of similar sediment texture was selected.

The results indicated that sediments from disposal areas were more contaminated with PCBs than naturally accumulating sediments, but not significantly more; and there was a marked PCB gradient in surficial sediments which increased from east to west. The level of PCBs in the surficial sediments do not seem to be controlled by sediment characteristics but by sources of PCB contamination (i.e. disposal areas, industrial activity). The sediment PCB concentration was poorly correlated to the total organic carbon, and the clay and silt contents. PCBs were partitioned among the different sediment size fractions. PCBs tended to accumulate more in the clay fraction than in the silt and sand fractions. The vertical distribution of PCBs in sediments showed that the concentration of PCBs increased and decreased randomly with depth. This was probably the result of bioturbation. Unlike the sediments, deposit-feeding polychaetes contained only trace amounts of PCBs (< 1 ppb); and thus polychaetes are probably an insignificant route in transferring PCBs from the sediments to the estuarine food web. The samples were analysed for PCBs by gas chromatography. Aroclors 1254 and 1016 were identified.

307 - Clark, B. D. et al. 1971. The barged ocean disposal of waste. A review of current practice and methods of evaluation. U.S. E.P.A. Corvallis, OR. EPA 16073FGY 07/71. 120p.

308 - Clark, John R. 1970. Living resources. In: Alan Johnson (editor), Water pollution in the Greater New York Area. Gordon and Breach Science Publishers. New York, NY. pp. 101-106.

309 - Clausen, J. F., H. J. Fisher, R. J. Johnson, E. L. Moon, C. C. Shih, R. F. Tobias, and C. A. Zee. 1977. At-sea incineration of organochlorine wastes onboard the M/T Vulcanus. U.S. Environmental Protection Agency. Office of Research and Develoment. Research Triangle Park, N.C. (Environmental Protection Technology Series EPA-600/2-77-196). 87p.

310 - De Clerck, R., and J. Van de Velde. 1974. Biological description of a dumping area for industrial waste derived from a titanium dioxide process. ICES. C.M. 74/E:29. [See Ref. #'s 358, 359].

311 - Cochran, J. K., and R. C. Aller. 1979. Particle reworking in sediments from the New York Bight Apex: Evidence from 234 Th/238 Disequilibrium. Estuar. and Coast. Mar. Sci. 9:739-747.

Abstract: Two diver-collected cores from mud-rich areas of the New York Bight Apex have been analyzed for ^{238}U and ^{232}Th decay series nuclides with emphasis on $^{234}Th/^{238}U$ disequilibrium. Excess ^{234}Th is present in both cores and showed exponential decrease in the top 4 cm. Biogenic reworking by a deposit-feeding community characterized by a Nucula proxima - Nephtys incisa assemblage is apparently responsible for the form of the ^{234} Th profiles. Mixing coefficients of 0.3-0.6 x 10^{-6} cm^2/sec are calculated. No decrease of ^{210}Pb with depth (0-11 cm) is found nor is there any vertical structure in profiles of ^{228}Th. Episodic deposition followed by periods of stability cause the observed homogeneity of the longer lived nuclides as well as laminated horizons in the sediment. $^{234}U/^{238}U$ isotope ratios reflect addition of either sea water or sewage source uranium to the collection area.

312 - Cocoros, G., P. H. Cahn, and W. Siler. 1973. Mercury concentrations in fish, plankton, and water from three western Atlantic estuaries. Jour. of Fish Biology. 5:641-647.

Abstract: Twenty fish samples, plankton, and water were obtained from three western Atlantic estuaries located off the coasts of North Carolina, Maryland, and New York. All fish samples were collected at high tide, at approximately 25°C and 31 ppt salinity. Plankton and water samples were taken concurrently at each sample site. Total mercury concentrations were determined by flameless atomic absorption. No dangerously high mercury levels were found in any of the samples. Although there was some indication that the food chain is a likely source of mercury contamination in these fish, since levels in viscera were very much higher than those in the rest of the fish, no evidence of strong food-chain intensification was found. Slow mercury accumulation by the fish was evident. Accumulation of mercury by the fish took place through the digestive system. Fish mercury levels were twice that for the plankton. Phytoplankton had higher mercury levels than zooplankton.

313 - Cohen, Daniel M., and David L. Pawson. 1977. Observations from the DSRV Alvin on populations of benthic fishes and selected larger invertebrates in and near DWD-106. In: Baseline Report of Environmental Conditions in Deepwater Dumpsite 106. NOAA/ODP Dumpsite Evaluation Report 77-1. Rockville, MD. Vol. II. pp. 423-450.

314 - Cohen, Philip M. 1977. Ocean dumping - Research and monitoring of ocean disposal effects. Oceans 77. MTS-IEEE. 23B-4p.

315 - Cohn, Myra S., and Donna van de Sande. 1977. Red tides in the New York-New Jersey coastal area. Underwater Naturalist. American Littoral Society. 8(3):12-21.

316 - Cok, A. E., D. J. P. Swift, E. Shepard, R. J. Reynolds. 1973. 'Sediment' unmixing in the New York Bight area. Geol. Soc. Amer. Abstr. Prog. 5(2):150.

Abstract: In January of 1972 a five year National Oceanographic and Atmospheric Administration Continental Margin SEDimentation program was initiated by the Atlantic Oceanographic and Meteolorologic Laboratory and the Marine Geology and Geophysics Office to investigate the impact of dumping in excess of 500,000 cubic yards/year of sewage sludge and an undetermined amount of other debris in the New York Bight. Grab sampling at 350 stations, vibra-coring, seismic and side-scan sonar profiling plus data from rack-mounted Geodyne current meters indicate that the distribution and movement of "normal" sediment in the area has been greatly influenced by the dumping of solid wastes. Fine-grained "sediment" believed to be reworked sewage sludge drape areas of moderately coarse sands and gravels. Although bottom currents attain a velocity greater than 1/2 knot in the vicinity of Ambrose light, an anomalous sediment distribution coupled with highly vained topography over much of the New York Bight dump site indicates recovery of this area will be a very long-term process.

A detailed geologic map of the area reveals three avenues for sludge dispersal: 1) In toward New York Harbor under the influence of estuarine circulation, 2) down the Hudson canyon, and 3) in a north-eastern direction toward the south shore of Long Island. Sediment color has proved to be a valuable tool in this area for rapid identification of over forty distinct sediment types.

317 - Collins, J. C., (editor) 1960. Radioactive Wastes. John Wiley & Sons, Inc. New York, NY.

318 - Colony, R. J. 1932. Source of Long Island and New Jersey sands. J. Sediment Petrol. 2:150-159.

319 - Colwell, R. R., J. D. Walker, G. S. Sayler, P. A. Seesman, and B. F. Conrad. 1976. Microbiological studies of the Atlantic continental shelf. In: M. G. Gross (editor), Amer. Soc. Limnol. and Oceanogr. Spec. Symp. No. 2. pp. 275-280.

Abstract: Microbial degradation of petroleum, polychlorinated biphenyls, and similar substances was investigated in the laboratory and at stations in the Atlantic Ocean near offshore drilling sites. Microorganisms were analyzed as to type, including total viable, aerobic, heterotrophic, and petroleum-degrading bacteria, yeasts and fungi, actinomycetes, and chitin-hydrolyzing bacteria.

Mixed hydrocarbon substrate generally yielded higher counts than media made up with South Louisiana crude oil. The inorganic nutrient (PO_4, NO_3, NO_2, NH_4) content of the seawater medium influenced the amount of petroleum hydrocarbons degraded by microorganisms present in sediment and in the water column. Sediment bacteria from the deep ocean environment (3,500 m) were able to degrade crude oil; microorgansism present in surface seawater samples collected at stations along the southeast Atlantic coast degraded crude oil, but the microbial potential for degradation of oil was limited.

Polychlorinated biphenyl (PCB) and PCB-degrading bacteria were recovered from surface water and sediment samples collected along the southeast Atlantic coast from Miami to Cape Hatteras. Oil and PCB-degrading bacteria are potential indices for oil and PCB contamination.

320 - Compaan, H. 1980. Incineration of chemical wastes at sea - a short review. ICES. C.M. 80/ACMP:4.

321 - Conlan, K. E. 1979. The biological effects of ocean dumping: A selected, annotated bibliography. Contractor Rep. Ser., Inst. Ocean Sci. Patricia Bay, Canada. No. 79-2. 208p.

322 - Convention on the Prevention of Marine Pollution by Dumping of Waste and other Matter; P.R. No. 287. London, England. November 20, 1972.

323 - Convention on the Prevention of Marine Pollution by Dumping from Ships and Aircraft. Miscellaneous No. 21 (1972). Oslo, Norway; February 1972.

324 - Conway, H. Lee, and Terry E. Whitledge. 1979. Distribution, fluxes and biological utilization of inorganic nitrogen during a spring bloom in the New York Bight. Jour. Mar. Res. 37(4):657-668.

Abstract: The utilization of ammonium (NH_4) and nitrate (CO_3) by natural phytoplankton communities was measured during a Ceratium tripos-dominated spring bloom in the New York Bight. During 6-8 hour uptake experiments, the inshore phytoplankton communities that were typically located at depths >25 m were characterized by NH_4 and NO_3 uptake rates that were less light-dependent than those of the more uniformly distributed communities at the shelf break. Measurements of NH_4 utilization, as a percentage of NH_4 plus NO_3 utilization, yielded values of 59% inshore and 70% at the shelf break. High biomass of zooplankton at the shelf break suggested they could account for 62% of the regenerated production (i.e., based on NH_4), while inshore, estimates of regenerated NH_4 from zooplankton, benthos, and bacterioplankton could account for 38 to 53% of the phytoplankton requirements. In the cross-shelf distribution, the ambient concentrations of NH_4 and NO_3 in the surface water were approx. 0.4 and approx. 0.1 umol L^{-1}, respectively; however at depths between 25 and 50 m, concentrations of approx. 2 and approx. 4 umol L^{-1} were observed.

325 - Cook, David O. 1979. Mathematical modeling predictions of the geological effects of sewage sludge dumping on the continental shelf. In: H. D. Palmer and M. G. Gross (editors), Ocean Dumping and Marine Pollution. Dowden, Hutchinson and Ross, Inc. Stroudsburg, PA. pp. 185-203.

Abstract: A far-field (beyond the initial mixing zone) water quality math model has been adapted to predict geological implications of ocean dumping. The example simulated is ongoing discharge of digested sewage sludge by the City of Philadelphia at a disposal site located at mid-shelf in the Middle Atlantic Bight. After calibration of the model using hydrographic field data, predictions were made of the horizontal distributions of suspended solids and deposition following an individual dumping incident. The distributions shift south-southwest of the discharge line in response to non-tidal currents and disperse over a larger area with time. Following a 1.3 x 10^5 kg discharge of sludge solids, the maximum predicted far-field suspended solids concentration is 2.06 mg/1 (after initial mixing), and the maximum depositional rate is 30.9 x 10^{-5} g/cm^2. The predictions are highly dependent upon assumed settling characteristics of the sludge.

The hydrodynamic simulation of the study area appears to be reasonably accurate. The geological predictions, in particular those for deposition, represent a first order estimate because of uncertainties in sludge solid settling behavior and pre/post depositional sludge fate. Regarding this fate, the effects of sludge recycling by zooplankton in the water column and detritus feeders on the sea floor are unknown. Also, the presence of physical structures in bottom photographs suggests that settled sludge solids would be subject to periodic resuspension. Until these uncertainties are resolved and qualified, the primary use of predicted despositional patterns is to indicate those areas of the sea floor where the benthic community may potentially be affected by sludge disposal.

326 - Cordle, Frank. 1981. Mercury levels in fish. Oceanus. 24(1):42-43.

Abstract: The FDA response to Officer and Ryther (Oceanus. 24(1):34-41). Discussion of legislation, data, and development of the standard level.

327 - Council on Environmental Quality. 1970. Ocean dumping--a national policy. U.S. Govt. Print. Off. Washington, DC. 45p.

328 - Cox, G. V. 1975. The Davy Jones garbage dump. Envir. Sci. Technol. 9(2):108-111.

329 - Cox, G. V. 1976. New York Bight--an environmental baseline study. Eighth Annual Offshore Technology Conference. pp. 665-674.

330 - Cox, Geraldine V. (Senior editor). 1977. Environmental survey of a proposed alternate dumpsite in the outer New York Bight. Raytheon Company. Portsmouth, RI.

Abstract: Based on water quality degradation at the existing sewage sludge disposal site, the Environmental Protection Agency (USEPA) asked the National Oceanic & Atmospheric Administration (NOAA) to select two alternative ocean disposal sites for further study on March 8, 1974. NOAA designated two areas, as candidate sites, Figure 1, for EPA on May 8, 1974. NOAA based their selection on the following criteria (MESA Report 1976).

o The location should minimize the chance of contaminants reaching shorelines and beaches;

o The location should minimize, to the extent possible, adverse effects on living marine resources;

o The location should (a) exclude the no-fishing zone defined under international bilateral agreement[1] and (b) avoid areas most intensively fished for surf clams.

o The location should be within 65 nmi (120 km) of the harbor entrance. This criterion was prescribed on the basis of economic considerations by EPA and does not have marine environmental significance.

By applying these criteria, and existing information on water movements, topographic controls, biological processes, and marine resources, areas 1-A and 2-A were recommended as alternative dumpsite areas because they were:

o Not immediately adjacent to the Hudson Shelf Valley. It is extremely difficult to predict the ultimate fate of material dumped into the Valley. There are indications of both deposition and erosion in the Valley and transport both up and down the Valley. The Valley also is a migration route for certain fish and shellfish, supports active fisheries, and serves as a winter aggregation zone for some fishes.

o Seaward of the clockwise circulation gyre in the inner portion of the Bight. This gyre, thought to have its western edge aligned with the Hudson Shelf Valley, could serve to transport material back into the Bight Apex.

o Beyond the areas most intensively fished for surf clams. The fished areas are generally shoreward of the 20-fm (36m) isobaths along N.J. and Long Island. The greatest concentrations are between the 10- and 20 fm (18- and 37-m) isobaths.

331 - Crawford, Ronald W., Charles F. Powers, and Richard H. Backus. 1951. The depth of unconsolidated sediments in the New York Harbor area and its approaches. In: John C. Ayers et al. The Hydrography of the New York Area. Status Report No. 9. (Contract N6 ONR 264, task 15). Cornell Univ. Ithaca, NY.

332 - Crawford, Ronald W. 1952. The distribution of oxygen in the waters of New York Bight, Block Island Sound, and Newport Bight; Cruise Stirni I, July-September, 1951. In: John C. Ayers et al., The Hydrography of the New York Area. Status Report No. 17. (Contract N6 DNR 264, task 15). Cornell Univ. Ithaca, NY.

333 - Crickmore, M. J. 1972a. Initial behaviour of sludge. In: Out of sight - out of mind. U.K. Department of the Environment. London. 2:45-54.

334 - Crickmore, M. J. 1972b. Sludge dispersion on the sea bed. Out of sight - out of mind. U.K. Department of the Environment. London. 2:171-208.

335 - Cripps, T. 1973. Some computer applications in marine disposal work. C.I.E.S.M. Journ. d'Etudes Poll. Mar. (Athenes, 1972). p. 69.

336 - Crist, R. W., and J. L. Chamberlin. 1978. Bottom temperatures on the continental shelf and slope south of New England during 1977. (Unpublished manuscript).

337 - Cronin, L. Eugene. 1967. Interim report on gross physical and biological effects of overboard disposal. Chesapeake Biological Laboratory, Univ. Maryland. Special report to the Maryland Board of Public Works and Commission on Submerged Land. 93p. (mimeo).

338 - Csanady, G. T. 1981. An analysis of dumpsite diffusion experiments. In: B. H. Ketchum, D. R. Kester and P. K. Park (editors), Ocean Dumping of Industrial Wastes. Plenum Press. New York, NY. pp. 109-129.

Abstract: The oceanic dispersion of industrial waste barged to Deep Water Dumpsite 106 is effected in the first instance by barge-bound eddies ("wake dispersion"), then by naturally occurring mean shear and turbulence. Wake dispersion is very efficient, producing initial dilution up to a factor of 10^4, and is controlled by the dimensions and the forward speed of the barge. Subsequent dispersion by natural oceanic processes is slow under stratified summer conditions, describable by an effective diffusivity of order 300 cm^2 sec^{-1} or a diffusion velocity in the neighborhood of 0.2 cm sec^{-1}

339 - Csanady, G. T. (In Press). Long-term mixing processes in slopewater. In: I. W. Duedall, B. H. Ketchum, P. K. Park and D. R. Kester (editors), Wastes in the Ocean, Vol. 1: Industrial and Sewage Wastes in the Ocean. John Wiley and Sons, Inc. New York, NY.

Abstract: The fate of industrial waste barged to the deepwater dumpsite 106 Mile site in New York Bight can be followed in diffusion experiments for a maximum period of about three days only, after which the waste becomes too dilute and its advection by currents too erratic for successful detection. However, a natural tracer is available to demonstrate the principal mechanisms of long-term mixing in this area: the freshwater originating from land runoff.

Potentially important transport processes for the freshwater runoff are advection by a large cyclonic gyre between the Gulf Stream and the North American continent, an upwelling-like circulation in the vertical plane, a random process of parcel separation at the shelf-edge front, and a similar random process of warm core ring shedding by the Gulf Stream. The random parcel and ring shedding processes are akin to mixing due to mechanical turbulence and are amenable to similar statistical treatment. The landward salt transport by the latter processes constitutes a large fraction of the total salt transport, which compensates for the freshwater flow.

An estimate of the rate at which parcels of anomalous water break down in the slope water region can be reached by considering the balance of mean square salinity fluctuation. The estimated time scale of decay may be supposed to parameterize the long-term diffusion process of barged waste, after the latter has formed a cloud comparable in size to the typical parcels of anomalous water. This decay time scale is estimated to be of the order of five months, on the basis of historical data of mean square salinity fluctuations and the known rate of freshwater inflow.

340 - Curran, P. 1979. Preliminary investigation into the application of oblique multispectra photography for the monitoring of sewage outfalls. Jour. of Envir. Manag. 9(1):41-48.

Abstract: A preliminary flight using oblique multispectral and false color photography was performed to evaluate the potential of such a system for the monitoring of sewage movement from an outfall at Weston-Super-Mare, Somerset. In a laboratory spectrophotometric study, the difference between turbid seawater and sewage was small. This result was confirmed by aerial photography, where qualitative analysis off the imagery proved disappointing. Nevertheless, quantitative information determined via image densitometry demonstrated the ability of this system to discriminate sewage-laden seawater. For remote sensing to be used economically in an operational system, there is a need for narrower spectral bands, ground markers, image enhancement, color compensating filters for the false color film, and polarized visible light photography.

341 - Dagg, Michael. 1978. Estimated _in situ_, rates of egg production for the copepod _Centropages typicus_ (Kroyer) in the New York Bight. Jour. of Exper. Mar. Biol. and Ecol. 34(3):183-190.

Abstract: _Centropages typicus_ (Kroyer) and _Pseudocalanus_ sp. are the two predominant copepods in the Continental Shelf waters immediately south of Long Island. New York. The estimated, in situ, rate of egg production for _Centropages typicus_ ranged from 5 to 230 eggs female^{-1} day^{-1} during this study. Variability was partly attributed to seasonal variation in water temperature and partly to variations in the physiological condition of individual females. It could not be shown that the ability to produce eggs varied seasonally due to factors related to food. Egg production by _Pseudocalanus_ sp. is probably from 1-10 eggs female^{-1} day^{-1}. The two reproductive behaviours result in average abundances of adults that are approximately the same, although the peaks are at different times of the year, indicating that _Centropages typicus_ has a higher mortality between the egg and adult stages than _Pseudocalanus_ sp.

342 - Dallaire, E. E. 1971. Ocean dumping: What and where, if at all? Civil Engineering-ASCE. 41(11):58-62.

343 - Dainty, S. H., R. F. Pearson, L. H. Thompson, and E. V. Finn. 1970. A regional scheme for the disposal of London's sludge to sea. Symp. Water Pollution Control in Coastal Areas. Bournemouth. 19-21 May, 1970. Institute of Water Pollution Control. London. Paper No. 17. 12p.

344 - Dames and Moore, Inc. Ecology Consultants. 1979. Site selection study for open water disposal of dredged materials: Long Island Sound, Block Island Sound and ajacent open waters. New England Division Corps of Engineers. 39p.

Abstract: The EIR includes a site selection phase which will identify areas suitable for the deposition of dredged materials considering all significant consequences to environments and resources. The EIR will define and discuss the environmental and economic impacts of open water disposal on both a regional and disposal site basis.

The EIR will also consist of a management analysis wherein procedures will be proposed to mitigate potentially adverse impacts of the deposition and to increase the positive productive effects of that deposition.

Personal interviews have been conducted with those individuals or groups considered knowlegeable of the study region who may have professional interests with regard to that environment and/or management of dredged material disposal in those waters; and, those individuals or groups with concerns about the environmental or economic consequences of utilizing these open waters for deposition of dredged sediments.

345 - Davies, C. M. 1974. Location of spoil grounds in sedimentary systems. Dock Harb. Auth. 55:120-124.

346 - Davies, I. M. 1976. The atmospheric deposition of some heavy metals in the Firth of Forth. ICES. C.M. 76/E:41.

347 - Davies, I. M. 1980. The role of suspended solids in the distribution and speciation of mercury in the Forth Estuary. ICES. C.M. 80/E:43.

348 - Davies, I. M., and J. M. Pirie. 1980. Evaluation a "Mussel Watch" project for heavy metals in Scottish coastal waters. Mar. Bio. 57:87-93.

349 - Davies, I. M. 1981. Survey of trace elements in fish and shellfish landed at Scottish Ports 1975-76. Scott. Fish. Res. Rep. No. 19. 28p.

350 - Davies, J. M. 1979. A direct method for studying the effects of pollutants on the sea bed. Scott. Fish. Bull. 45:31-32.

351 - Dawson, R. and J. P. Riley. 1977. Chlorine-containing pesticides and polychlorinated biphenyls in British coastal waters. Est. Coast. Mar. Sci. 5(1):55.

352 - Dayal, R., S. A. Oakley, and I. W. Duedall. 1978. Sediment geochemical studies of the 2800 m Atlantic Nuclear Waste Disposal site. Marine Science Research Center. State Univ. of New York. Stony Brook, NY. Final Report EPA Office of Radiation Programs. 98p.

Abstract: Sediment geochemical investigation of the abandoned 2800 m Atlantic nuclear waste disposal site reveals that the surrounding sediments can serve as an effective geochemical barrier to lateral or vertical migration of released radioactivity from the waste containers. The presence, concentration levels and distribution patterns of both ^{137}Cs and ^{134}Cs in sediments in the immediate vicinity of a waste container indicate sediment contamination by released radioactivity. ^{137}Cs distribution profile in a sediment core collected in the vicinity of a waste container suggests bioturbation as an effective mechanism for redistributing vertically the released radioactivity. In comparison with bioturbation, migration of radionuclides via molecular diffusion through pore waters is of minor significance. Considering bioturbation as an eddy diffusional process, a sediment mixing rate of 350 cm^2 is calculated.

353 - Dayal, R., M. G. Heaton, M. Fuhrmann, I. W. Duedall. 1979a. A geochemical and sedimentological study of the dredged spoil deposit in the New York Bight. State University of New York. Stony Brook, NY. 265p.

Abstract: Geochemical and sedimentological investigations of the dredged spoil deposit reveal that the metal concentrations in dredged spoil sediments are highly variable and considerably elevated over concentrations observed in sediment outside the deposit and in underlying natural sediment. Compared to metal enrichments reported for other coastal deposits, the enrichment observed in dredged spoil sediments are orders of magnitude greater. Based on the sedimentary record, the fluxes of sediment and associated metals to the dumpsite have been estimated for the dumping periods 1936-73 and 1973-78. The calculated rates and magnitudes of sediment and metal fluxes to the New York Bight, via dredged material dumping, are found to orders of magnitude higher than those reported for other coastal deposits.

354 - Dayal, R., A. Okubo, I. W. Duedall, and A. Ramamoorthy. 1979b. Radionuclide redistribution mechanisms at the 3800-m Atlantic nuclear waste disposal site. Deep-Sea Res. 26(12A):1329-1345.

Abstract: As part of an investigation of the abandoned site located off the coast of New Jersey. Several box cores were taken in Aug. 1976. The D/R/V Alvin was used to locate and examine waste canisters and to sample surrounding sediment. Both 135Cs and 134Cs were present in significant concentrations in sediment sampled close to a waste canister. The observed 137Cs concentrations are orders of magnitude higher than fallout background concentration levels in the area. The presence, concentration levels, and distribution patterns of the Cs isotopes indicate that the waste canister is the source of the observed radionuclides and that these elevated radionuclide concentration levels are highly localized around the canister. Considering the fine-fraction mineralogy and exchange capacity of sediment and the adsorption characteristics of the Cs ion, sediment at the study site apparently serves as a barrier to radionuclide migration. Because of the sediment profile of 137Cs in a submersible-collected sediment core in the immediate vicinity of a canister, bioturbation may be the dominant mechanism by which Cs is redistributed. Interaction of adsorption and bioturbation may accelerate interstitial migration of Cs+. A sediment mixing rate of 1.140 cm^2/yr is calculated using a mathematical model developed quantitatively to describe the observed 137Cs profile. The same model and the calculated mixing coefficient was applied to account for 134Cs depth profile. The total 137Cs activity released into the overlying water column as a result of bioturbation is a small fraction (0.3%) of the 137Cs released from the canister.

355 - Dean, David, and Harold H. Haskin. 1964. Benthic repopulation of the Raritan River estuary follows pollution abatement. Limnol. and Oceanogr. 9(4):551-563.

356 - Dean, R. B., and J. E. Smith, Jr. 1974. The properties and composition of sludges. In: Proc. Seminar on Methodology for Monitoring of the Marine Environment. U.S. Environmental Protection Agency. Office of Monitoring Systems. Washington, DC. EPA-600/4-74-004. pp. 259-274.

357 - Deb, P. K., A. J. Rubin, A. W. Launder, and K. H. Manuy. 1966. Removal of COD from wastewater by fly ash. Proc. 21st Ind. Waste Conference. Purdue University. Ext. Service. 21:848.

358 - De Clerck, R., J. Van de Velde, and W. Vyncke. 1973. On the effects of dumped organic industrial waste deriving from the production of proteolytic enzymes on density, distribution and quality of fish and shrimps. Aquaculture. 2:323-335. [See Ref. # 310].

359 - De Clerck, R., and J. Van de Velde. 1974. Biological description of a dumping area for industrial waste derived from a titanium oxide process. ICES. C.M. 74/E:29.

360 - De Fraja Frangipane, E., F. Mosetti, and F. Ramponi. 1975. Feasibility of marine disposal of sewage from the city of Trieste. In: E. A. Pearson, and E. de Fraja Frangipane (editors), Marine Pollution and Marine Waste Disposal. (Supplement to Progress in Water Technology.) Proceedings of the 2nd International Congress. San Remo. 17-21 December 1973. Pergamon Press. pp. 423-439. [See Ref. # 478].

361 - De Fraja Frangipane, E. 1979. Present state of knowledge on marine wastewater disposal. In: S. H. Jenkins, (editor). Marine Municipal and Industrial Wastewater Disposal. Proceedings of a conference, Sorrento, Italy 23-27 June 1975. (Progress in Water Technology Volume 4). Pergamon Press. Oxford. pp. 11-17.

362 - DeFalco, Paul, Jr. 1965. Hudson-Champlain and Metropolitan coastal comprehensive water pollution control project. In: Proceedings Conference in the matter of pollution of the interstate waters of the Hudson River and its tributaries - New York and New Jersey. U.S. Health, Education and Welfare. Metuchen, NJ. Vol. I:177-257.

363 - Department of Fisheries and Environment Canada (author, G. K. Sato). 1977. A summary of ocean dumping permits issued by Fisheries and Environment Canada during 1975 and 1976. Ottawa, Canada. 184p.

364 - Department of Fisheries and Environment Canada (authors, Sato and Hassan). 1978. A summary of ocean dumping permits issued by Fisheries and Environment Canada during 1977. Ottawa, Canada. 215p.

365 - Department of Fisheries and Environment Canada (authors, Marine Sciences and Information Directorate, Ocean Aquatic Sciences). 1979. A summary of permits issued under the ocean dumping control act 1978. Ottawa, Canada. 400p.

366 - DeSylva, Donald P. 1952a. The transparency, by black and white secchi disc, of the waters of Long Island Sound, Block Island Sound and Newport Bight; Cruise Stirni III, January-February 1952. In: John C. Ayers et al., The Hydrography of the New York Area. Status Report No. 18. (Contract N6 DNR 264, task 15). Cornell Univ. Ithaca, NY.

367 - DeSylva, Donald P. 1952b. The distribution of water color in Long Island Sound, Block Island Sound and Newport Bight. In: John C. Ayers et al., The Hydrography of the New York Area. Status Report No. 19. (Contract N6 DNR 264, task 15). Cornell Univ. Ithaca, NY.

368 - Dethlefsen, V. 1972. On the effects of red mud on marine organisms. ICES. C.M. 72/E:7.

369 - Dethlefsen, V. 1973. Zur Frage des Fischvorkommens im Dunnsaureverklappungs-gebiet nordwestlich Helgolands. Arch. FischWiss. 24:64-75.

370 - Dethlefsen, V. 1974. Review of scientific aspects of sea disposal of red mud. ICES. C.M. 74/E:10.

371 - Dethlefsen, V. 1978. Occurrence and abundance of some skeletal deformities diseases and parasites and major fish species in dumping areas off the German coast. ICES. C.M. 78/E:8.

372 - DeTurk, Richard, and Robert Wood. 1970. Managing the national environment: a regional plan for water, sewage, air and refuse. Tri-state Transportation Commission. New York, NY. 38p.

373 - Deutsche Forschungsgemeinschaft. 1973. Litoralforschung abwasser in Küstennahe. [Shore Study of waste water in the coastal area]. Bad Godesburg, Germany. 93p.

374 - Devanas, M. A., C. D. Litchfield, C. McClean and J. Gianni. 1980. Coincidence of cadmium and antibiotic resistance in New York Bight Apex benthic microorganisms. Mar. Pollut. Bull. 11(9):264-269.

Abstract: Microorganisms resistant to 1-500 ppm cadmium were isolated over a 11-month period from sediments at the sewage sludge, dredge spoils and industrial acid waste disposal sites, as well as at an estuarine outflow to the New York Bight Apex. Tests for antibiotic resistance in these isolates revealed that 94% of the isolates were resistant to one or more antibiotics and that 91% of the original isolates could be characterized by multiple drug resistance. Different selective pressures may account for the various genera and antimicrobial resistance patterns observed. The possibility of extrachromosomal linkage of cadmium and streptomycin resistance is discussed.

375 - Devine, Michael, Edward R. Meyer, and Tom O'Connor. 1981. Site description and dumping history and study plan. In: Assessment Report on the Effects of Waste Dumping in 106-Mile Ocean Waste Disposal Site. NOAA/ODP Dumpsite Evaluation Report 81-1. Rockville, MD. pp. 3-10.

376 - Dewling, R. 1973. Oil spills, hazardous materials, vessel protection, and ocean dumping. In: American Ordance Association Proceedings of 3rd Meeting on Environmental Protection. Fort McNair, Washington, DC. pp. 99-106.

377 - Dewling, R. T., and P. W. Anderson. 1976. New York Bight I: Ocean dumping policies. Oceanus. 19(4):2-10.

Abstract: 80% of all ocean dumping taking place off N.Y. and N.J. coasts. Examining the disposal of municipal sludges, industrial wastes, and acid wastes; dumping permits, environmental impacts and a look at possible alternatives.

378 - Dewling, R. T., and P. W. Anderson. 1979. Sludge dumping: Meeting the 1981 deadline. (sewage; industrial waste; dredging spoil) In: 15th Annual Conference, Marine Technology 1979 on Ocean Energy, New Orleans, October 10-12, 1979. Marine Technology Society. Washington, DC. pp. 34-39.

379 - Diaz, Henry. 1979. Atmospheric conditions and comparison with past records. In: R. L. Swanson and C. J. Sinderman (editors), Oxygen Depletion and Associated Benthic Mortalities in New York Bight, 1976. NOAA Professional Paper No. 11. U.S. Dept. of Commerce. Rockville, MD. pp. 51-77.

Abstract: Some of the relevant atmospheric forcing fields affecting the surface environment in New York Bight were examined for February through August 1976, the period before and during the oxygen-depletion event. The same kinds of physical conditions in the historical record were compared to identify possible similarities.

Three salient points relating to atmospheric conditions emerged. First, sea-surface temperatures throughout the Bight were relatively high early in the year when compared with the record for the past 100 years, but there were other similar occurrences in the record. Also, February and March air temperatures over the northeast were near their warmest levels in the past century. The unusually warm temperatures may have aided in the early development and subsequent strengthening of stratification of Bight waters. Second, the monthly surface wind patterns showed persistent south to southwesterly flow during the entire period, blowing with above-normal constancy during May and June. Steady wind conditions could have a pronounced effect on Bight circulation and exchange processes. The general atmospheric circulation patterns over eastern North America departed considerably from the norm during the February through June months. This resulted in a minimum of storm activity over the Bight, and presumably less mixing of the water column, compared with the record for the past 25 years. And third, vertical motion calculations using both open-ocean and coastal upwelling estimates indicate that upwelling/onwelling may have been prevalent during most of the analysis period.

380 - Dick, R. I. 1975. Alternative to marine disposal of sewage sludge. ACS Symp. Ser. 18:453-466.

381 - Ditsworth, G. R., A. M. Teeter, R. J. Callaway. 1978. New York Bight suspended matter and oceanographic data: 1973-1974 total suspended matter: traverse stations June 1974, and prior cruises total suspended matter and physical oceanographic data June-July, 1974 cruise total suspended matter, December, 1974 cruise. EPA Environmental Research Laboratory. Corvallis, OR. 66p.

Abstract: The concentration and size of particulate matter suspended in the water column were determined as part of an overall study of sludge dispersion in New York Bight. Sampling points were established along a traverse which extended from New York Harbor, through the dump zone, to buoy NB. Data obtained from surveys in 1973 and 1974 are given in this report along with other oceanographic data obtained during a June 27-July 1, 1974 cruise.

382 - Dixon, T. R., and T. J. Dixon. 1979. Munitions in British coastal waters. Mar. Pollut. Bull. 10(12):352-357.

383 - Documentation Associates. 1977. Deep ocean mining environmental study (DOMES) literature survey. Documentation Associates Information Services. Incorporated. Los Angeles, CA. 231p.

384 - Domenowske, R. S., and R. I. Matsuda. 1969. Sludge disposal and the marine environment. Jour. Water Poll. Control Fed. 41(9):1613-1624.

385 - Dominik, J., V. Forstner, A. Magini, and Hans-Erich Reineck. 1978. ^{210}Pb and ^{173}Cs chronology of heavy metal pollution in a sediment core from the German Bight (North Sea). Senckenbergiana marit. 10(4/6):213-227.

386 - Dooley, H. D., R. Johnston, and D. C. Moore. 1979. Report on pre-dumping conditions at St. Abbs and Bell Rock Sewage Sludge Dumping Grounds (Firth of Clyde). Dept. of Agriculture and Fisheries for Scotland. Aberdeen Marine Laboratory Working Paper. No. 79/5. Aberdeen, Scotland. 12 pp. + Appendices.

387 - Dooley, H. D., and G. Slesser. 1981. Distribution of cryolite sludge in the Inner Moray Firth as inferred from hydrographic measurements. Dept. of Agriculture and Fisheries for Scotland. Aberdeen Marine Laboratory Working Paper. No. 81/2. Aberdeen, Scotland. 13p.

388 - Drake, D. E., and P. G. Hatcher. 1974. Dispersal of waste solids in New York Bight. Abstracts with Programs. GSA. 6(7):715.

Abstract: The distribution and composition of particulate suspended matter in the New York Bight indicate that much of the solid wastes disposed of in the Bight Apex are initially transported northward and partially deposited in the Christiaensen Basin (the headward portion of the shelf channel). Suspended sediment distributions and current meter measurements indicate a general northward drift of subsurface water. This advective current forms the western limb of a clockwise gyre in the Bight Apex. The gyre appears to be strongly influenced by surface winds and estuarine flow. The carbohydratge/TOC ratio (a possible indicator of waste materials) and the suspended sediment data show that a portion of the waste materials transported into the basin are moved northward toward the shoreline, then eastward (seaward) by the gyre. Desposition of some of this sediment occurs beneath the southward-moving limb of the gyre on the eastern flank of Cholera Bank. Additionally, some of the waste materials appear to complete the circuit and are deposited in the Hudson shelf channel 15-20 km south of the disposal sites. Based on suspended sediment concentrations above the dredge spoil and sewage sludge sites and current meter data, we estimate that at least 10-20% of the sewage sludge site, no sludge at all is found in bottom sediments because of currents and the relative topographic high at the site.

389 - Drake, David E. 1974. Suspended Particulate Matter in the New York Bight Apex: Sept.-Nov., 1973. NOAA Technical Report ERL 318 - MESA 1. Boulder, CO. 53p. NTIS No. COM 75-10913/AS.

Abstract: The distribution of suspended particulate matter in the New York Bight Apex was studied during the fall of 1973 as part of the Marine Ecosystems Analysis program of NOAA.

Five surveys from September through November revealed consistent suspended matter distributions that reflect the Bight Apex water circulation. Two major currents dominate during the fall season of limited river flow and gradually weakening water column stratification: (1) relatively fresh surface water, containing between 1 to 4 mg/l of suspended particles, flows from Hudson estuary and down the New Jersey coast within 5 to 10 km from shore; and (2) northward flow along Hudson shelf channel occurred during all surveys. Low concentrations of suspended matter and ashed-weight fractions dominated by diatoms indicate a central shelf origin for the shelf-channel current. Surface winds have a strong influence on the currents at all depths in the Bight Apex. During the fall season, winds are predominantly from western quadrants and tend to sweep through upwelling and northward flow along Hudson shelf channel. Current meter records appear to show that the shelf channel flow waxes and wanes in response to the strength and direction of surface winds.

Total suspended matter distributions and dispersion patterns of iron particles dumped at the acid-waste dumpsite support the existence of a clockwise gyre in the central portion of the area during the fall season; the shelf-channel current forms the western limb of this gyre. Dredge spoil and sewage sludge dumped near the head of the shelf channel settle within the shelf channel to form organic-rich mud lenses. However, some of this material is entrained by the northward valley current and transported from the shelf channel to the northward over Cholera Bank. The bank crest does not accumulate fine sediment owing to relatively vigorous wave surge. However, as the turbid current turns south along the east side of Cholera Bank, fine-grained, organic-rich material begins to settle. The result is two disconnected lenses of organic-rich sediment on either side of the sand bank. Contaminated sediments are widely dispersed in the Bight Apex water column. However, major dispersion appears to be centered on the Hudson shelf channel with good evidence that material dumped at the valley head is transported both up and down the channel in substantial quantities.

One of our surveys coincided with a moderate storm which produced steep, 1.0- to 3.0-m seas. Resuspension of clay and silt-sized material occurred throughout the area, and, owing to the near absence of water column stratification, this sediment was rapidly mixed toward the sea surface. A minimum of 10^4 metric tons of sediment was entrained during this storm, and it is clear that such high energy events strongly influence the fate of natural and artificial sediments in the New York Bight.

390 - Drake, David E. 1977. Suspended particulate matter in the New York Bight Apex, fall 1973. Jour. of Sedimentary Petrology. 47(1):209-228.

Abstract: Four surveys of suspended particulate matter in the New York Bight Apex from September through November 1973 revealed consistent distributions that reflect the water circulation. Two major currents were dominant during the fall season of limited river flow and gradually weakening stratification of the water column: (1) low-salinity (29 to 31 ppt) surface water, containing between 1 to 4 mg/1 of suspended particles, flows from Hudson estuary southward along the New Jersey coast within 5 to 10 km from shore; and (2) northward flow along Hudson shelf channel occurred during all surveys. Low concentrations of total suspended matter and ash fractions dominated by diatoms indicate a central shelf origin for the shelf-channel current.

The distribution of total suspended matter and the dispersion of patterns of iron particles released at the acid-waste dumpsite support the existence of a clockwise gyre in the central of the area during autumn; the shelf-channel current forms the western limb of this gyre. Dredge spoil and sewage sludge dumped near the head of the shelf channel settle into this depression to form mud lenses rich in organic matter. Some of this material is entrained into a turbid bottom layer by the northward current and transported from the depression to the northeast over Cholera Bank. The bank crest does not accumulate fine sediment owing to relatively vigorous wave surge. As the turbid current turns south along the east side of Cholera Bank, fine grained, mineral and organic material begins to settle, forming two disconnected lenses of mud on either side of the sand bank.

Contaminated sediments are widely dispersed throughout the water column in the Bight Apex. Major dispersion appears to be centered on the Hudson shelf channel, and there is good evidence that material dumped near the channel head is transported up and down the channel in substantial quantities.

391 - Draxler, A. F. J. 1979. Transient effects of ocean waste water sludge dumping. Jour. Water Poll. Control Fed. 51(4):741-748.

Abstract: Wastewater sludge (3.5 x 10^6 annually) from the New York metropolitan area is barged to a site approximately 22-km equidistant between Long Island, New York, and Sandy Hook, New Jersey, where it is dumped in water 29-m deep. The sludge contains approximately 5% solids, which are 55% organic material, and has a large nutrient and heavy metal content. The spatial and seasonal distribution of nutrients in the water and metals in the sediment of the Bight and the contribution of ammonium nitrogen by waste dumping have both been previously investigated. In contrast to those studies, the purpose of these experiments was to investigate the dynamic processes that follow a single sludge dump: the settling of particulates, the dilution of nutrients, and the changes in oxygen concentration of the receiving water.

392 - Drucker, Barry S. 1978. Distribution and economic analysis of selected heavy minerals in the inner New York Bight. Master's Thesis. Long Island University. Brookville, NY.

Abstract: Ninety-four sand samples from the inner New York Bight have been collected and analyzed with respect to the heavy mineral concentration of selected economic mineral species. Three general regions have been differentiated according to total heavy mineral weight percentages.

The regional variations include: a greater concentration of kyanite, sillimanite, zircon, and monazite in the northern area with diminishing concentrations toward the south. Staurolite is common throughout the sampling area and rutile is relatively rare. Ilmenite and leucoxene are concentrated in the southern area.

These variations suggest that the sediments of the inner New York Bight have several sources, and that this material has undergone selective sorting and dispersal by several different continental and oceanographic processes. The distribution of the individual minerals are secondarily affected by differences in particulates size, specific gravity and morphology.

Several sample sites from two major areas of heavy mineral concentration have been analyzed with respect to economic value. The data are preliminary and in no way indicate expected recoverable reserves, but they do show that vast tonnages of "heavies" exist, and that these deposits could be termed "potential future resources" until more detailed vertical and lateral sample selections are completed.

393 - Duca, C., and A. Corda. 1969. Evacuation des eaux residuaires dans la mer Noire. (Emptying waste waters into the Black Sea.) Rapp. P.-V. Reun. Commn Int. Explor. Scient. Mer Mediterr. 19:917-918.

394 - Duce, R. A., G. T. Wallace, and B. J. Ray. 1976a. Atmospheric trace metals over the N.Y. Bight. Graduate School of Oceanography. University of Rhode Island. Kingston, RI. 44p.

Abstract: Iron, aluminum, zinc, cadmium, lead, and sodium were determined in a series of near surface atmospheric particulate samples collected form R/V WESTWARD from several areas of the New York Bight. The atmospheric concentrations were generally about 10-20% of the mean concentrations observed over a two-year period at several locations in New York City. This was not the case for atmospheric sodium, whose primary source is the sea. Atmospheric concentrations of the other elements over the Bight showed a dependence on distance from the coast and surface wind direction.

395 - Duce, R. A., G. T. Wallace, and B. J. Ray. 1976b. Atmospheric trace metals over the New York Bight. NOAA TR ERL 361-MESA 4. Boulder, CO. 17p. NTIS No. PB 263817/AS.

396 - Duebler, Earl E., Jr. 1952. Water color distribution in the Newport Bight and New York Bight areas, cruise Stirni I, July-September 1951. In: John C. Ayers et al., The Hydrography of the New York Area. Status Report No. 14. (Contact N6 DNR 264, task 15). Cornell Univ. Ithaca, NY.

397 - Duedall, I. W., M. J. Bowman, and H. B. O'Connors. 1974a. Sewage sludge and ammonia concentrations in the New York Bight Apex. State Univ. of New York. Stony Brook, NY. 37p.

Abstract: Water column ammonia concentrations were determined at several stations at or near the sewage sludge dump site in the New York Bight Apex and also at several background stations located on a perimeter surrounding the dump area. There areas are suspected to be grossly affected by the input of sewage sludge barged from the New York metropolitan region. Parachute drogues were used to track the movement of water in the study area. Within the dump site, ammonia concentrations were patchy and usually increased from surface to bottom, with surface concentrations less than 1 uM and bottom and bottom concentrations in the range 1-9 uM. However, at one station that had recently received sludge input, the surface ammonia concentration was about 200 uM. Except for two background stations west of the dump site, all background stations showed no significant vertical variation in ammonia. The observations suggest that the effects of sludge dumping manifest themselves throughout the water column within a short time, but that such high levels probably do not persist for long periods.

398 - Duedall, I. W., H. B. O'Connors, J. H. Parker, W. Miloski, and G. Hulse. 1974b. On the tidal variations of physical and chemical properties measured on transects between Sandy Hook, New Jersey, and Rockaway Point (Long Island), New York. State Univ. of New York. Stony Brook, NY. 28p. 102 graphs. 58 data tables.

Abstract: On November 5 and 8, 1973, two preliminary surveys were carried out to examine the distribution and tidal variation in physical and chemical properties along two transects between Sandy Hook (New Jersey) and Rockaway Point (Long Island). The properties measured were: salinity, temperature, dissolved oxygen, chlorophyll fluorescence, turbidity, phosphate (dissolved), ammonia, nitrite, and nitrate. We found that the largest tidal variation in properties occurred on the Sandy Hook side of the transects, where the water contains a greater portion of river water. The optimum time between resampling any given station is somewhere between 30 and 90 minutes. For future transport measurements, we suggest the use of six strings of current meters to be moored within the transect region; each string should contain at least three current meters.

399 - Duedall, I. W., H. B. O'Connors, and B. Irwin. 1975a. Fate of wastewater sludge in the New York Bight Apex. Jour. of Water Poll. Control Fed. 47(11):2702-2706.

Abstract: The fate of wastewater sludge dumped into coastal water is largely unknown. In the New York Bight Apex (hereafter called the Apex), investigations concerning the impact of dumped sludge are presently being pursued by a number of scientists, federal agencies, and Congressional Committees. Large quantities of sludge from the New York metropolitan region are dumped almost daily in the Apex at a location about 24 km from the entrance to New York Harbor (Figure 1). The depth of water at the dump site is about 30 m. Pararas-Carayannis[8] estimated that the average annual input of sludge to the Apex was 3.2×10^6 cu m during the period 1965-70. Sludge contains significant quantities of organic carbon and nitrogen compounds; therefore, it is reasonable to assume that such material when dumped in the ocean, may act as a metabolic substrate for marine microorganisms.

The objectives of this report are (a) to report some values for the total organic carbon (TOC), organic nitrogen (ON), total carbohydrate (TCH, and calorie content of dried sludge, and (b) to make some inferences as to the fate of ocean-dumped sludges based on the sludge analyses, some preliminary carbon and nitrogen analyses of sediment from the Apex, and recently reported in situ oxygen utilization observations in sediments underlying a sludge disposal area.

400 - Duedall, I. W., M. J. Bowman, and H. B. O'Connors, Jr. 1975b. Sewage sludge ammonium concentrations in the New York Bight Apex. Estuar. and Coast. Mar. Sci. 3:457-463.

Abstract: Water column ammonium concentrations were determined at several stations at or near the sewage sludge dump site in the New York Bight Apex on 30 and 31 July, 1973, and also at several stations located on a perimeter surrounding the dump area. Parachute drogues were used to track the movement of water over a 31-hour period. Within the dump site, ammonium concentrations were patchy and usually increased from surface to bottom with concentrations in the ranges of less than 1 uM and 1-9 uM, respectively. However, at one station that had recently received sludge input, the surface ammonium concentration was over 500 uM and the bottom concentration was about 200 uM. All except two background stations showed no significant vertical variation in ammonium. The observations suggest that sludge dumping increases ammonium content of the water column, but that high levels probably do not persist for long periods.

401 - Duedall, I. W., R. Dayal, A. Okubo, K. W. Jones, H. W. Hobart, and, R. E. Shroy. 1975c. Short-term variability in the composition and distribution of sewage sludge in the ocean after dumping. Trans. Amer. Geophys. Union, Abstr. Publ. EOS. 59(4):244-245.

402 - Duedall, I. W., H. B. O'Connors, and S. A. Oakley. 1976. Sewage sludge disposal and water column properties in the New York Bight Apex. State Univ. of New York. Stony Brook, NY. 10p.

Abstract: An oceanographic study in the N.Y. Bight Apex designed to monitor water properties after a sewage sludge in the ocean produces large abundances of NH_4^+, PO_3^- and suspended solids, some sludge components were observed to reach the bottom (21 m) in less than 40 minutes, after about 200 minutes surface and intermediate depth concentrations of dissolved and particulate phases are back to baseline conditions. Bottom concentrations of both dissolved and particulate phases, however, remained elevated for the duration of our experiment. It is not known for how long the concentrations remained high.

403 - Duedall, I. W., and H. B. O'Connors. 1976. Final Report Part I: The abundances, distribution and flux of nutrients and chlorophyll a in the New York Bight Apex. Part II: Sandy Hook/Rockaway Point transect study: Data Report of Cruises from November 1973 to June 1974. NOAA DR ERL MESA-20. Boulder, CO.

Abstract: Part I: Five cruises were made between Nov. 1973 and June 1974 to measure tidal, spatial, and seasonal changes in salinity, temperature, and the concentrations of ammonium, nitrite, nitrate, phosphate, silicic acid, chlorophyll a and suspended matter in the waters between Sandy Hook, N. J. and Rockaway Point, N.Y. Concentrations of nutrients and chlorophyll a were much greater than those in adjacent coastal waters. Sewage effluent is the main source of chemicals in the waters near the metropolitan region, and effects from the Hudson river are also noted. Part II. This report presents tables of data along with surface current, weather and sea conditions during the cruises made in the N.Y. Bight area. Computed values of Sigma-t and degree of dissolved oxygen saturation are also included.

404 - Duedall, I. W., H. B. O'Connors, J. H. Parker, R. E. Wilson and A. S. Robbins. 1977a. The abundances, distribution and flux of nutrients and chlorophyll a in the New York Bight Apex. Estuar. and Coast. Mar. Sci. 5:81-105.

Abstract: Tidal, spatial, and seasonal changes in salinity, temperature, and the concentrations of ammonium, nitrite, nitrate, phosphate, silicic acid, chlorophyll a and suspended matter in the waters between Sandy Hook, New Jersey, and Rockaway Point, New York, were measured during five cruises which took place between November 1973 and June 1974. Over this period concentrations of nutrients and chlorophyll a were much greater than those found in the adjacent coastal waters. The main source of the ammonium, nitrite, and phosphate is sewage effluent which is discharged into the waters surrounding the New York metropolitan region; nitrate comes mainly from the Hudson River and silici acid is discharged in large amounts from river and sewage sources.

The largest tidal variation in salinity and nutrient and chlorophyll a concentrations occurs near Sandy Hook where the Hudson River discharge has the greatest influence. Near Rockaway Point, nutrient and chlorophyll a concentrations are generally lower and salinities higher than those observed near Sandy Hook because of the inflow of Bight water by non-tidal currents. During the spring freshet nutrient concentrations, especially ammonium, are low along the transect due to (1) dilution by the spring freshet and (2) utilization by the abundant phytoplankton. Flux calculations for the June observations indicate that most of the nutrients and chlorophyll a are being transported from the lower Hudson Estuary into the New York Bight Apex.

405 - Duedall, I. W., H. B. O'Connors, S. A. Oakley, and H. M. Stanford. 1977b. Short-term water column perturbations caused by waste water sludge dumping in the New York Bight Apex. Jour. Water Poll. Control Fed. 49(10):2074-2080.

Abstract: Water column properties in the New York Bight Apex were monitored before and after a controlled spot dump of wastewater sludge. Ammonium, phosphate, and suspended solids in the discharged wastewater sludge dispersion plume caused the greatest water column pertubations. Temperature, salinity, pH, and the concentrations of dissolved oxygen and chlorophyll a in the water column were not significantly affected by the waste water dumping, thus suggesting, for the hydrographic and discharge conditions of the experiment, a relatively rapid descent rate for the particular wastewater sludge. By 110 minutes after dumping, the water column above the thermocline (17 m) had nearly returned to its background condition with respect to the variables measured. Below the thermocline, however, concentrations of ammonium, phosphate, and suspended solids remained high for a period greater than 2.5 hours.

406 - Duedall, I. W., R. Dayal, J. H. Parker, H. W. Kramer, K. W. Jones, and R. E. Shroy. 1978. Distribution, composition and morphology of suspended solids in the New York Bight Apex. In: M. Wiley (editor), Estuarine Interactions. Vol. 3. Academic Press. New York, NY. pp. 533-563.

407 - Duedall, I. W., H. B. O'Connors, R. E. Wilson, and J. H. Parker. 1979. The lower bay complex. MESA New York Bight atlas monograph 29. MESA New York Bight Project and New York Sea Grant Institute. Albany, NY. 47p.

Abstract: The Lower Bay complex is the seaward part of New York Harbor and includes Raritan, Sandy Hook, and Lower bays. It connects with Upper Bay through a narrow constriction between Staten Island and Brooklyn. The bay complex is relatively shallow (5 to 20 m or 16 to 66 ft) but has an irregular topography due mainly to the numerous ship channels in Lower and Raritan bays. The weather in the bay complex is typical of a midlatitude coastal region, with the adjacent Atlantic Ocean acting as a buffer.

A dynamic and complex estuarine system, the bay complex receives a large, seasonally variable inflow of fresh water originating mainly from the Hudson River with lesser amounts from the Raritan and Passaic rivers. Sewage effluent is also a significant source of fresh water. The nontidal inflow of salt water through the Sandy Hook-Rockaway Point transect is confined to Ambrose and Sandy Hook channels and also through the entire water column near Rockaway Point. Because of the variable inflow of fresh water, the distribution of water properties (salinity, nutrients, chlorophyll *a*) also varies seasonally).

Tides and tidal currents in the bay complex are semidiurnal; their patterns are complicated because of the shape of the bay, the variation in freshwater discharge, Coriolis acceleration, and the intricate connection of waterways. Tidal variations in water properties are large and can be perturbed significantly by storms.

Sewage effluent from the New York metropolitan area is the principal source for the high concentrations of nutrients observed in the bay complex. A large fraction of these nutrients are consumed in biological processes occurring within the bay complex. There is a net transport of nutrients and chlorophyll *a* to the Apex of New York Bight. The nutrients originating from the bay complex and transported seaward have been implicated as an important factor leading to the decline of oxygen in bottom waters of the Bight during summer periods.

408 - Duedall, Iver W., Frank J. Roethel, James D. Seligman, Harold B. O'Connors, Jeffrey H. Parker, Peter M. J. Woodhead, Ramesh Dayal, Bart Chezar, Beverly K. Roberts and Hugh Mullen. 1981. Stabilized power plant scrubber sludge and fly ash in the marine environment. In: B. H. Ketchum, D. R. Kester and P. K. Park (editors), Ocean Dumping of Industrial Wastes. Plenum Press. New York, NY. pp. 315-346.

Abstract: Solid (stabilized), brick-like forms of scrubber sludge and fly ash from coal combustion were tested for their physical properties, leaching behavior in seawater, effect on organisms, and long term stability in a marine environment. Laboratory results showed that trace elements did not leach significantly from test samples. In the field, colonization and overgrowth of test blocks progressed very rapidly, and block integrity, as determined by compressive strength, was maintained for over 500 days.

These preliminary results and other results from ongoing studies suggest that marine disposal of scrubber sludge and fly ash, in the stabilized form, may be environmentally acceptable.

409 - Duedall, Iver W., B. H. Ketchum and P. Kilho Park. (In Press). Overview: Global Inputs, Characteristics, and Fates of Ocean-Dumped Industrial and Sewage Wastes. In: I. W. Duedall, B. H. Ketchum, P. K. Park, and D. R. Kester (editors), Wastes in the Ocean, Vol. I: Industrial and Sewage Wastes in the Ocean. John Wiley and Sons, Inc. New York, NY.

Abstract: Records of ocean dumping permits issued by member countries of the London Dumping Convention are published annually by Inter-Governmental Maritime Consultative Organization (IMCO). The reports from 1976 through 1979 were analyzed for annual trends in the permit activity (that is, number of permits and kinds and quantities of permits). During the 1976-1979 period over 2000 permits were issued, mainly for the disposal of dredged material, industrial wastes, and sewage sludge. For these wastes, total annual tonnages ranged from 35 x 10^6 and 245 x 10^6 metric tons for 1976 and 1978, respectively. Estimated tonnages of industrial wastes ranged from 8 x 10^6 in 1976 to 18 x 10^6 tons y^{-1} in 1979, with U.S.A., France, and the United Kingdom leading all other countries as dumping of the majority of wastes. Annual tonnages of sewage sludge, essentially all from U.S.A., and the United Kingdom during the 1976-1979 period remained nearly constant at 16 x 10^6 tons y^{-1}.

Some physical and chemical properties, taken mainly from the scientific literature, of the following industrial and sewage wastes were summarized: Acid-iron, pharmaceutical, coal fly ash (fly ash alone and consolidated fly ash and scrubber sludge) and sewage sludge. Abundances of trace elements in these wastes varied by orders of magnitude when compared on an element-to-element basis. Fly ash and sewage sludge have the highest concentrations of elements of environmental concern (arsenic, cadmium, lead).

The distribution and behavior, effects, and fate of ocean-disposed wastes are complex requiring knowledge of physical, chemical and geochemical, biologic and sedimentary processes. More research is required in all the disciplines to understand the important scientific aspects of industrial and sewage wastes in the ocean.

410 - Duedall, I. W. et al. (In Press). Diffusion of calcium and sulfate ions in stabilized coal wastes. In: I. W. Duedall, B. H. Ketchum, P. K. Park and D. R. Kester (editors), Wastes in the Ocean, Vol. I: Industrial and Sewage Wastes in the Ocean. John Wiley and Sons, Inc. New York, NY.

Abstract: Fly ash and scrubber sludge wastes (mainly calcium sulfite hemihydrate and gypsum) from coal burning power plants were stabilized with lime to produce solid, brick-like forms. The flux of calcium (Ca^{2+}) and sulfate (SO_4^{2-}) ions from blocks placed in test tanks containing seawater or estuarine water was measured for 168 days for one waste and 147 for another. Initially, fluxes of Ca^{2+} and SO_4^{2-} leaving the blocks were about 10^{-7} mole mm^{-2} day^{-1} for two different wastes with approximately the same fly ash: sludge ratio of 1:1; at the end of the experiment, fluxes had decreased to 2×10^{-8} to 3×10^{-8} mole mm^{-2} day^{-1}. A one-dimensional model based on diffusion was developed to describe the fluxes. The model predicts diffusivities of 1.2×10^{-9} to 3×10^{-9} cm^2 s^{-1} and the depth (x_c) of penetration of the diffusion process: in 10 days, $x_c = 0.06$ to 0.10 cm; in 365 days x_c - 0.39 to 0.61 cm; and in 30 years, $x_c = 2.1$ to 3.3 cm. The results of the model suggest that stabilized coal waste blocks will have a long life in seawater if erosion and biological processes do not have a major effect on block properties.

411 - Duedall, I. W., B. H. Ketchum, P. K. Park and D. R. Kester (editors). (In Press). Wastes in the Ocean, Vol. I: Industrial and Sewage Wastes in the Ocean. John Wiley and Sons Inc. New York, NY.

412 - Duinker, J. C. and R. F. Nolting. 1974. Trace metals in the Southern Bight (copper, zinc, manganese and iron). ICES. CM. 74/E:26.

413 - Duinker, J. C., and R. F. Nolting. 1977. Dissolved, and particulate trace metals in the Southern Bight. Mar. Pollut. Bull. 8(3):65-71.

414 - Duinker, J. C. 1980. Processes affecting the behaviour of contaminants (metals and organochlorines) during estuarine mixing and in coastal areas, with particular reference to the Southern North Sea. ICES. C.M. 80/E:34.

415 - Duncan, J. W. K. 1969. Sewage sludge disposal by pumping and barging into sea. Nigerian J. Sci. 3(1):69-81.

416 - Duncan, R. N. 1974. The 1972 convention on the prevention of marine pollution by dumping of wastes at sea. Jour. of Maritime Law and Commerce. 5(2):299-315.

417 - Dunster, H. J. 1969. United Kingdom studies on radioactive materials released in the marine environment. Biological Implications of the Nuclear Age. Symposium. Proceedings. (Held in Livermore, Calif., March 5-7, 1969). Sponsored by University of California. Lawrence Radiation Laboratory. Bio-medical division. Livermore and U.S. Atomic Energy Commission. Division of Technical Information. Oak Ridge, TN. pp. 145-155.

418 - Eagle, R. A. (Undated). Assessment techniques for examining the biolgical effects of waste disposal at sea. U.K. Min. of Agr., Fisheries and Food (MAFF). Burnham-on-Crouch. Internal Rep. 152p. Mimeo.

419 - Eagle, R. A. 1973. Benthic studies in South East of Liverpool Bay. Estuar. and Coast. Mar. Sci. 1:285-299.

420 - Eagle, R. A., M. G. Norton, R. S. Nunny, and M. S. Rolfe. 1978a. The field assessment of effects of dumping wastes at sea: 2. Methods. U.K. Min. of Agr., Fisheries and Food (MAFF). Lowestoft. Fisheries Res. Tech. Rept. No. 47. 24p.

Conclusion: The methods employed in MAFF's surveys of waste disposal areas to date have been described. However, further work continues on the development of new or improved monitoring methods, for application in the field, in the laboratory and in the work-up of the resulting data. Developments of significance will continue to be reported in this series of reports.

Although it was intended to provide sufficient detail of the procedures adopted for most purposes, details of a practical nature as well as detailed descriptions of computer programs etc. have had to be omitted.

421 - Eagle R. A., P. A. Hardiman, M. G. Norton, and R. S. Nunny. 1978b. The field assessment of effects of dumping wastes at sea: 3. A survey of the sewage sludge disposal area in Lynne Bay. U.K. Min. of Agr., Fisheries and Food (MAFF). Lowestoft. Fisheries Res. Tech. Rept. No. 49. 22p.

Abstract: The survey of the area in Lyme Bay used for the dumping of sewage sludge shows dispersion processes to be ineffective in preventing accumulations of dumped material in the sediments. Despite the small quantity of sludge dumped (62000 m^3/year) levels of carbon, nitrogen and trace metals such as mercury, copper, zinc, lead, nickel and chromium are significantly elevated in the area most commonly used for dumping.

The addition of organic materials by dumping does not appear to have been utilized by any of the benthic species and there is no evidence of enrichment of numbers of individual species or of the number of species. In the zone most contaminated by dumping, however, there appeared to be a slight reduction in the number of species found in other parts of the survey area.

422 - Eagle, R. A., P. A. Hardiman, M. G. Norton and R. S. Nunny. 1979a. The field assessment of effects of dumping wastes at sea: 4. A survey of the sewage sludge disposal area off Plymouth. U.K. Min. of Agr., Fisheries and Food (MAFF). Lowestoft. Fish. Res. Tech. Rept. No. 50. 24p.

Conclusions: This survey, of the dumping ground south of Plymouth and the surrounding area, has revealed only minor effects attributable to the dumping of sewage sludge.

Zones to enrichment by organic detritus and trace metals from sewage sludge have been identified, but in all cases the concentrations are less than those inshore of the dumping site which are associated with coastal discharges and the deposit of dredge spoil. The area is one of high dispersion due to the resuspending action of winter storms, and thus significantly increased accumulations are unlikely to result at the current rate of dumping.

The investigation of the macrobenthos has shown the area to support a diverse fauna. Effects of sludge dumping on the benthos cannot be detected.

423 - Eagle, R. A., P. A. Hardiman, M. G. Norton, R. S. Nunny, and M. S. Rolfe. 1979b. The field assessment of the effects of dumping wastes at sea: 5. The disposal of solid wastes off the north-east coast of England. U.K. Min. of Agr., Fisheries and Food (MAFF). Lowestoft. Fish. Res. Tech. Rept. No. 51. 34p.

Conclusions: Surveys have allowed the distribution of dumped colliery waste and fly ash to be determined in an area extending from south of the Wear dumping ground to north of the Blyth dumping ground.

The primary areas affected by tipping are those where deposition of the waste occurs immediately after dumping. These are found, in most cases, inshore of the areas designated for dumping.

Although the waste reaches the sea bed very soon after dumping, subsequently some dispersion of a proportion of it takes place by movement in suspension and as bedload. Thus sediments surrounding the zones of active tipping also contain particles of waste origin. The most readily defined dispersion pathways for these wastes are those arising from the southward movement of sand-sized particles as bedload.

The effects of waste disposal on the benthos have been investigated in two of the four dumping areas. Fly ash and colliery waste dumping has caused severe depletion of the benthos (both in terms of species and abundance) in the areas where the waste is tipped with greatest intensity. There is evidence that some recovery has taken place on older and weathered waste deposits. Thus, the productivity of the areas currently affected could increase following a cessation in tipping.

424 - Eberman, J. W. 1956. Sewage and Indutrial Wastes. 28(11):1365-1370.

425 - EEC. 1976. Marine pollution: EEC draft directive on dumping at sea. Europe Environment No. 15. Jan. 10. pp. 11-13.

426 - Egan, W. G., and J. M. Cassin. 1973. Correlation of in situ fluorescence and bioluminescence with biota in the New York Bight. Biological Bull. 144(2):262-275.

Abstract: A series of experiments is described which emphasizes the in-situ aspects of bioluminescence and fluorescence photometric observations. The fundamental question of isolating a particular organism and determining whether it fluoresces or bioluminesces was not answered, but a phenomenon was observed and the probably causative organisms deduced. The causative organisms may be inferred by a comparison of the in-situ bioluminescence, short-wavelength UV produced fluorescence and long-wavelength UV-produced fluorescence with laboratory determined biota distribution. Strong bioluminescence occurs above the thermo/halocline and appears to be caused mainly by Peridiniaceae and Gymnodiniaceae. The long-wavelength UV fluorescence is correlated with the total chlorophyll and consequently with the total biomass.

427 - EG&G, Environmental Consultants. 1974. Currents observed in New Jersey coastal waters near 39°28'N latitude and 74°15'W longitude during February 1974. EG&G Environmental Consultants. Waltham, MA. 32p.

Abstract: Currents in the vicinity of the proposed site showed considerable directional variability during February 1974. Rapidly changing winds caused numerous reversals in the alongshore component from a southerly to a northerly direction.

Tidal effects were observed in the onshore-offshore direction. These effects were most pronounced in the upper level currents.

428 - EG&G, Environmental Consultants. 1977a. Plan for demonstrating compliance with ocean dumping final revision of regulations and criteria. Report to Allied Chemical Corp. Morristown, NJ.

429 - EG&G, Environmental Consultants. 1977b. Dispersion in waters of the New York Bight acid dump grounds of by-product hydrochloric acid discharged from a towed barge. Report to Allied Chemical Corp. Morristown, New Jersey. EG&G, Environmental Consultants. Waltham, MA. 97p.

Abstract: On 13 August 1977, an 8-hour study was made of dispersion in the wake of barge AC No. 5 which was discharging by-product hydrochloric acid (dyed to enhance identification) into the New York Bight Acid Dump Grounds. Twenty-eight stations were occupied. For each station, waste concentration, pH, fluoride concentration, and salinity were determined at six levels. Temperature structure was measured at hourly intervals.

Although a shallow density gradient at 10 m confined waste to the surface mixed layer, waste dilution as indicated by dye, fluoride, and pH, was rapid. At 1 min, it was 6,500:1; at 4 min. 15,000:1. At 4 hr, dilution was approximately 100,000:1. Values of pH returned to within 0.2 pH units of ambient at 4 hr; this observation is consistent with the pH depression projected for waste dilutions observed after 4 hr.

Substantial distoration and translation of the plume was observed. An apparent wind-induced current shear caused the surface manifestation of the plume to move eastward and separate from the subsurface core. Tidal currents translated the plume approximately 3.6 km west of the barge track.

430 - EG&G Environmental Consultants. 1977c. Dispersion in waters of the New York Bight and acid dump grounds of acid iron wastes discharged from a towed barge. Report to NL Industries, Inc. Highstown, New Jersey. EG&G Environmental Consultants. Waltham, MA. 112p.

Abstract: On 12 August 1977, a 5-hour study was made of dispersion in the wake of barge Moran 108 which was discharging acid-iron waste into the New York Bight Acid Dump Grounds. A single station was taken in the waste on the following day. In all, 22 stations were occupied. For each station, iron concentration, waste concentration, pH, and salinity were determined at six levels. Temperature structure was measured at hourly intervals.

Although a density gradient just below 10 m confined waste to the surface mixed layer, rapid lateral dispersion of acid-iron waste was indicated by iron and pH data. Thirty-nine minutes after discharge, mixing in the barge's wake and by naturally occurring turbulence diluted acid-iron waste 9,400:1. At 4 hr, waste in the plume was diluted further to approximately 90,000:1. A single station, approximately 18 hr after discharge, showed additional dilution to 116,000:1, pH measured in the waste wake returned to within 0.2 pH unit of ambient within 1 hr of discharge. Observed pH values are consistent with the pH depression predicted from seawater buffering equations for waste concentrations observed during that first hour.

Substantial distortion and translation of the plume cross-section was observed. Winds of 10-15 kts out of the west caused intense surface mixing which rapidly spread the plume to 2 km wide in 4 hr. In addition, the wind stress on the surface appears to have slowed the westward tidal current at the sea surface compared to the bottom of the surface mixed layer so that the farthest westward extension was at 10 m and the farthest eastward extension was at 0 to 5 m. Tidal currents carried the axis of the plume approximately 1.6 km west of the barge track on the day of the study.

431 - EG&G Environmental Consultants. 1977d. Summer 1977 chemical oceanographic monitoring cruise New York Bight acid dump grounds: Cruise report. EG&G Environmental Consultants. Waltham, MA. 38p.

Abstract: This report presents results for the first phase of a chemical oceanographic monitoring program being conducted in the vicinity of acid-waste disposal site in New York Bight. The program, which is designed according to guidelines promulgated in the Ocean Dumping Final Revision of Regulations and Criteria and methodologies used in the Marine Ecosystem

Analysis (MESA) New York Bight Project, is based on seasonal sampling (summer, fall, and winter 1977-1978) at five permanently established disposal site stations, two seasonally established "waste transport" stations, and two permanent reference stations. During 15, 16, and 18 August 1977, all nine stations were sampled at four depths with replicate casts for temperature, suspended particulate matter, pH, dissolved oxygen, salinity, alkalinity, fluoride, chlorophyll a, and iron (dissolved and particulate). Other dissolved and particulate heavy metals (As, Cd, Cr, Cu, Pb, Hg, Ni, Ti, V, and Zn) were determined in samples from replicate casts at the centrally located disposal site station and one reference station. The replicate sediment samples from these two stations and from the waste transport station were analyzed for iron and the other heavy metals.

432 - EG&G Environmental Consultants. 1977e. Measurements of the dispersion of barged waste near $38^{\circ}33'$N latitude and $74^{\circ}20'$W longitude. Report to E. I. DuPont de Nemours and Co. Waltham, MA.

433 - EG&G Environmental Consultants. 1977f. Measurements of the dispersion of barged waste near $38^{\circ}50'$N latitude and $72^{\circ}15'$W longitude at the "106" dumpsite. Report to E. I. DuPont de Nemours and Co. Waltham, MA.

434 - EG&G Environmental Consultants. 1978. Fall 1977 chemical oceanographic monitoring cruise-New York Bight acid dump grounds cruise report. EG&G Environmental Consultants. Waltham, MA. 43p.

Abstract: This report presents results for the second phase of a chemical oceanographic monitoring program being conducted in the vicinity of the acid-waste disposal site in New York Bight. The program, which was designed according to guidelines promulgated in the Ocean Dumping Final Revision of Regulations and Criteria and methodologies used in the Marine Ecosystem Analysis (MESA) New York Bight Project, is based on seasonal sampling (summer, fall, and winter 1977-1978) at five permanently established disposal site stations, two seasonally established "waste transport" stations, and two permanent reference stations. During 14, 15, and 16 November 1977, nine stations were sampled at three depths for temperature, suspended particulate matter, pH, dissolved oxygen, salinity, alkalinity, fluoride, chlorophyll a, and iron (dissolved and particulate). Other dissolved and particulate heavy metals (As, Cd, Cr, Cu, Pb, Hg, Ni, Ti, V, and Zn) were determined in samples from replicate casts at the centrally located disposal site station and one reference station. The replicate sediment samples from these two stations and from the waste transport station under an old waste plume were analyzed for iron and the other heavy metals.

435 - E. I. du Pont de Nemours and Company. 1972. Waste dispersion characteristics in an oceanic environment. University of Delaware College of Marine Studies and Hydroscience, Inc. Draft of a report to the Water Quality Program, U.S. Environmental Protection Agency. Edgemor, DE. 393p.

436 - Eisen, Paul A., Arleamon Sadler Jr., and Marilyn E. Schenffler. 1978. Data administration for Marine Ecosystems Analysis. NOAA TM ERL MESA-36. Boulder, CO. 79p. NTIS No. PB 293911/AS.

Abstract: The data administration system for storing and retrieving oceanographic data used by MESA New York Bight Project is presented. The main objective of the system is to make Project data and results readily available to users. In this regard, three commonly used strategies for organizing data, two utilizing data files, and one a data-base, are described. This leads to the introduction of the Project's method for handling data which meets objectives through use of a total organization of the data system, a flexible file formatting scheme and data catalogue report generation. The total organization of the data system is an amalgam of two commonly used strategies (traditional/flexible and data-base/key-task). The flexible file formatting scheme is based on a network structure. A specific example, demonstrating STD data retrieval shows the ease with which data may be retrieved using an interface system for special applications. The data catalogue is a tool to inventory, track, and report the status of Project data files.

437 - Eisen, Paul A., and Joel S. O'Connor. 1979. MESA contribution to sampling in marine environments. In: G. P. Patil and M. Rosenzweig (editors), Contemporary Quantitative Ecology and Related Ecometrics. Statistical International Ecology Series Vol. 12. Co-operative Publishing House. Fairland, MD. pp. 545-567.

Summary: Four MESA-sponsored studies are described that use statistical methods to address improved sampling in the New York Bight. The first study finds that metals in sediments can be sampled more efficiently if their observed log normal spatial distribution is considered. The proposed strategy concentrates sampling in regions of highest, nonuniform contamination. The second contribution outlines a method to monitor the influence of waste disposal on benthic faunal assemblages. If sediment strata are first identified by statistically significant differences in benthic species, abundances among strata then can be used to monitor waste methods for estimating tidal datum planes at locations where only a short series of observations is available. A t-test is used to determine if the estimated datum calculated by each method is acceptable. Although the standard method of calculating datums is generally acceptable, the alternative method can be used to advantage on the West Coast of the United States. Finally, weather data from the East Coast have been subjected to cross-correlation analyses to study the feasibility of interpolating analyses to study the forcing in data-sparse marine areas. Sufficient correlation is shown to exist in space and time to proceed with experiments in hindcasting.

438 - Elder, J. A. et al. 1974. Conceptual model of the New York Bight. Water Resources Engineers. Walnut Creek, CA.

Abstract: The changes that may occur in the physical, chemical and biological characteristics of the Bight, as results of man's activities in the years to come, are difficult to predict. Yet, it is essential that we seek to manage these changes--to preserve, to protect, even to enhance, the intrinsic values identified with this vast natural water resource. Management of such a resource demands the capability to assess in quantitative terms the possible consequences of alternative strategies or physical works for regulating or modifying the Bight's environment. One attractive approach to providing such a capability is through the design of a "model" that can simulate the behavior of the Bight under imposed stresses. Such a model, if it can be devised to represent reasonably and economically the complex behavior of this piece of the oceanic environment, could be of enormous value to decision makers, allowing them to test alternative strategies before the fact, to observe probable changes, and to make adjustments as may be required to comply with stipulated objectives.

The National Oceanographic and Atmospheric Administration (NOAA) has commissioned Water Resources Engineers (WRE) of Walnut Creek, California, Springfield, Virginia, and Austin, Texas, to develop a comprehensive conceptual ecologic model of New York Bight. This report constitutes the completion document for the model's conceptual development under Phase I of Contract No. 3-35423 between NOAA and WRE, dated June 29, 1973.

439 - Eleftheriou, A., D. C. Moore, D. J. Basford, and M. R. Robertson. 1981. Experimental studies on the effects of sewage sludge on a benthic community. ICES. C.M. 81/E:43. 9p.

440 - Ellis, D. 1929. The dilution of sewage in the sea. Jl R. Tech. Coll. Glasgow, 2:698-707.

441 - Elridge, Richard W. 1970. Research and development for better solid waste management. Bioscience. 19(11):984-988.

442 - Engler, R. M. 1974. Discussion of regulatory criteria for ocean disposal of dredged materials: Elutriate test rationale and implementation guidelines. Government Reports Announcements. 74(10):131. May 17.

443 - Erb, R. A. 1969. A wasteline to the sea--a study toward regional disposal of concentrated wastes. Proc. Water Resources Seminar. Univ. Del. Newark, DE.

444 - ERCO. 1981. New York Bight benthic sampling survey: Coprostinol, polychlorinated biphenyl aromatic hydrocarbon measurements in sediments. Final report for contract NA-80-FA-C-00038. Submitted to NOAA, NMFS Sandy Hook Laboratory. ERCO. Cambridge, MA. 20p.

445 - Erickson, L. E. 1972. Simultaneous dispersion and decay of radioactive wastes. Health Physics. Jour. 22(2):177-186.

446 - Esser, Stephen C. (In Press). Long term changes in some finfishes of The Hudson-Raritan estuary. In: G. F. Mayer, A. Calabrese, F. A. Cross, D. C. Malins, D. W. Menzel, J. S. O'Connor and F. J. Vernberg

(editors), Ecological Stress in the New York Bight: Science and Management. Estuarine Research Federation. Columbia, SC.

Abstract: Long-term changes in abundance as expressed in catch statistics have occurred for most finfish species of the Hudson-Raritan estuary. The commercial catch of American shad within the Hudson-Raritan estuary has experienced significant fluctuations and decline since the end of the nineteenth century. Possible causes, including physical changes, pollution, and overfishing, are reviewed. An extensive sturgeon fishery flourished in the Hudson River during the 1800's but declined in the early 1900's. A similar decline occurred in the smelt fishery, which prospered in the Raritan River in the early 1800's. Only the striped bass population appears to have increased slightly in recent years. Other species, including bluefish, scup, and flounders, are caught both in the estuary and in the ocean, but the combined statistics make it impossible to determine the estuarine harvest alone. Most of these species use the estuary as a spawning or nursery area. Their populations are influenced by water quality of the Hudson and its tributaries.

447 - Essink, K. 1972. Biologische aspecten van afvalwaterlozing in the mariene milieu. (Biological aspects of waste water disposal in the marine environment). Landbouwkundig Tijdschrift. 84(7):250-255.

448 - Essink, K., and W. J. Wolff. 1978a. Pollution of the Wadden Sea area. Rep. Wadden Sea Working Group. No. 8. 61p.

449 - Essink, K., and P. de Wolf. 1978b. Pollution by organic waste in the Dutch Wadden Sea and the Ems-Dollard estuary. Rep. Wadden Sea Working Group. No. 8. pp. 39-45.

450 - Ewing, M., J. L. Worzel, N. C. Steenland and F. Press. Geophysical investigations in the emerged and submerged Atlantic coastal plain. Part V: Woods Hole, New York and Cape May sections. Bull. Geol. Soc. Amer. 61:877-892.

451 - Fader, S. W. 1972. Barging industrial liquid wastes to sea. Jour. Water Poll. Control Fed. 44:314-318.

452 - Falk, L. L. 1975. Industrial viewpoint on ocean disposal. In: T. M. Church, (editor). Marine Chemistry in Coastal Environments. American Chemical Society: 169th National Meeting: Special Symposium on Marine Chemistry in the Coastal Environment. ACS Ser. 18:406-409.

453 - Falk, L. L., T. D. Myers, and R. V. Thomann. 1977. Waste dispersion characteristics and effects in an oceanic environment. NTIS No. PB 268 157. 306p.

454 - Farrell, Joseph B., and B. Vincent Salotto. 1973. Effect of incineration on metals, pesticides, and polychlorinated biphenyls in sewage sludge. Proc. Nat. Symp. Ultimate Disposal Wastewaters and their residuals. p. 186-198.

455 - Fasoli, U., and P. Melli. 1973. Self-purifying capacity of sea water. C.I.E.S.M. Journ. d'Etudes Poll. Mar. (Athenes) p. 119.

456 - Fellman, M. H. 1970. Trace materials in wastes disposed to coastal waters: Fates, mechanisms and ecological guidance and control. Federal Water Quality Administration, Northwest Region. Corvallis, OR. Working Paper No. 78. 102p.

457 - Ferguson, T. L. 1975. Determination of incinerator operating conditions necessary for safe disposal of pesticides. Municipal Environmental Research Lab. Cincinnati, OH. (EPA 600/2-75-041). 400p.

458 - Feuerstein, D. L. 1975. Predesign surveys and monitoring of waste disposal systems. In: E. A. Pearson, and E. de Fraja Frangipane (editors), Marine Pollution and Marine Waste Disposal. (Supplement to Progress in Water Technology.) Proceedings of the 2nd International Congress. San Remo. 17-21 December 1973. Pergamon Press. New York, NY. pp. 359-362.

459 - Figley, William, Bruce Pyle, and Bruce Halgren. 1979. Socioeconomic impacts. In: R. L. Swanson and C. J. Sinderman (editors), Oxygen Depletion and Associated Benthic Mortalities in New York Bight, 1976. NOAA Professional Paper No. 11. U.S. Dept of Commerce. Rockville, MD. pp. 315-322.

Summary: Table 14-7 summarizes the estimated, actual, and potential economic losses to New Jersey's commercial and recreational marine fisheries as a result of oxygen depletion during summer 1976. The surf clam was by far the hardest hit commercial resource, followed by the ocean quahog, finfish, lobster, and sea scallop. During 1976 actual commercial losses sustained to finfish and lobster fisheries were an estimated $7.9 million, and potential losses to the entire commercial fishing industry, based upon estimates of resource losses, were estimated at $62.5 million. For commercial fisheries, total actual and resource losses until recruitment begins to replenish affected stocks are estimated at $569 million.

Resources losses cannot be directly converted into losses for the fishing industry. Fish prices usually fluctuate inversely with fish availability, and the fisherman is compensated for reduced catches by increased unit prices. Also, most fishermen do not depend upon a single species and are able to take advantage of other species that may be abundant. Despite not being able to determine industry losses directly, the tremendous losses experienced by several of New Jersey's fishery resources have caused significant financial losses to many fishermen and businesses. In fisheries that had significant mortalities, especially surf clam, the decline in harvest and economic losses is expected to continue in future years until recruitment begins to replenish stocks. In addition, the impact of the disruption of the food chain by the elimination of much of the benthic community may have far-reaching consequences that will hamper commercial fishermen in future years.

Losses to the recreational fisheries during 1976 were $3.7 million. The party, charter, and dive boat fleets absorbed most of the financial losses. Losses by the reduced number of private-boat ocean fishing trips were minimal and were distributed over a large number of recipients--tackle shops, marinas, restaurants, motels, gas stations.

460 - First, Melvin, N. 1969. Waste incineration at sea and ocean disposal of non-floating residues. Paper presented at the 62nd annual meeting of the Air Pollution Control Association. New York, NY. June 8-12, 1969. Paper No. 69-33. 22p. (mimeo).

461 - First, M. W. (editor). 1972a. Municipal Waste Disposal by Shipborne Incineration and Sea Disposal of Residues. Report. Harvard University School of Public Health. Boston, MA. 592p.

462 - First, M. W. 1972b. Composition of incinerator residues and analyses of marine organisms. In: M. W. First (editor). Municipal Waste Disposal by Shipborne Incineration and Sea Disposal of Residues. Report. Harvard University School of Public Health. Boston, MA. Section 3. pp. 1-24.

463 - First, M. W. 1972c. Engineering feasibility. In: M. W. First (editor). Municipal Waste Disposal by Shipborne Incineration and Sea Disposal of Residues. Report. Harvard University School of Public Health. Boston, MA. Section 3. 33p.

464 - First, Melvin, W. 1972d. Solid waste management. In: M. W. First (editor). Municipal Waste Disposal by Shipborne Incineration and Sea Disposal of Residues. Report. Harvard University School of Public Health. Boston, MA. 22p.

465 - First, M. W. 1973. Municipal waste disposal by shipborne incineration and sea disposal of residues. Harvard School of Public Health. Boston, MA. EPA 670 2-73-017. 603p.

466 - Fisher, H. J. 1976. Status of ocean incineration of organic chlorides aboard Matthias III as of 29 June, 1976. TRW Report to EPA.

467 - Flierl, G. R. 1980. Modelling the effects of wastes dumped in deep ocean gyres. Report to NOAA Ocean Dumping Program. MIT. Cambridge, MA.

468 - Flierl, Glenn R. (In Press). Simple models of waste disposal in a gyre circulation. In: I. W. Duedall, B. H. Ketchum, P. k. Park, and D. R. Kester (editors), Wastes in the Ocean, Vol. I: Industrial and Sewage Sludge in the Ocean. John Wiley and Sons, Inc. New York, NY.

Abstract: We discuss several simple models of the distribution of waste and impact of waste dumping in a region where there is a circulating flow. First, the distribution in the absence of biological feedback effects is considered and the dependence upon the flow rate, the scale of the gyre, the decay time of the waste, and the dumping rate are described. Second, a biological model with a growth rate dependent on the density of organisms and a death rate varying with the waste concentration is used to examine what the impact on organisms within the gyre might be. Finally, various situations with biological feedback, introduced by making the decay rate of the wastes dependent on the level of biological activity, are considered. When the death rate varies nonlinearly with concentration, "catastrophes" (in the mathematical sense) can occur: the population may suddenly die out with only a small increase in the dumping rate. The models here are intended to illustrate possible behaviors when the physics, chemistry, and biology of a polluted ecosystem are all considered together; no attempts are made to apply these models in detail.

469 - Florida Ocean Science Institute. 1971. Limitations and effects of waste disposal on an ocean shelf. U.S. EPA. Water Pollut. Cont. Res. Series. 16070EF12/71.

470 - Folger, D. W. 1976. Direct observation of the sea floor in mid-shelf dumpsites, Middle Atlantic Bight. In: EOS: Trans. Amer. Geophy. Union. 56(6):370.

471 - Folger, D. W., H. D. Palmer, and R. A. Slater. 1979. Two waste disposal sites on the continental shelf off the middle Atlantic states: Observations made from submersibles. In: H. D. Palmer and M. G. Gross (editors), Ocean Dumping and Marine Pollution. Dowden, Hutchinson and Ross, Inc. Stroudsburg, PA. pp. 163-184.

Abstract: Nineteen dives were conducted in August 1974 and July 1975 to survey the Philadephia sewage and DuPont acid dumpsites about 65 km southeast of the mouth of Delaware Bay. Though 1974 dives in the sewage dumpsite were carried out the day after 2 million gallons of sewage was released, little evidence of sewage remained in the water column or on the bottom within the area of the dumpsite. Bottom currents were typically weak (2-5 cm/s) and flowed in an easterly or northerly direction. The bottom was characterized by broad rounded ripples (30-100 cm wavelength; 2-5 cm wave height) or small hummocks and pits (15 cm diameter). Sand dollars, starfish, scallops, and crabs were most common on the bottom together with lesser numbers of hake, flounder, shrimp, skates, and snails. Oscillatory wave motion, tidal currents, and benthic organisms apparently were the main mechanisms redistributing sediment on the bottom during the period of the dives. In 1974, at the center of the acid waste site, where a 1-million-gallon release of acid (pH 0.06) had just taken place, visibility in the water column 6-9 m deep was reduced to zero. Below this opaque layer, suspended matter was composed mainly of large reddish-yellow flocks 1-3 cm long. The visibility of reduction and the flocks were apparently due to precipitates from the acid waste that were concentrated at the top of the thermocline. Microtopography and epibenthic organisms in the acid site were similar to those in the sewage site, but the bottom was covered by a thin (2 cm) yellowish to reddish, granular, flaky material. During dives conducted in 1975, bottom samples were collected by both the mother ship and the submersible. At five sites, in both dumpsites, gravel (including shell fragments) averaged 6.1%; sand, 88%; and silt plus clay, 2.1% of the sediment. Both coarsest and finest materials were accumulating in ripple troughs. Fe, Ti, and Cr in sediments were all higher at the acid site than in surrounding areas. Thus, sewage sludge apparently was being oxidized, eaten, or tansported away from the dumpsite and accumulating elsewhere, whereas some acid waste was accumulating on the bottom in the disposal area.

472 - Fontaine, M., and J. C. Lacaze. 1973. Considerations sur la methodologie des recherches concernant les effects ecologiques et biologiques des rejects en mer. (Consideration on the methodology of research concerning the ecological and biological effects of discharge to the sea). CIESMM. Journees d'etudes sur les Pollutions Marines. Commission Internationale pour l'Exploration Scientifique de la Mer Mediterranee. Monaco. pp. 107-108.

473 - Forns, J. M. 1977. Phytoplankton and zooplankton taxonomic investigations of two interim ocean dumpsites. EPA-68-01-3211. U.S. Environmental Protection Agency. Region III. Philadelphia, PA. 40p.

474 - Foster, D. N., and D. M. Stone. 1963. Ocean disposal of ash. University of New South Wales, Water Research Laboratory. Manly Vale, New South Wales. (Report submitted to Electricity Commission of New South Wales.) 130p.

475 - Foxworthy, J. E., and H. R. Kneeling. 1969. Final Report - Eddy diffusion and bacterial reduction in waste fields in the ocean. Report 69-1. Allan Hancock Foundation. University of Southern California. Los Angeles, CA. 176p.

476 - Foyn, E. 1964. Disposal of waste in the marine environment and the pollution of the sea. Oceanogr. Mar. Biol. Annu. Rev. 3. pp. 95-114.

477 - Foyn, E. 1971. Municipal wastes. Impingement of man on the ocean. Wiley-Interscience. New York, NY. pp. 445-459.

478 - Frangipane, E. D. 1979. Principles and guidelines for the discharge of wastes into the marine environment. WHO/UNEP. [See Ref. #'s 360, 361].

479 - Franz, David R. 1976. Distribution and abundance of inshore populations of the surf clams Spisula solidissima. In: M. G. Grant (editor), Amer. Soc. Limnol. and Oceanogr. Spec. Symp. 2:44-413.

Abstract: The distribution and abundance of juvenile and adult surf clams were determined inshore of the 3-mi contour in July 1974 at 141 sites off Long Island, New York. Transects were located every 2 mi from Montauk to Rockaway Point, and samples were taken at 0.5, 1.5, and 2.5 mi offshore. Selected stations were revisited in July 1975.

East of Shinnecock Inlet adult clams occurred at densities between 0.5 and 3.0 bushels per dredge haul. West of Shinnecock adults declined gradually, reaching a minimum west of Jones Inlet to East Rockaway Inlet. Abundances increased precipitously off Rockaway Beach.

Highest densities of juveniles occurred at 0.5 mi and decreased rapidly farther offshore. Juveniles were more abundant at the west end of Long Island, particularly inshore at 0.5 mi. Higher densities of junveniles here may result from accumulation of larvae produced farther east and transported westward via longshore currents. The convergence of tidal and longshore currents may effectively "trap" larvae off western Long Island.

Clam stocks off eastern Long Island probably are older than 9 years and are mostly composed of 1-3 age classes. Clam stocks off the Rockaways appear younger. The commercial fishery is apparently dependent on massive settlements of larvae occurring irregularly and infrequently.

480 - Franz, David R. (In Press). An historical perspective on molluscs in Lower New York Harbor, with emphasis on oysters. In: G. F. Mayer, A. Calabrese, F. A. Cross, D. C. Malins, D. W. Menzel, J. S. O'Connor, and F. J. Vernberg (editors), Ecological Stress in the New York Bight: Science and, Management. Estuarine Research Federation. Columbia, SC.

Abstract: Information on benthic organisms and communities in the lower Hudson estuary prior to the 1950's is scarce. Records from the late 1800's indicate that species richness in the shallow bay habitats along Staten Island was comparable to that presently characteristic of Great South Bay and northeastern Long Island Sound. By the 1920's many species had disappeared from Staten Island, and the number of species remaining approached that which presently exists. These data indicate that major environmental deterioration occurred during the years 1890-1920. Supporting evidence is provided by historical accounts of New York's oyster industry. Before 1900, oysters were found throughout much of the lower estuary and north of Ossining, New York, including Newark bay, Arhtur Kill, Kill Van Kull, Jamaica Bay, Raritan Bay, and the New Jersey shore of the Hudson. By the turn of the century, the shellfish industry was limited to water south of the Narrows; by 1920 it largely had disappeared from harbor waters.

Destruction of shellfish beds was caused by a combination of factors including sewage pollution, harbor dredging, and industrial pollution. Although it is difficult to determine the relative importance of individual factors, sewage-related pollution appears to have played a major role in bringing about the general degradation of the estuarine environment. This is suggested by records of dissolved oxygen (DO) levels in the harbor, which were collected beginning in 1909. These data clearly indicate that by the 1920's, DO levels over much of the harbor had declined to critical levels (0-20% saturation in summer).

While the lower Hudson estuary is seriously impacted, communities of benthic organisms exist throughout. The degree of environmental deterioration and the recovery of the estuary probably can be measured best by comparing these benthic communities with those of nearby, less polluted estuarine systems.

481 - Freeland, George L., and Donald J. P. Swift: 1975. New York alternative dumpsite assessment - reconnaissance. Study of surficial sediment. In: TX International Congress of Sedimentology, Nice, France. Theme 10. pp. 13-18. Also appeared in Offshore Technology Conference, Paper No. OTC 2385, 505-511.

Abstract: Evaluation of offshore areas as potential dumpsites requires assessment of bottom sediment character and transport. Preliminary evaluation of the potential for deposition of dumped materials, primarily sewage sludge, were made at two sites 60 nm from New York Harbor in 20- to 30- fm water depths. Since sludge particle density is barely over 1.0, geological data were analysed for potential deposition and transport of fines in particular in addition to the sand-sized fraction. Results suggest a net southwestward bottom sediment transport, intensifying during winter storms.

482 - Freeland, G. L., and G. F. Merrill. 1976a. Deposition and erosion in the dredge spoil and other New York Bight dumping areas. Proc. Amer. Soc. of Civil Engineers Specialty Conference on Dredging and Its Environmental Effects. Mobile, AL. January 26-28, 1976. pp. 936-946.

Abstract: The disposal of solid-wastes from the New York City metropolitan area is the cause of considerable environmental concern, as most of these wastes are dumped in marine waters outside of the harbor mouth. Dredge spoil and sewage sludge constitute over 94% of the volume of material dumped containing solids.

483 - Freeland, G. L., and G. F. Merrill. 1976b. Sedimentological aspects of bathymetric changes over a 37-year period in the New York Bight Apex. Geol. Soc. of Amer. Abs. 8(2):177-178.

Abstract: Comparison of data from a detailed 1973 bathymetric survey in the New York Bight Apex, done as part of NOAA's Marine Ecosystems Analysis (MESA) Program in the New York Bight, with results of the previous survey in 1936 reveal net changes of significance to sedimentologists studying sediment transport and to environmental managers attempting to determine accumulation of dispersal of dumped materials. A contoured net change map permits measurements of areas of net erosion and deposition and calculation of volumes for each. Assuming a preliminary estimate of 0.09 m of sea level rise in the intervening 37 years, the total area of 715 km^2 has undergone sediment erosion of 258 x 10^6 m^3 and deposition of 214 x 10^6 m^3 for a net removal of 44 x 10^6 m^3 equal to a layer 0.06 m thick over the entire area. However, 75 x 10^6 m^3 of dredge materials and 5.3 x 10^6 m^3 of cellar dirt (construction debris) have been dumped by man from sources outside the Apex. The dredge spoil dumping alone has caused over 10 m of shoaling. Subtracting the anthropogenic sediment results in a net removal of 143 x 10^6 m^3 of sediment, or 0.24 m over the area outside of dumpsites and dredged channels. Examination of the Christiaensen Basin, the natural amphitheater at the head of the Hudson Shelf Valley and a natural sink for fine sediment, including sewage sludge dumped nearby, reveals that 75% of the area has undergone erosion, mostly in the center of the basin, while deposition of up to 2/3 m has occurred on the northwest side of the Basin.

These data, combined with surficial sediment and side-scan sonar data, point to active sediment transport in the area, even to the extent to scouring fine sediment from topographic lows.

484 - Freeland, George L. Donald J. P. Swift, and William L. Stubblefield. 1976c. Surficial sediments of the NOAA-MESA study areas in the New York Bight. In: M. G. Gross (editor), Amer. Soc. Limnol. and Oceanogr. Spec. Symp. 2:290-101.

Abstract: In the New York Bight Apex, extensive sedimentological studies and a 1973 bathymetric survey reveal that the only significant change in bottom topography since 1936 occurred at the dredge spoil dumpsite where the dumping of 98 x 10^6 m^3 of dredged material has caused up to 10 m of shoaling. The center of the Christiaensen Basin, a natural collecting area for fine-grained sediment, is not doubt contaminated with sludge but shows no apparent sediment buildup during the intervening 37 years. The Apex outside of the Christiaensen Basin is floored primarily by sand ranging from silty fine to coarse, with small areas of sandy gravel, artifact (anthropogenic) gravel, and mud. Nearshore mud patches appear to be covered at times with sand and occasionally scoured out. Sidescan sonar records show linear bedforms, indicative of sand movement, over most of the Apex area.

Two midshelf areas have been proposed as interim alternative dumping areas. The northern area is in a tributary valley of the ancestral Long Island river system. Fine sands cover the northeast part and medium sands predominate to the west and south. Bottom photographs show a smooth, slightly undulatory, mounded, or rippled sea floor.

In the southern alternative dumping area coarse sand and gravel deposits lie on the crest and east flank of the Hudson divide, while medium and fine sand occurs in the ridge and swale topography to the west. These distributions suggest fine sediment is winnowed from the crest and east flank of the divide and deposited to the west. Veatch and Smith Trough contains a veneer of shelly, pebble sand with large, angular clay pebbles and occasional oyster shells derived from exposed early Holocene lagoonal clay. These studies suggest that if sewage sludge were dumped, widespread dispersion, mostly to the southwest, could be expected, with winter resuspension and transport of fine material on the bottom. Possible permanent buildup on the bottom could be expected if dredged material were dumped.

485 - Freeland, George L., and George F. Merrill. 1977a. The 1973 bathymetric survey in the New York Bight Apex: Maps and geological implications. NOAA TM ERL-MESA-19. Boulder, CO. 22p.

Abstract: A hydrographic survey of the New York Bight Apex was undertaken by the New York District of the Corps of Engineers under contract to NOAA as part of the MESA Program. A bathymetric map was prepared and a comparison was made between the 1973 data and hydrographic data from the most recent previous survey of the area, H-6190, done in 1936 by the U.S. Coast and Geodetic Survey. A resulting contoured net bathymetric change map shows that the most significant change has occurred in the dredge-spoil dumpsite, where there has been up to 10 m of shoaling. Calculations of volumes of eroded and deposited sediment indicate that the area has generally eroded and that, except at the dredge-spoil dumpsite and the now abandoned dumpsites near Ambrose and Sandy Hook Channels, dumping is not causing significant changes in water depths.

486 - Freeland, G. L., J. W. Lavelle, D. J. P. Swift, and R. A. Young. 1977b. NOAA's waste-disposal studies in New York Bight. Amer. Assoc. of Petroleum Geologists Bull. 61(5)786.

Abstract: Waste-disposal studies in the New York Bight center around three programs: the effects of dumping in the Bight Apex, mid-shelf potential dumpsites, and the new Inner Shelf Sediment Transport Experiment (INSTEP) which will attempt to resolve the pattern (rates and directions) of deposition, transport, and erosion on the inner Long Island Shelf for both cohesive and noncohesive sediment.

In the Bight Apex, natural sedimentation, affecting 82% of the area, has resulted in net erosion of 58 x 10^6 m^3 of sediment over the 37-year period from 1936 to 1973. Dumping over the same period has resulted in a net deposition of 81 x 10^6 m^3 at dredge spoil and cellar dirt (construction rubble) dumpsites. Sewage sludge, constituting only 3.3% of dumped solids, is nearly neutrally buoyant, highly mobile, and does not form a deposit at the sewage-sludge dumpsite, but mixes into nearby deposits of natural mud.

Two mid-shelf areas, north and south of the Hudson Shelf Valley, were found to be pristine and not recommended for use. Material dumped would have a net southwestward movement as a result of storm conditions, thus potentially contaminating a much larger area than is affected at present.

INSTEP will examine the temporal and spatial variance of sediments flux on the bottom and through the water column. It will assess the pattern of sediment transport on straight shoreface, in an inner shelf ridge topography, and in an area of nearshore mud patches. Near-bottom sediment fluxes will be measured by electromagnetic current sensors coupled to forward-scattering nephelometers, plus additional vertical arrays of EM meters coupled with an acoustic meter which incrementally measures the suspended-sediment concentration in the first meter above the bottom. Bottom sediments will be mapped and monitored quarterly for a year. Completion of the program is expected in 1980.

487 - Freeland, G. L., and D. J. P. Swift. 1978. Surficial sediments. MESA New York Bight Atlas Monograph 10. MESA New York Bight Project and New York Sea Grant Institute. Albany, NY. 93p.

Abstract: ATLAS MONOGRAPH 10 describes what is known about surficial and suspended sediment in New York Bight. The surficial sediment cover of the continental shelf in the Bight consists of a sheet of sand up to 10 m thick with small areas of gravel and muddy sand. Off the shelf edge, mud increases to become the dominant sediment. Modification of the natural pattern of sedimentation has occurred in the Bight Apex where man has dumped his wastes for over 60 years. The Apex is now a "worst case" example of ocean pollution. With proper management of municipal and industrial wastes, the Apex may eventually return to a near-natural setting, still in use by man, but without harmful ecological effects.

488 - Freeland, George L., Donald J. P. Swift, and Robert A. Young. 1979. Mud deposits near the New York Bight dumpsites: Origin and behavior. In: H. D. Palmer and M. G. Gross (editors), Ocean Dumping and Marine Pollution. Dowden, Hutchinson and Ross, Inc. Stroudsburg, PA. pp. 73-95.

Abstract: A detailed investigation of the floor of the New York Bight Apex reveals that muddy sediment occurs in the Christiaensen Basin in the fringing lagoons and marshes near the mouths of tidal inlets, and in ephermeral patches on the Long Island shoreface, where mud settles into the troughs of sandwave-like bedforms. The appearance and disappearance of these nearshore mud patches has led to an erroneous concept of a sewage-sludge "front" advancing toward the Long Island beaches from the sewage-sludge dumpsite. While Bight Apex mud deposits do contain sewage sludge, this contamination is usually not apparent in gross physical properties or composition; sophisticated chemical techniques are required to detect the contamination. Fine-grained deposits near the New York Bight sewage-sludge dumpsite are primarily of natural origin and should, therefore, be called mud, not sludge. Fine-grained sediment at the dredge-spoil dumpsite is also mostly natural sediment, but it is anthropogenic in that it is dredged harbor and estuary sediment.

The impact of sewage-sludge dumping in the New York Bight Apex depends to a large extent on the pathways and rates of the natural sediment transport system. The rate of sewage-sludge input is higher than the estimated rate of fine sediment input from land runoff. However, the input of suspended fine sediment into the Bight Apex by coastwise shelf currents, though not precisely known, may be much higher than the input from either dumping or land runoff. The rate at which organic matter is cycled through the suspended-sediment system via biological production and consumption of organic suspended sediment is estimated to be almost an order of magnitude higher than the rate of suspended-sediment input from either land runoff or dumping. Such limited information as is available suggests that the rate of sewage-sludge dispersal is high relative to the rate of input. Our model leads us to infer that 1) sewage sludge is rapidly diluted in the Bight Apex by natural sediment, and 2) the more stable sewage-sludge components are widely dispersed through the natural mud deposits of the New York Bight system.

489 - Freeland, G. L., D. I. Stanley, D. J. P. Swift, D. N. Lambert. 1981. Hudson shelf valley, North American Atlantic Shelf: Its role in modern shelf sediment transport. Mar. Geol. 42:399-427.

490 - French Ministry of Quality of Environmental Life. 1976. Burning of industrial chlorine wastes on the high seas. In: Final environmental impact statement. Maritime Administration Chemical Waste Incineration Ship Project. U.S. Department of Commerce. Maritime Administration. Washington, DC. Vol. 1. 62p.

491 - Fuhrmann, Mark. 1979. Sedimentology of the New York Bight dredged material dumpsite deposit. Master's Thesis. Adelphi University. Garden City, NY. 56p.

Abstract: The New York Bight dredge spoil dumpsite, located outside of the New York harbor, is now a major geological feature of the Bight. Ten vibracores, ranging in length from 3.1 to 8.4 meters, were taken along two transects across the study area. Each of the cores is described and the sediment types observed are discussed. Volume and mass estimates as well as bathymetric comparisons are made for the three survey years: 1936, 1973 and 1978. From the dumping a deposit has evolved for which stratigraphic correlations were made, some of which can be dated. The deposit itself is dynamic. Sand, discrete bedding and laminated sediments are predominant at the center of the deposit while massive bedding and fine grained material are common on the downslope fringes. Statistical analysis confirms that the cores, at the periphery of the deposits, contain significantly more mud than do the cores near the center of dumping activity. Grain size, bedding features and stratigraphy all strongly indicate that there is selective movement and deposition of fine grained sediments to the sides of the deposit that are below wave base. Incursion of natural, fine grained, glauconitic sands from erosion windows near the site was seen in the cores as flaser-like and lenticular beds of often laminated sands. These were found primarily on the flanks of the site and comprise approximately 7% of the total volume of the deposit. Anthropogenic materials found within the dumped sediment include: coal, cinders, metallic and glassy fly ash spherules, metal/rust flakes, wood, concrete, glass, a nail and a swatch of cloth.

492 - Fujioka, R. S., H. H. Hashimoto, E. B. Siwak, R. H. F. Young. 1981. Effect of sunlight on survival of indicator bacteria in seawater. Appl. Envir. Microbiol. 41: 690-696.

493 - Fulk, R., Gruber, D., and Wullschelger, R. 1975. Laboratory study of the release of pesticide and PCB materials to the water column during dredging and disposal operations. Contract Report D-75-6 prepared by Envirex, Inc. Milwaukee, Wisconsin, to the U.S. Army Engineer Waterways Experiment Station. Vicksburg, MS.

494 - Furr, A. K., A. W. Lawrence, S. S. C. Tong, M. C. Grandolla, R. A. Hofstader, C. A. Bache, W. H. Gutenmann, and D. J. Lisk. 1976. Multi-element and chlorinated hydrocarbon analysis of municipal sewage sludges of American cities. Environ. Sci. Technol. 10:683-687.

Abstract: An analytical survey of 68 elements, dieldrin and PCBs was conducted in sewage sludges of 16 American cities during 1972-73. Unusually elevated concentrations of certain elements were found in specific areas. High PCB levels were in sludge from Schnectady, possibly from the predominant manufacture of electrical equipment there.

495 - Fyn, Ernst. 1965. Disposal of waste in the marine environment and the pollution of the sea. In: H. Barnes (editor), Oceanogr. Mar. Biol. Vol. 3. Hafner Publishing Company. New York, NY. pp. 95-114.

496 - Gadd, P. E., J. W. Lavelle, and D. J. P. Swift. 1978. Estimates of sand transport on the New York shelf near-bottom current meter observations. Jour. of Sed. Petr. 48(1):239-252.

Abstract: Calculations of cohesionless bottom-sediment movement within the New York Bight have been made by applying the transport formulae of Bagnold (1941, 1956, 1963), Einstein (1950), and Yalin (1963) to near-bottom current meter and surficial sand size observations. Current data were drawn from the records of eighteen long-term Savonious rotor current meter deployments at various locations within the Bight during Fall of 1973, and Spring of 1974.

The assumptions underlying the caculations are that wave activity was minimal at recording sites, that a drag coefficient of 3×10^{-3} reflecting small scale roughness and large boundary layer Reynolds number is suitable to convert measured currents to friction velocities, and that laboratory threshold velocities apply in the marine environment.

The calculations suggest that oceanic bottom sediment movement reaches maximum intensity during the fall and winter due to the added energy input from strong meteorological events. Calculated transport quantities on the inner shelf tend to decrease as depth and distance from shore increase. However, during the Fall of 1973, the deep waters near the head of the Hudson Shelf Valley exhibited current flows directed to the north in excess of 40 cm/sec. These up-channel flow events appear to be in response to strong, sustained westerly winds. The maximum sediment transport rate caused by these current velocities is two orders of magnitude greater than that occuring at much shallower depths along the New Jersey coast during the same measurement period. The Spring 1974 current velocity field yields transport rates of lesser magnitudes relative to the previous fall, with a net down-channel sediment flux at the head of the Hudson Shelf Valley. The New York Harbor entrance appears to be the site of near-continental sediment transport generated by swift tidal currents.

497 - Gaertner, Michael Peter. 1976. Seasonal migration of fishes of importance to New York State. Master's Thesis Marine Sciences Research Center. State Univ. of New York. Stony Brook, NY.

Abstract: It has long been known that many species of marine fishes undertake extensive seasonal migrations, especially along coasts with a north south orientation. These migrations are northerly and inshore during spring and summer and southerly and offshore during fall and winter. These seasonal movements have been verified by tagging and tag recovery studies and by exploratory fishing.

Statistics of commercial landings for the states from Maine to North Carolina are published monthly by the National Marine Fisheries Service in cooperation with the individual states. These statistics support the results of the tagging and fishing studies mentioned above. Therefore, it appears that these data for commercial monthly landings of fishes along the eastern coast of the United States represent a useful method for confirming seasonal movements of these fishes.

498 - Gaines, Jack L. 1976. Evaluation of the present shellfish harvesting closure lines in the New York Bight. U.S. Public Health Service. Northeast Technical Services Unit. Davisville, RI. 10p.

Abstract: Study of the validity of present closure lines and water quality of areas now harvesting shellfish, bacteriological analysis of water and sediments finds that inshore area off Long Beach does <u>not</u> meet current standards for approved harvesting and should be closed. Southern closure line is valid and New Jersey water quality is acceptable.

499 - Gallay, J. J. 1975. Incineration of hazardous wastewater residues at sea. In: Proc. Natl. Conf. on management and disposal of residues from treatment of industrial wastewater. Information Transfer Inc. Washington, DC. pp. 187-190.

500 - Galloway, J. N. 1979. Alternation of trace metal geochemical cycles due to the marine discharge of wastewater. Geochmi. Cosmochim. Acta. 43(2):207-218.

501 - Gameson, A. L. H. 1975 (editor). Discharge of Sewage from Sea Outfalls. In: Proceedings of an International Symposium held at Church House, London, 27 August to 2 September 1974. Progr. Water Technol. Suppl. XIV. 455p.

502 - Gardner, D., and J. P. Riley. 1973. Distribution of dissolved mercury in the Irish Sea. Nature. London. 241(5391):526.

503 - Garside, C., T. C. Malone, O. A. Roels, and B. A. Sharfstein. 1976. An evaluation of sewage-derived nutrients and their influence on the Hudson Estuary and New York Bight. Estuar. and Coast. Mar. Sci. 4:281-289.

Abstract: Sewage wastes discharged into estuaries contain oxidizable organic matter and inorganic nutrients from which further organic matter may be formed. This paper presents an evaluation of the total discharge of nutrients into the lower Hudson Estuary and estimates the quantities taken up by photosynthetic organisms under summer and winter conditions. Primary production is shown not to be nutrient-limited at any time in the Lower Estuary and in a seasonally variable area of the Apex of the New York Bight. Primary production contributes a significant part of the oxidizable organic matter to the lower Estuary. The general approach requires few input data and can be of value in assessing the impact of sewage wastes on receiving waters.

504 - Garside, C., and T. C. Malone. 1978. Monthly oxygen and carbon budgets of the New York Bight Apex. Estuar. and Coast. Mar. Sci. 6:93-104.

Abstract: A relationship between surface oxygen flux and particulate organic carbon as a means of estimating mean benthic respiration is proposed and examined using data from the Apex of the New York Bight for 1973-1974. A mean water column mass balance for oxygen involving mean surface oxygen flux, mean benthic and water column respiration, and photosynthetic oxygen production is used to give a first approximation of the carbon and oxygen budgets of the Apex annually. The carbon budget is compared with known organic carbon inputs to the Apex and found to be balanced. The major source of particulate organic carbon is demonstrated to be in situ primary production, with no other single source of POC providing more than 10% of the annual total. High production of POC by primary production is always accompanied by high evasive surface flux and generally by higher levels of water column respiration. Benthic respiration is controlled primarily by temperature, but at times by rapid supply of new substrate. It is postulated that one such source of supply is resuspended estuarine sediment as a result of high runoff. The possibility of anoxic conditions in the Apex is examined, and it is concluded that major organic loading events, as a result of high river runoff, could generate small anoxic regions near the estuary. However, more wide scale anoxic conditions could occur only if mechanisms other than those responsible for the property distribution reflected in the data used in this paper were operative.

505 - Garside, C. 1981. Nitrate and ammonia uptake in the Apex of the New York Bight. Limnol. and Oceanogr. 26(4):731-739.

506 - GEC (General Fisheries Council for the Mediterranean). 1972. The state of marine pollution in the Mediterranean and legislative controls. Stud. Rev. Gen. Fish. Coun. Mediterr. No. 51. 68p.

507 - Gerlach, S. 1981. Marine Pollution. Springer-Verlag, New York, NY. ISBN 3-540-10940-4. 250p.

Contents: Introduction. - Domestic Effluents. - Industrial Effluents, Pollution of the Sea By Ships. - Oil Pollution. - Radioactivity. - General Problems of Harmful Substances in the Sea. - Global Contamination of the

Oceans by Heavy Metals. - Global Pollution of the Oceans with Chlorinated Hydrocarbons. - Laws Against the Pollution of the Oceans. - Diagnosis and Therapy. - References. - Subject Index.

508 - GESAMP (Joint Group of Experts on the Scientific Aspects of Marine Pollution). 1975. Scientific criteria for the selection of the sites for dumping of wastes into the sea. Reports and Studies No. 3. [Report of the GESAMP Working Group on the Scientific Basis for Disposal of Waste into the Sea which met 4-8 February 1974, in Rome and 5-11 October 1974, in Copenhagen.] IMCO/FAO/UNESCO/WMO/WHO/IAEA/UN. 21p.

509 - Giblin, A. E., A. Bourg, I. Valiela, and J. M. Teal. 1980. Uptake and losses of heavy metals in sludge by a New England salt marsh. Amer. Jour. Bot. 67(7):1059-1068.

510 - Gienapp, H. 1978. Determination of the optical properties of the sea in the German Bight (Meeresoptische untersuchungen in der deutschen bucth). In: Tech. Hochschule On Meas. from Aircraft. pp. 471-492.

Abstract: The spectral distribution, transmission scattering, and CIE coordinate characteristics of the sea were investigated in the framework of the 1976 FMP program. The region between Helgoland and Sylt and the Elb estuary were studied. The sea on the southern boundary of the Verklappung's region was shown to be very dirty. Relations between color metric magnitudes Gelbstoff, Chlorophyll, and suspended matter contents were shown to exist. Comparisons were made with ERTS satellite photographs. A high degree of correlation in some of these relations was seen for the Elb estuary. The DIBIAS system is used in the calculations performed.

511 - Glendenning, K. A., and W. J. North. 1960. Effects of wastes on the giant kelp, Macrocystis pyrifera. In: E. A. Pearson (editor), Proceedings of the First International Conference on Waste Disposal in the Marine Environment. University of California. Berkeley, CA. July 22-25, 1959. Pergamon Press. New York, N.Y. pp. 82-92.

512 - Goldberg, Alexander S. 1971. A procedure for treatment and disposal of wastewater sludge. Jour. Water Poll. Control Fed. 43(9):1912-1920.

513 - Goldberg, E. D. 1975. Marine pollution. In: J. P. Riley and G. Shirrow (editors), Chemical Oceanography, 2nd edition. Vol. 3. Academic Press. New York, NY. pp. 39-89.

514 - Goldberg, E. D. 1976. Strategies for Marine Pollution Monitoring. John Wiley Interscience. New York, NY. 310p.

515 - Goldberg, Edward D. (editor). 1979a. Proceedings of a Workshop on Scientific Problems Relating to Ocean Pollution. Estes Park, Colorado. July 10-14, 1978. U.S. Dept. of Commerce. NOAA. Envir. Res. Labs. Boulder, CO. U.S. GPO. 677-034/2. 225p.

Summary: Review of specific catagories of wastes: synthetic organics, chlorination products, dredging and disposal, litter, mircoorganisms, trace metals, biostimulants, fossil fuel compounds, and biological effects.

516 - Goldberg, Edward D. (editor). 1979b. Proceedings of a Workshop on Assimilative Capacity of U.S. Coastal Waters for Pollutants. Crystal Mountain, Washington. July 29-Aug. 4, 1979. U.S. Dept. of Commerce. NOAA. Envir. Res. Labs. Boulder, CO. U.S. GPO. 677-096/1256. 284p.

Summary: Sources and specific case studies: Estuaries, coastal and open ocean, DWD-106, N.Y. Bight. Southern California Bight and Puget Sound.

517 - Goldberg, Edward D. 1981. The oceans as waste space: The argument. Oceanus. 24(1):2-9.

Abstract: Management of toxic wastes, DDT, Mercury and description of the concept of Assimilative Capacity with titration of polluting substances and endpoints, future needs.

518 - Golden, Paul C., and Michael A. Champ. 1974. Monitoring ocean disposal sites. Published in the Proceedings of the 10th Annual Conference of the Marine Technology Society, National Needs and Ocean Solutions. September 23-25. Washington, DC. pp. 107-113.

519 - Goodfellow, R. M., J. Cardoso, G. Eglinton, J. P. Dawson, and G. A. Best. 1970. A faecal sterol survey in the Clyde Estuary. Mar. Pollut. Bull. 8(12):272-276.

520 - Gopalan, U. K., and J. S. Young. 1975. Incidence of shell disease in shrimp in the New York Bight. Mar. Pollut. Bull. 6(10):149-153.

Abstract: Incidence of shell disease in shrimp is as high as 30% in certain localities in the New York Bight. Gross examination and histopathological preparations revealed cracking and pitting of the exoskeleton, common characteristics of crustacean shell disease, followed by necrosis of underlying tissue. Aquarium studies, which included tests of the effects of an antibiotic, indicated a possible infectious etiology. The disease may be related to pollution of the habitat by organic wastes including cellulose.

521 - Gordon, A. L., A. F. Amos, and R. D. Gerard. New York Bight water stratification-October 1974. In: M. G. Gross (editor), Amer. Soc. of Limnol. and Oceanogr. Spec. Symp. 2:45-57.

Abstract: Thermohaline stratification of New York Bight continental shelf water during October 1974 is basically of the summer regime. Salinity increases markedly with increased distance from the coast; yet a basic vertical structure is maintained: an upper isohaline layer; a salinity maximum at the top of the thermocline; a salinity minimum at the base of the thermocline; a deep isohaline layer associated with the cold near-bottom winter residual stratum; and (over the outer shelf) a bottom intrusion of relatively saline and warm slope water. Inversions in temperature and salinity are common within the thermocline.

The pycnocline is continuous over the shelf and slope, though some weakening and deepening occurs over the shelf break. Over the shelf it is mainly supported by the thermocline and over the slope by the halocline. The pycnocline may not be an effective barrier to isopycnal interchange of surface and deep layers in view of the relative slope of isopycnals to pycnocline.

In October 1974 oxygen distribution of the continental shelf was primarily two-layered, with a sharp division at the pycnoline. The lower cold layer has an oxygen concentration of near 60% of full saturation, with values near 3.6 ml/liter. This is low; if the deeper layer is principally a residue of the winter homogeneous condition with initial saturated oxygen values, it would represent oxygen consumption at a rate of 2.6 ml/liter during the six summer months after accounting for the low oxygen influx of slope water.

522 - Gordon, R. B., et al. (In Press). Summary report of the panel on the management of dredged material. In: G. F. Mayer, A. Calabrese, F. A. Cross, D. C. Malins, D. W. Menzel, J. S. O'Connor, and F. J. Vernberg (editors), Ecological Stress in the New York Bight: Science and Management. Estuarine Research Federation. Columbia, SC.

Abstract: New York Harbor is in a naturally shallow estuary. Left alone, the greater water depth in the Harbor would become about 6 m, too shallow to allow most vessels engaged in international trade to enter. Sediment reaches the Harbor from both the sea and the rivers. Appropriate engineering works, if taken in hand now could, over a period of many years, reduce the amount of sediment that accumulates in the Harbor. But there is no doubt that periodic maintenance dredging will be required as long as New York Harbor remains a major seaport. Much of the muddy sediment that is deposited in the Harbor is contaminated with a wide range of undesirable and toxic substances; more contamination is added while this sediment remains in the Harbor. The most serious problem associated with maintenance dredging is disposal of the contaminated sediment. For many years the accepted practice has been to discharge this at a site, the "mud dump", just off the Harbor entrance in the New York Bight. As dredging continues, the size of the mud dump continues to increase. Concern about the ecological consequences of the continued use of the mud dump has led to exploration of alternative disposal methods. All are more expensive, some, such as containerized ocean disposal, enormously more expensive. No one scheme will solve the problem of disposing of contaminated dredged material, but the main elements of a Regional Dredging Plan that will provide for continued economic operation of the Port while minimizing environmental damage are presented in this report.

523 - Gorsline, D. S. 1979. Shelf-sediment dynamics and solid-waste disposal. In: H. D. Palmer and M. G. Gross (editors), Ocean Dumping and Marine Pollution. Dowden, Hutchinson and Ross, Inc. Stroudsburg, PA. pp. 9-16.

Abstract: Dredged material is commonly natural material that has been deposited under riverine, estuarine, or deltaic conditions. It includes natural terrigenous and biogenic detritus, organic matter and debris, and adsorbed metal ions and organic compounds of natural or man-generated origins. Because the great bulk of the material is natural particles, its dispersion is identical to dispersion for naturally contributed sedimentary materials and the particles will follow the same paths. Other than the effect of the overwhelming mass dumped in a locality, the major problem environmentally is the smaller fraction of dredge spoil that contains pollutants. Since a large proportion of these contaminants is adsorbed on the clay fraction which moves as a suspended load, the major environmental impact is the disposal of the fine fraction.

Most harbors are located in estuaries or the lower reaches of major rivers, and so the most complex problems of control of spoil disposal are due to the inherent characteristics of estuarine circulation and particularly to the fact that estuaries are sediment traps. They typically draw from both river input and from adjacent shelves, and thus the problem of finding suitable dump sites outside a given system may be economically and technically difficult.

We need more information about such problems as the rates of benthic boundary layer processes, the paths of suspended sediment transport over shelves, and the net contributions of estuaries to shelves in normal and storm periods.

Deep water sites should be avoided until we know more about the biology of those environments.

524 - Gould, D. J. 1976. Ecological effects of sewage discharges to the sea: an assessment of research needs. Tech. Rept. TR 26. Water Research Centre. (U.K.).

525 - Goulet, Jr. Julien R., and Keith A. Hausknech. 1977. Physical oceanography of Deepwater Dumpsite 106, Update: July 1975. In: Baseline Report of Environmental Conditions in Deepwater Dumpsite 106. NOAA/ODP Dumpsite Evaluation Report 77-1. Rockville, MD. Vol. 1. pp. 55-86.

526 - Grabow, W. O. K., O. W. Prozesky, and L. S. Smith. 1974. Drug resistant coliforms call for review of water quality standards. Water Res. 8:1-9.

527 - Graikoski, J. T., R. A. Greig, and J. A. Babinchack. 1974a. Coliform and metal concentrations in sediments from the Atlantic and Long Beach area - New York Bight, January 15-17, 1974. NOAA, Middle Atlantic Coastal Fisheries Center. Woods Hole, MA. Informal Report No. 22. 5p.

528 - Graikoski, J., R. Greig, D. Wenzloff, B. Nelson and A. Adams. 1974b. Trace metals in marine biota and sediments collected from offshore waters of New York Bight. NOAA, Middle Atlantic Coastal Fisheries Center. Sandy Hook Laboratory, NJ. Informal Report No. 38. 9p.

529 - Grant, W. D., and O. S. Madsen. 1978. Bottom friction under waves in the presence of a weak current. NOAA TM ERL MESA-29. Boulder, CO. 138p. NTIS No. PB287426/AS.

530 - Grasshoff, K. 1978. Globale und regionale probleme der meeresverschmutzung.. Schriftenreihe des Agrarwissenschaflichen Fachbereichs der Univ. Kiel. 53, 1976 (1978). Coll. Repr. Inst. Meeresk. Univ. Kiel. No. 21. pp. 18-26.

531 - Gray, J. S. (Chairman), D. Boesch, C. Heip, A. Jones, J. Lassig, R. Vangerhorst, and D. Wolfe. 1980. Ecological working group report to monitoring of biological effects of pollution in the sea. In: International Council for the Exploration of the Sea. Rapports et Proces-Verbaux des Reunions. Vol. 179. Charlottenlund Slot, Denmark.

532 - Great Britain-Parliment. 1974. Dumping at Sea Act, 1974. Her Majesty's Stationery Office. London. 13p.

533 - Greig, R., et. al. 1974. Distribution of five metals in sediments from the New York Bight. U.S. Department of Commerce. NOAA. Middle Atlantic Coastal Fisheries Center. Milford, CT. 36p.

Abstract: Sediment samples from 107 stations located in the dumping areas of the New York Bight were analyzed for the metals Cu, Cr, Pb, Zn and Ni. Sampling at most of the stations was also done on a quarterly interval for possible variations of metal concentrations with time. The range of concentrations of the metals found in the area sampled (expressed as ppm on dry weight of the sediment) was as follows: Cu - 215-0,8, Cr - 300-1.6, Pb - 266-0.9, Zn - 520-1.3, Ni - 38-0.8. The highest concentrations of metals were found in sediments at stations located in the dumping areas. With a few notable exceptions, the concentrations of the metals did not vary greatly over the quarterly year sampling period.

534 - Greig, R. A. 1975. Comparison of atomic absorption and neutron activation analyses for the determination of silver, chromium, and zinc in various marine organisms. Anal. Chem. 47:1682-1684.

535 - Greig, R. A., and J. Jones. 1976. Nondestructive neutron activation analysis of marine organisms collected from ocean dump sites of the middle eastern United States. Arch. Envir. Contam. and Toxicol. 4(4):420-434.

Abstract: The concentrations of eight metals were determined by a nondestructive neutron activation technique for eleven species of fish and shellfish. The marine organisms were collected from ocean dump sites off New York City, off New Haven, Connecticut, and off Delaware Bay.

Antimony was not detected in most of the organisms examined in this study; the detection limit was about 0.02 to 0.05 ppm. Antimony levels ranged from 0.01 to 0.129 ppm in fish that had detectable levels. Cobalt levels were low in all samples with most levels in the range of 0.1 to 0.3 ppm. Chromium concentrations at 0.3 to 1.0 ppm were only roughly quantitated by the procedure employed. Most marine organisms examined had chromium levels at or below these values. Nickel was not detected in any of the organisms examined; the detection limit was in the 3- to 6-ppm range.

Rubidium concentrations were 0.6 to 1.5 ppm for most organisms; only rough quantitative measurement was possible at these levels. Selenium levels ranged from about 0.3 to 3.8 ppm in all samples. Silver concentrations were below 0.3 ppm in most organisms. Silver concentrations as high as 10 to 30 ppm, however, were found in the digestive gland of rock crab. Zinc levels in windowpane flounder liver were about 6 to 9 times greater than the 4- to 10-ppm levels found in muscle. Zinc concentations in rock crab muscle, on the other hand, were only slightly higher than the 15- to 32-ppm concentration found in the digestive gland. Fish other than windowpane flounder had zinc levels that ranged from 4 to 9 ppm in the muscle and 14 to 42 ppm in the liver. Shellfish other than rock crab had zinc levels of 15 to 30 ppm in muscle and 17 to 40 ppm in the digestive gland.

536 - Greig, R. A., D. R. Wenzloff, and J. B. Pearce. 1976. Distribution and abundance of heavy metals in finfish, inverterbrates and sediments collected at a deepwater disposal site. Mar. Pollut. Bull. 7(10):185-187.

Abstract: Analyses of Deep Water Finfish for several heavy metals indicate that deep-water fish have less metal in muscle tissue than do fish taken from the continental shelf. Liver tissue from deep-water fish also, generally, contain less metals. Sediments and a single inverterbrate collected at deep-water stations have been analyzed for heavy metals.

537 - Greig, R. A., D. R. Wenzloff, A. Adams, B. Nelson, and C. Shelpuk. 1977a. Trace metals in organisms from ocean disposal sites of the Middle Atlantic Eastern United States. Arch. Environm. Contam. and Toxicol. 6:395-409.

Abstract: Concentrations of silver, arsenic, cadmium, chromium, mercury, manganese, lead, and zinc were determined for selected marine fish and shellfish collected at or near three ocean disposal sites, a control site, and three inshore areas of the middle eastern United States. The disposal sites were off New Haven, Connecticut; New York City, and Delaware Bay. The control site was at Chincoteague Inlet, Virginia, and other areas were in Long Island Sound (near the New Haven disposal site); Barnegat Bay, New Jersey, and a second area off New York City near the Bight Apex disposal site.

With the following exceptions, the amounts of trace metals in biological samples did not vary substantially among the geographic areas: Silver in rock crab flesh averaged 0.79 ppm for the New York Bight disposal site compared with 0.24 to 0.38 ppm for four other areas. Cadmium, manganese, and zinc concentrations were greatest in rock crab flesh collected from Long Island Sound area 2 (not a known disposal site) compared to three other areas; mean levels were 1 vs. 0.1 ppm, 29 vs. 0.8 to 1 ppm, and 64 vs. 32 to 36 ppm. The manganese content in gills of rock crabs from the same area in the Sound was 22 ppm compared with 6 ppm for Chincoteague Inlet. Digestive glands of channeled whelk collected from a disposal site in Long Island Sound contained the greatest amount of silver, cadmium, and zinc compared to the control area, Chincoteague Inlet, and a site outside the disposal area in the Sound. Levels were 20 vs. 6 to 7 ppm of silver, 24 vs. 16 and 17 ppm of cadmium, and 2650 vs. 1,025 and 405 ppm of zinc, respectively. In addition, whelk digestive gland from the two areas in the Sound contained about 1,100 ppm of copper compared to only 32 ppm for Chincoteague Inlet.

Trace metal concentrations in the organisms used during this study were of the same order of magnitude as those reported by various investigations for a variety of organisms collected from waters in the U.S. Atlantic and Pacific as well as in British waters of the Atlantic.

538 - Greig, R. A., A. Adams, and D. R. Wenzloff. 1977b. Trace metal content of plankton and zooplankton collected from the New York Bight and Long Island Sound. Bull. Environ. Contam. Toxicol. 18:3-8.

Abstract: This study provides baseline data on trace metals in zooplankton from Long Island Sound and New York Bight and compares any differences in geographical locations. The levels of Ag, Cd, and Ni did not markedly differ in various locations with ranges from 0.5 to 4.6 ppm for all samples. Cr varied considerably for different samples because of different sample sizes with the range being 7 ppm to 35.2 ppm. Cu, Pb, and Zn levels varied greatly in the New York Bight as a function of species or geographic location. Interspecific differences could not be studied since different plankton species were collected at different stations, but individuals of the same species collected at different Long Island Sound stations were not identified because of their poor condition after freezing. Cu levels ranged from 2-14 ppm, Pb being uniform at 11-23 ppm, Zn varying greatly at 7-120 ppm.

539 - Greig, Richard A., and Wenzloff, Douglas R. 1977c. Trace metals in finfish from the New York Bight and Long Island Sound. Mar. Pollut. Bull. 8(9):198-200.

Abstract: Concentrations of silver, cadmium, chromium, copper, mercury, manganese, nickel, lead, and zinc were measured in five species of finfish collected from the New York Bight and two from Long Island Sound. With few exceptions, the trace metal contents were similar for the various species examined and also for a single species when comparing catch locations. Metal levels also were similar to concentrations reported by two other investigators who studied fish from the North Atlantic.

540 - Greig, R. A., R. N. Reid, and D. R. Wenzloff. 1977d. Trace metal concentrations in sediments from Long Island Sound. Mar. Pollut. Bull. 8(8):183-188.

541 - Greig, Richard, and Douglas Wenzloff. 1977e. Final report on heavy metals in small pelagic finfish, euphausid crustaceans and Apex predators, including sharks, as well as on heavy metals and hydrocarbons (C_{15}+) in sediments collected at stations in and near Deepwater Dumpsite 106. In: Baseline Report of Environmetnal Conditions in Deepwater Dumpsite 106. NOAA/ODP Dumpsite Evaluation Report 77-1. Rockville, MD. Vol. III. pp. 547-564.

542 - Greig, R. A., and R. A. McGrath. 1977f. Trace metals in sediments of Raritan Bay. Mar. Pollut. Bull. 8(8):188-192.

543 - Greig, R. A., D. Wenzloff, C. Shelpuk, and A. Adams. 1977g. Mercury concentrations in three species of fish from North Atlantic offshore waters. Arch. Environm. Contam. Toxicol. 5:315-323.

544 - Greig, R. A., D. R. Wenzloff, and J. B. Pearce. 1978a. Levels of heavy metals in selected finfish and shellfish from the New York Bight. ICES. C.M. 70/E:19.

545 - Greig, R. A., D. R. Wenzloff, C. L. MacKenzie, Jr., A. S. Merrill, and V. S. Zdonowiez. 1978b. Trace metals in sea scallops, Placopecten magellanicus, from eastern United States. Bull. of Environ. Contam. and Toxicol. 19:326-334. p. 326-334.

546 - Greve, P. A. 1971. Chemical wastes in the sea: New forms of marine pollution. Science. 173:1021-1022.

547 - Grew, G. W. 1978. Signature extraction of ocean pollutants by eigenvector transformation of remote spectra. In: Proceedings Joint Conference on Sensing of Environmental Pollutants, 4th. New Orleans, LA. November 6-11. American Chemical Society. Washington, DC.

Abstract: Spectral signatures of suspended matter in the ocean are being extracted through characteristics vector analysis of remote ocean color data collected with MOCS (Multichannel Ocean Color Sensor). Spectral signatures appear to be obtainable through analyses of 'linear' clusters that appear on scatter diagrams associated with eigenvectors. Signatures associated with acid waste, sewage sludge, oil, and algae are presented. The application of vector analysis to two acid waste dumps overflown two years apart is examined in some detail. The relationships between eigenvectors and spectral signatures for these examples are analyzed. These cases demonstrate the value of characteristic vector analysis in remotely identifying pollutants in the ocean and in determining the consistency of their spectral signatures.

548 - Grice, George D., Peter H. Wiebe, and Elaine Hoagland. 1973. Acid-iron waste as a factor affecting the distribution and abundance of zooplankton in the New York Bight. I. Laboratory studies on the effects of acid waste on copepods. Estuar. and Coast. Mar. Sci. 1:45-50.

Abstract: Laboratory experiments were conducted using three species of copepods to study the effects acid-iron wastes might have on zooplankton. Individuals of *Calanus finmarchicus*, *Temora longicornis* and *Pseudocalanus sp*. were maintained in bottles containing varying dilutions of acid wastes in sea water or control sea water solutions for 24 to 48 hours to examine survival of adults exposed to high concentrations of waste. They were also maintained for time periods up to 18 days to ascertain effects on reproduction and survival of young. In addition, individuals of *C. finmarchicus* were transferred through a series of increasing dilutions of acid wastes and into filtered sea water to simulate the short-term effects of acid waste concentrations in the wake of a discharging barge.

Substantial mortality of adults of the above three copepods occurred at concentrations of acid waste producing pH's of approximately 6.5 and lower. However, this mortality is not indicative of mortality in the fields as these concentrations and pH values exist for only a short time (less than 3 min) due to rapid mixing of the acid waste with sea water. Individuals maintained in buffered acid wastes of comparable dilutions showed no mortality, while individuals maintained in test media using sulfuric acid in place of acid waste showed high mortalities at pH's of 5.5 or less. Thus, acidity of test solutions may be a principal cause of copepod mortality rather than some toxic component of the waste material. Inhibition of reproduction and deleterious effects on survival of young were observed in experiments of 18 days duration at concentrations of acid wastes which do not in fact persist for such periods on the acid grounds. No mortality was observed when *C. finmarchicus* was transferred through acid waste dilutions with pH values and time periods comparable to those they would encounter on the acid grounds during discharge of acid wastes. Acid waste discharges, therefore, did not appear to be responsible for the large variations in zooplankton biomass previously observed in the survey area, since mortality of species due to short term exposure to high concentrations of waste appears small and little effect on adults or larval forms at great dilutions could be demonstrated.

549 - Grigg, Richard W., and Robert S. Kiwala. 1970. Some ecological effects of discharged wastes on marine life. Calif. Fish Game. 56(3):145-155.

550 - Grommers, F., J. Mahoney, and M. First. 1972. A simulation model for solid waste management. In: M. W. First (editor), Municipal Waste Disposal by Shipborne Incineration and Sea Disposal of Residues. Report. Harvard University of Public Health. Boston, MA. Section 7. p. 7:1-7:38.

551 - Gross, M. Grant. 1967. Sinking rates of radioactive fallout particles in the north east Pacific Ocean, 1961-62. Nature. London. 216(5116):670-672.

552 - Gross, M. G. 1969. New York City - A major source of marine sediment. Marine Science Research Center. State Univ. of New York. Stony Brook, NY. Tech. Rep. No. 2. 24p.

553 - Gross, M. G. 1970a. Preliminary analysis of urban wastes. New York, Metropolitan Region. Marine Sciences Research Center. State Univ. of New York. Stony Brook, NY. Tech. Rept. 5. 35p. NTIS No. AD 746959.

554 - Gross, M. G. 1970b. Analysis of dredged wastes, fly ash, and waste chemicals - New York, Metropolitan Region. Marine Sciences Research Center. State University of New York. Stony Brook, NY. Tech. Rept. 7. 33 pp. NTIS No. AD 734 337. (Reprinted in Congressional Record and in Hearings before the Senate Subcommittee on Public Works, Resource Recovery Act of 1969. Part 5p. 2347-2358.)

555 - Gross, M. G. 1970c. Analysis of dredged wastes, fly ash and waste deposits, New York metropolitan region. GSA Bull. 83:3163-3176.

556 - Gross, M. G. 1970d. New York metropolitan region -- a major sediment source. Water Resources Research 6(3):927-931.

557 - Gross, M. Grant, J. A. Black, R. J. Kalin, J. R. Schranel, and R. N. Smith. 1971. Survey of marine waste deposits, New York Metropolitan Region. Mar. Sci. Res. Center Tech. Rep. Series No. 8. State Univ. of New York. Stony Brook, NY. 72p. NTIS No. AD 723431.

558 - Gross, M. Grant. 1972. Marine waste deposits near New York. Mar. Pollut. Bull. 3(4):61-63.

559 - Gross, M. G. 1972b. Geologic aspects of waste solids and marine waste deposits, New York metropolitan region. Bull. Geol. Soc. Amer. 83:3163-3176.

560 - Gross, M. Grant. 1974a. Disposal of water-borne and barged waste-solids and environmental effects in the New York Bight. Chesapeake Bay Institute. John Hopkins University. Baltimore, MD.

Abstract: Discharges of waste solids to NY Bight (1963-1968) were

Dredged wastes	3.5 million metric tons of dry solids
Rubble	.59 million metric tons of dry solids
Sewage sludges	.2 million metric tons of dry solids
Industrial sludges	.35 million metric tons of dry solids

Wastes can be detected by their black color, odor, human artifacts, high carbon contents, and anomalous metal contents (high Ag,Cu,Cr, Pb). Waste deposits in Hudson Channel locally exceed 12 m in thickness and cover 150 km^2.

Physical alteration of the bottom and associated biological effects have been documented. Low dissolved oxygen concentrations can occur in the disposal areas during late summer.

561 - Gross, M. Grant. 1974b. Sediment and waste deposition in New York harbor. In: O. A. Roels (editor), The Hudson River Colloquium. Annals N.Y. Academy of Sciences. 250:112-128.

Abstract: In this paper the physical alterations (by man) of the Hudson River estuary are discussed. Particular attention is paid to the sediments and waste deposits that covered much of the harbor bottom and large areas of the New York Bight.

The paper includes sections on maintenance dredging, sand and gravel mining, plus quantitative data on flux of sediment into N.Y. outer harbor.

562 - Gross, M. Grant. 1975a. Trends in waste disposal in U.S. coastal waters, 1968-1974. In: Thomas M. Church (editor), Mar. Chem. in the Coastal Environ. Amer. Chem. Soc. Washington, DC. pp. 594-404.

Abstract: Estuaries and coastal oceans receive large volumes of waste solids dredged from harbors or removed from sewage treatment plants or industrial plants and placed on the adjacent continental shelf, which often receives little sediment derived from other sources. Legislation, implemented in 1973, has caused a significant reduction in the number of sites used; from 140+ in the late 1960s to 110 in 1973. Dredged spoil disposal apparently decreased from 56 million tons in 1968 to 37 million tons in 1973. Dumping of industrial wastes, sewage sludges, and construction debris has increased at annual rates of +2.9%, +3.9%, and +14.2%. The present trend is to operate fewer disposal sites located farther from shore.

563 - Gross, M. G. 1975b. Scientific aspects of ocean disposal. American Chemical Society. 169th National Meeting. Special Symposium on Marine Chemistry in the Coastal Environment. ACS Ser. No. 26.

564 - Gross, M. G. (editor). 1976a. Proceedings of the Symposium on the Middle Atlantic Continental Shelf and the New York Bight. Amer. Soc. Limnol. and Oceanogr. Spec. Symp. 2. 441p.

565 - Gross, M. G., R. L. Swanson, and H. M. Stanford. 1976b. Man's impact on the middle Atlantic continental shelf and the New York Bight -- Symposium Summary. In: M. G. Gross (editor), Amer. Soc. Limnol. and Oceanogr. Spec. Symp. 2:1-12.

Abstract: The primary contributor of contaminants to the New York Bight is not sewage sludge dumping, but wastewater discharges, river runoff from the Hudson and barge discharges.

Swimmers in barely acceptable waters of Coney Island Beach reported gastrointestinal symptoms.

Paper presents an overview of waste disposal, NY Bight environmental characteristics, productivity, fish diseases, and health problems.

566 - Gross, M. G. 1976c. Sources of urban wastes. In: M. G. Gross (editor), Amer. Soc. Limnol. and Oceanogr. Spec. Symp. 2:150-161.

Abstract: The coastal ocean has long been used by cities along the Middle Atlantic Bight-particularly in the New York-New Jersey metropolitan region-to dispose of municipal, industrial, and dredged wastes; the volumes increased by more than 5% per year in the early 1970s. Sediment eroded from agricultural land and from construction sites must be dredged from navigation channels after deposition by rivers. Solids from sewage (treated and untreated) mix with the riverborne sediments so that large volumes of dredged materials must be handled as wastes. Riverborne sediment load and littoral drift must be dredged from navigation channels; that material is now dumped at sea. Industrial wastes, such as from titanium dioxide production and coal ash, have been dumped at sea. Others, such as steel-making slag, have been used for landfill. Construction and demolition debris have been dumped at sea when no landfill sites were available. Refuse, garbage, and incinerator ash are commonly disposed of in coastal wetlands.

Alternative disposal strategies and sites will be needed to supplement present regulations to reduce the urban wastes now dumped at sea.

567 - Gross, M. Grant. 1976d. New York Bight II: Problems of research. Oceanus 19(4):11-18.

Abstract: Summary of research on the effects of waste disposal in the New York Bight: Discusses effects of pollutants on bottom sediments, benthic biological processes, water column, finfish and fisheries, and public health.

568 - Gross, M. G. 1976e. Waste disposal. MESA New York Bight Atlas Monograph 26. MESA New York Bight Project and New York Sea Grant Institute. Albany, NY. 32p.

Abstract: MONOGRAPH 26 discusses wastes dumped in the major disposal sites in New York Bight. Much more of the waste solids comes from dredge spoil and rubble than from industrial sludges and sewage sludge; the entire sediment load from littoral drift and river discharge mixes with wastes and must be handled as dredge spoil. Accumulations of waste deposits not only create hills on the ocean bottom but also affect bottom-dwelling marine life, pointing to finrot in flounder and shell erosion in lobster and crab. So little is known about the effects of ocean-dumped materials, even from recent studies, that the situation demands continued extensive research. Meanwhile, dumping in the Bight goes on because it is cheaper and there is more room than on land.

569 - Gross, M. Grant. 1978. Effects of waste disposal operations in estuaries and the coastal ocean. Annual Review of Earth and Planetary Science. 6:127-143.

570 - Gross, M. Grant, and W. B. Cronin. 1979. Dredging and disposal in Chesapeake Bay, 1975-2025. In: H. D. Palmer and M. G. Gross (editors), Ocean Dumping and Marine Pollution. Dowden, Hutchinson and Ross, Inc. Stroudsburg, PA. pp. 131-145.

Abstract: In Chesapeake Bay frequent dredging is required in three locations: Bay entrance, harbors, and the turbidity maxima at the fresh-/salt-water interfaces. Between 1975 and 2025, 560 x 10^6 m^3 (740 x 10^6 yd^3) are scheduled to be dredged to maintain existing channels and to deepen them to accommodate larger vessels. One-third of the dredging is associated with new construction projects; two thirds comes from maintenance dredging. Approximately half the maintenance dredging will be done in port areas where much of the dredged material may be mixed with industrial wastes and sewage discharges.

Sources of deposits in channels and port facilities in Chesapeake Bay are poorly understood. River sediment discharge to Chesapeake Bay is probably 5 to 10 x 10^6 tons (15 to 30 x 10^6 m^3/yr) per year, in years without major floods. Smaller but unknown amounts of sediment are contributed by sewage treatment plants, urban runoff, industrial processes, and shoreline erosion.

Available data suggest mostly localized and transitory biological effects for those disposal operations involving unpolluted materials that do not alter the physical or chemical character of the bottom. Long-term effects of dredged material disposal in the Bay such as altered current regimes or sediment transport can not be predicted with confidence.

Because of shortages of wetland disposal sites, opposition to open-water disposal, and high costs of diked disposal areas, new disposal sites and techniques will likely be needed in the next 50 years to handle the disposal of materials from needed dredging in Chesapeake Bay. Dredging and disposal of dredged materials will continue to be major geological processes in Cheapeake Bay.

571 - Grosslein, M. D. 1976. Some results of fish surveys in the mid-Atlantic important for assessing environmental impacts. In: M. G. Gross (editor), Amer. Soc. Limnol. and Oceanogr. Spec. Symp. 2:312-328.

Abstract: Knowledge of the distribution, biomass, and composition of the major biological communities is essential to the data base required for assessing the impact of environmental change on the marine ecosystem. Since 1967 the National Marine Fisheries Service has conducted regular trawl surveys in the mid-Atlantic, providing quantitative measures of density distributions of demersal fish populations. The surveys show that each species occupies wide areas of the shelf and that there is a high degree of species mixture, particularly in the New York Bight. Significant numbers of adult stages, planktonic eggs, and larvae can be found over the whole mid-Atlantic shelf all year. Thus there are no shelf areas free of risk from potential damage from waste disposal activities. Impacts of any change must consider a multispecies community and include all life stages.

Precision of large-scale surveys is relatively low, making it difficult to detect any but major changes. Broad-scale surveys must be augmented with detailed laboratory and field experiments on physiology and behavior (especially food chains) of selected organisms and communities to get insight into probable effects of sublethal factors. The complexity and scope of marine ecosystems require long-term but well-coordinated research programs to ensure proper integration of small-vs. large-scale studies and field vs. laboratory experiments.

572 - Grosslein, M. D., and T. Azarovitz. (In press). Fish distribution. MESA New York Bight Atlas Monograph 15. New York Sea Grant Institute. Albany, NY.

573 - Grunseich, Gary Steven. 1977. The decomposition of sewage sludge in sea water. Master's Thesis. Marine Science Research Center. State Univ. of New York. Stony Brook, NY.

Abstract: Laboratory experiments were conducted over a twelve-week period to follow the decomposition of sewage sludge in seawater and in sediment-seawater mixtures under aerobic and anaerobic conditions at 4 and 21°C. Results showed that the sewage sludge decomposed more rapidly in the presence of oxygen. Dissolved organic carbon, a major carbon source in sewage sludge, abruptly decreased to very low concentrations in 3-4 weeks in aerobic systems; concentrations of particulate carbon decreased only gradually during the twelve-week period. Aerobic conditions at the sewage sludge dump site in the New York Bight Apex can promote rapid decomposition of sewage sludge if it is distributed evenly over an area of at least two square miles.

574 - Grunseich, G. S., and I. W. Duedall. 1977. The decomposition of sewage sludge in seawater. Marine Science Research Center. State Univ. of New York. Stony Brook, NY. 30p.

Abstract: Laboratory experiments were conducted over a twelve-week period to follow the decomposition of sewage sludge in seawater and in sediment-seawater mixtures under aerobic and anaerobic conditions at 4° and 21°C. Results showed that the sewage sludge decomposed more rapidly in the presence of oxygen. Dissolved organic carbon, a major carbon source in sewage sludge, abruptly decreased to very low concentrations in 3-4 weeks in aerobic systems; concentrations of particulate carbon decreased only gradually during the twelve-week period. Aerobic conditions at the sewage sludge dumps site in the New York Bight Apex can promote rapid decomposition of sewage sludge if it is distributed evenly over an area of at least 5.2 km^2 (2 mi^2).

575 - Grunseich, G. S., and I. W. Duedall. 1978. The decomposition of sewage sludge in sea water. Water Res. 12:535-545.

Abstract: Laboratory experiments were conducted over a twelve-week period to follow the decomposition of sewage sludge in seawater and in sediment-seawater mixtures under aerobic and anaerobic conditions at 4^0 and 21^0C.

Results showed that the sewage sludge decomposed more rapidly in the presence of oxygen. Dissolved organic carbon, a major carbon source in sewage sludge, abruptly decreased to very low concentrations in 3-4 weeks in aerobic systems; concentrations of particulate carbon decreased only gradually during the 12-week period. Aerobic conditions at the sewage sludge dump site in the New York Bight Apex can promote rapid decomposition of sewage sludge if it is distributed evenly over an area of at least 5.2km^2 (2 miles).

576 - Guarino, C. F. 1967. Sludge disposal by barging to sea. Water, and Sewage Works. 114:126-127.

577 - Guarino, C. F., et al. 1975. Land and sea solids management alternatives in Philadelphia. Jour. Water Poll. Control Fed. 47(11)2551-2564.

578 - Guarino, C. F., M. D. Nelson, and S. Townsend. 1977. Philadelphia sludge disposal in coastal waters. Jour. Water Poll. Control Fed. 49(5):737-744.

579 - Guarino, C. F., M. D. Nelson, and S. S. Almeida. 1979. Ocean dispersal as an ultimate disposal method. [Includes heavy metals (Pb, Cr, Ni, Zn, Cu, Cd, Mn, Zr, V) and coliform levels]. Jour. Water Poll. Control Fed. 51:773-782.

580 - Guigres, F., and J. Perieres. 1978. Les rejets en mer des navires et leur controle. [The disharge to the sea of vessels and their control.] Ind. Pet. Monde Gaz-Chim. 497:65-77.

581 - Gulf Coast Waste Disposal Authority. 1978. Gulf coast waste disposal authority five year status report 1970-1975 (Texas). Houston, TX. 24p.

582 - Gunnerson, C. G., R. P. Brown, and D. D. Smith. 1969. Marine disposal of solid wastes. A paper presented at ASCE Ocean Engineering Conference. Miami Beach, FL.

583 - Gunnerson, C. G. 1975. Ocean Dumping in the New York Bight. Gov. Rpts. Announcements. 75(25):97 Dec. 1975. NOAA TR ERL-321. Boulder, CO. 83p.

Abstract: The New York Bight extends seaward over 15,000 m^2 from Long Island and New Jersey to the edge of the continental shelf, some 80-100 nm offshore. Wastes from 20 million people are discharged to the Bight. These wastes arrive by various routes--ocean dumping, outfall sewers, air pollution, river discharge, land runoff, thermal discharges, vessel wastes, and occasional spills. Although impacts of these waters on the marine environment are not clearly understood, there is evidence that the waters, bottom sediments, and living resources are under stress. An average of 7,400,000 m^3/yr of dredge spoils were dumped between 1965 and 1970, plus waste acid, and construction and demolition debris. The hazards of this dumping are not known, but above normal incidence of fin-rot disease in fish in the area and in closing of the area to shell-fishing indicate an endangered situation.

584 - Gunnerson, Charles G., et at. (In Press). Summary report of management panel on domestic or muncipal waste discharges. In: G. F. Mayer, A. Calabrese, F. A. Cross, D. C. Malins, D. W. Menzel, J. S. O'Connor, and F. J. Vernberg (editors), Ecological stress in the New York Bight: Science and Management. Estuarine Research Federation. Columbia, SC.

Abstract: Studies of municipal waste discharges and sewage sludge dumping reveal that the latter account more than half of the loading of the most serious pollutants in the New York Bight region. Of the total inputs, liquid municipal discharges account for 35% to 70% and municipal sludge dumping accounts for 5% to 15%, depending upon the particular contaminant. Increasing costs of conventional waste disposal practices and regulation have not been accompanied by apparent corresponding benefits. There are consequently increasing needs for determining willingness to pay for alternative policy, institutional, and technological options in New York metropolitan area waste generation and disposal practices.

585 - Gunnerson, C. G. 1981. The New York Bight ecosystem. In: R. A. Geyer (editor), Marine Environmental Pollution 2. Dumping and Mining. Elsevier Scientific Publ. Co. Amsterdam. pp. 313-378.

586 - Gusev, A. G. 1962. Effect of sewage from the coal-tar chemical industry on fisheries and aquatic organisms. Nauchn. Tekh. Byul. Gos. Nauchn. Issled. Inst. Ozernogo i Rechnogo Rybn. Khoz. 16:60-64 Referat Zhur. Biol. 1963, No. 20192 (Translation) Sport Fish Abstr. #7475.

587 - Haedrich, Richard. 1977. Neuston fish at DWD-106. In: Baseline Report of Environmental Conditions in Deepwater Dumpsite 106. NOAA/ODP Dumpsite Evaluation Report 77-1. Rockville, MD. Vol. II. pp. 481-485.

588 - Halcrow, W., D. W. Mackay, and I. Thornton. 1973. The distribution of trace metals and fauna in the Firth of Clyde in the relation to the disposal of sewage sludge. Jour. Mar. Biol. Ass. UK 53. pp. 721-739.

589 - Halcrow, W., D. W. Mackay, and J. Bogan. 1974. PCB levels in Clyde marine sediments and fauna. Mar. Pollut. Bull. 5(9): 134-136.

590 - Halebsky, Max. 1979. A study of the economics, and environmental viability of a U.S. flag toxic chemical incineration ship. MARAO Contract No. 7-38029. Prepared by Global Marine Development Inc. Newport Beach, CA. 18p.

591 - Hall, J. B., Jr., and A. O. Pearson. 1977. Results from the National Aeronautics and Space Administration remote sensing experiments in the New York Bight - April 7-17, 1975. NASA TM X-74032. 186p.

Abstract: A cooperative operation between the National Aeronautics and Space Administration (NASA) and the National Oceanic and Atmospheric Administration (NOAA) was conducted in the New York Bight during April 7-17, 1975, to evaluate the role of NASA remote sensing technology to monitor ocean dumping. Six NASA remote sensing experiments were flown on the C-54, U-2, and C-130 NASA-aircraft, while NOAA obtained concurrent sea truth information using helicopters and surface platforms. The experiments included 1) a Radiometer/Scatterometer (RADSCAT), 2) an Ocean Color Scanner (OCS), 3) a Multichannel Ocean Color Sensor (MOCS), 4) four Hasselblad cameras, 5) an Ebert spectrometer, and 6) a Reconafax IV infrared scanner and a Precision Radiation Thermometer (PRT-5).

The purpose of this report is to present the results of these experiments relative to the use of remote sensors to detect, quantify, and determine the dispersion of pollutants dumped into the New York Bight. Descriptions of the experiments, experimental methods, data analyses techniques, and significant results for each experiment are given in separate sections of the report.

592 - Hall, R. G., A. M. Meaburn, S. Chang, R. Greig, and J. B. Pearce. 1978. Retrieval and interpretation of data on trace metals in fish and shellfish from the New York Bight. Report to the NOAA/MESA New York Bight Project. NMFS Northeast and Southeast Fisheries Centers. 52p.

593 - Halsband, E. 1968. Physiologische Untersuchungsmethoden zur Bestimmung des Schadlichkeitsgrades von Abwassergiften in Sub-, Brack- und Salzwasser. Helgolander wiss. Meeresunters. 17:224-246.

594 - Halsband, E. 1971. Physiologische studien zur ermittlung des schaedlichkeitsgrades von "Rotschlamm" in seewasser. Wasser, Luft Und Betrieb. 15(7):268-273.

595 - Han, G., D. V. Hansen, and A. Cantillo. 1979. Diagnostic model of water and oxygen transport. In: R. L. Swanson and C. J. Sinderman (editors), Oxygen Depletion and Associated Benthic Mortalities in New York Bight, 1976. NOAA Professional Paper No. 11. U.S. Dept. of Commerce. Rockville, MD. pp. 165-192.

Conclusions: The first conclusion to be drawn from the use of the diagnostic model is that even during May and June 1976, the replacement time of water in New York Bight was relatively short compared to the general seasonal cycle of property changes for the Bight as a whole and for the inshore segments of principal interest. A second conclusion is that oxygen depletion in the critical region during the May-June period was due to oxygen utilizaiton about 3 times greater than in other regions of the inner shelf and nearly 10 times greater than that occurring over the outer shelf, rather than simply being due to the length of stratified season, stagnation, or advection of low-oxygen water. The most critically affected regions, in fact, had a relative advective oxygen input substantially greater than other areas.

The key problem in examining the anoxia episode is explaining the high consumption rates that occurred off New Jersey during spring and summer 1976. Numerous causes have been suggested, including ocean dumping, estuarine discharge, primary production and an anomalous bloom of dinoflagellate organisms. All these suggestions do not adequately address the questions of why 1976 and why the inner shelf off New Jersey. Although observations obtained and techniques used are deficient for explicit quantitative modeling of carbon/oxygen relationships, or even the distribution of particulate material or *Ceratium tripos* in the Bight, a strong qualitative case can be made from results of the diagnostic circulation model of the Bight, and the model study of particulate material transport by Festa and Hansen (1976), that the 1976 anoxia episode off New Jersey came about in the following way.

For reasons not well understood but probably related to the pattern of surface wind, and quite possibly to the heavy discharge from the Hudson River during early 1976, the circulation in and near the Apex was kinematically equivalent to that commonly occurring in coastal plain estuaries for the entire period from May 18 to June 29. A similar situation existed in the average along the entire inner New Jersey shelf over the last 26 days of the period.

Such circulations tend to trap and concentrate suspended particulate matter having a small sinking velocity. This concentration process functions irrespective of where such particulate materials are introduced into the circulation pattern. Hence, particulates from the Hudson River, the dumping activities, and plankton productivity would have concentrated in the Apex and over the inner shelf off New Jersey. This concentration of oxidizable material had an exorbitant biochemical oxygen demand that depleted the oxygen supply in spite of reasonable active ventilation of waters below the thermocline.

If *Ceratium tripos*, which are capable of relatively rapid movement, seek a position below the thermocline, this behavior will couple to the circulation just as do inanimate particle distributions. In the case of *Ceratium tripos* the concentration mechanism can be expected to function with particular efficiency because the organism can optimize its vertical movement relative to that of the water. A principal outstanding question is whether *Ceratium tripos* did in fact contribute in a dominant way to the oxygen demand off New Jersey, or whether the concentration of other oxidizable particulates would have led to anoxia even in the absence of *Ceratium tripos*.

Conditions in May and June 1975 have not been diagnosed. But the comparison of current meter data between 1975 and 1976 shows that the convergent circulation pattern which fostered inshore concentration of particulates did not exist during spring 1975.

Though we can calculate what occurred in 1976, the comparison of oxygen utilization rates to average rates suffers because we have a poor idea of what the average really is. Changes in both terms of the oxygen mass balance, ($QC+Q^{V}C$) and S, contributed to the large decrease in concentrations with time preceding the anoxic episode. A really illuminating result is that an order of magnitude change in these terms is not necessary to produce the observed conditions. Relatively small changes, smaller than a factor of two, can considerably influence the mass balance. The simplicity of the oxygen mass balance equation conceals many feedback loops between the water transport and oxygen production. Decreased transport will, for example, prevent nutrients from reaching an area, thus decreasing plankton growth rates. In the solution to these problems and others, such as nutrient supply, organic loadings, and planktonic growth rates, and the oxygen consumption rates of all these constituents, lie the answers to why the anoxia developed off New Jersey in summer 1976. The oxygen transport is only a piece of the puzzle.

596 - Hancock, J. A., R. P. Jacobs, M. R. Knapp, and J. S. Glascow. 1974. Waterborne debris in marine pollutin incidents. U.S. Coast Guard. Office of Research and Development. Washington, DC. (CG-D-108-74). 460p.

597 - Hand, T. D., A. W. Ford, P. G. Malone, D. W. Thompson and R. B. Mercer. 1978. A feasibility study of response techniques for discharges of hazardous chemicals that sink. Washington, DC. U.S. Coast Guard, Office of Research and Development. Report CG-D-56-78. 334p.

598 - Hann, R. W. 1975. Ocean discharge. U.S. Nat. Acad. Sci., Nat. Res. Coun., Study Panel. Assessing potential ocean pollutants. Washington, DC. pp. 228-299.

599 - Hann, R. W., Jr. 1976a. Evaluation of safety factors with respect to ocean disposal of waste materials. Texas A&M University. Sea Grant Publication. TAMU-SG-77-202. College Station, TX. 101p.

600 - Hann, R. W. Jr., T. F. Zapatka, M. C. Zapatka, and C. Baskin. 1976b. History of ocean dumping in the Gulf of Mexico. Texas A&M Univ. Civil Eng. Dept. College Station, TX. 87p.

601 - Hansen, D. V. 1977. Circulation. MESA New York Bight Atlas Monograph No. 3. MESA New York Bight Project and New York Sea Grant Institute. Albany, NY. 23p.

Abstract: Information sources on currents in New York Bight date back 65 years. Only during the last 15 years, however, has it been possible to collect sufficiently accurate data to enable quantitative statements. These more recent results indicate that the standard deviation of the currents in New York Bight, as in other oceanic areas, is several times greater than the mean flow. This temporal variation is due primarily to tides and wind. Because the wind effects are random in time, various observations are not readily combined to provide a composite description.

Data from systematic observation projects have revealed some major features of circulation in the Bight, but do not yet allow description of many details. The major feature of Bight circulation is a relatively slow flow to the southwest over most of the outer continental shelf with some indication of a clockwise eddy in the inner Bight. There is the expected exchange circulation, characterized by seaward flow of estuarine waters near the surface and landward flow of deeper waters between the Hudson/Raritan estuary and the offshore waters, and there is some indication that the landward flow may extend as far as 64 km (40 mi) offshore in the Hudson Shelf Valley. All of these features are masked primarily by stronger but variable wind-driven currents on a day-to-day basis, and may be drastically altered for periods of several weeks. This is especially so during summer in conjunction with sustained periods of little rainfall or strong southerly winds.

602 - Harbison, Richard, L. Madin, and V. McAlister. 1977. Gelatinous zooplankton at Deepwater Dumpsite 106. In: Baseline Report of Environmental Conditions in Deepwater Dumpsite 106. NOAA/ODP Dumpsite Evaluation Report 77-1. Rockville, MD. Vol. II. pp. 305-307.

603 - Hard, C. G., and H. D. Palmer. 1976. Sedimentation and ocean engineering: Ocean dumping. John Wiley Interscience. New York, NY. pp. 557-577.

604 - Hardy, C. D., and E. R. Baylor. 1975. Surface tension reductions and urban wastes in the New York Bight. Jour. of Geophy. Res. No. 18 (80):2696-2699.

Abstract: The distribution of surface tension in the New York Bight was from oil drop spreading measurements at sea. Reductions in surface tensions of greater than 30% (<50.9 dyn/cm) of that calculated for clean seawater were associated with sewage sludge dumping and urban waste discharges. The potential role of surface films as an agent in the transport of metallic and lipophilic pollutants is briefly discussed.

605 - Hardy, C. D., E. R. Baylor, and P. Moskowitz. 1976. Sea surface circulation in the northwest Apex of the New York Bight - with appendix: bottom drift over the continental shelf. Volumes I and II. NOAA TM ERL MESA-13. 132p. Boulder, CO. NTIS No. PB269028.

Abstract: Plastic interface drift cards, designed to move within the uppermost thin layer of the sea surface, were released within the New York Bight between January and August 1974 to simulate the movement of sea surface contaminants, particularly oil slicks, within that study area. It is concluded that the probability of sea surface contaminants reaching the costs of New York and New Jersey is maximal during the summer, primarily because of the increased frequency of winds blowing from the east to southwest. The appendix is concerned with a seabed drifter study of sewage sludge dumping grounds south of Long Island which shows that these dumping grounds are located within areas of substantial inshore movement. It is concluded that some fraction of this sludge is transported away from the dumpsite and towards the New York shorelines.

606 - Hardy, C. D., and E. R. Baylor. In Press. Surface tension reductions and urban wastes in the New York Bight. Jour. Geophys. Res.

607 - Harms, U. 1974. The levels of heavy metals (Mn, Fe, Co, Ni, Cu, Zn, Cd, Pb, Hg) in fish from onshore and offshore waters of the German Bight. ICES. C.M. 74/E:5.

608 - Harmsen, H. 1974. Schadlose vernichtung von gefahrlichen chemischen sondermull auf See. [Harmless destruction of dangerous dust-like chemicals in the sea]. Umwelthygiene. 25(9):201-204.

609 - Harold, L. C., and R. A. Baldwin. 1967. Ecological effects of antibiotics. FDA Papers. February.

610 - Harris, W. H. 1974a. Testimony given at the U.S. Environmental Protection Agency, Region II hearing held 11 February 1974 on renewal of permits for the ocean dumping of sewage sludge in the New York Bight. New York, NY.

611 - Harris, W. H. 1974b. Testimony given at the New York State Assembly Environmental Committee hearing held 22 March 1974 on sewage sludge dumping in the New York Bight. New York, NY.

612 - Harris, W. H. 1974c. Sewage sludge expansion, migration, accumulation, and identification. NASSAU Co. New York inner continental shelf. Technical Briefing Report No. 5. Presented at the congressional hearing on ocean sludge dumping conducted by the Senate Public Works Committee's Subcommittee on Environmental Pollution. Marine Sciences Research Group, Institute for Chronological and Geochemical Studies. Brooklyn College, New York, NY.

613 - Harris, W. H. 1975. Heavy metal ratios as indicators of dispersal and origin of waste solids in the New York Bight. Geol. Soc. of Amer. Abst. 7(1):72.

Abstract: The distribution and heavy metal composition of waste sediments in the New York Bight indicate that sewage sludge disposed of in the Bight Apex is initially transported northwestward and partially deposited in the Christiansen Basin (the headward portion of the Hudson Shelf Channel). A portion of the sewage sludge transported into this basin is moved northward toward Long Island and then eastward in the nearshore zone beneath the western and northern limbs of the clockwise gyre in the Bight Apex.

Ratios of selected heavy metals in waste sediments accumulating in the nearshore zone, e.g., Ag/Sn, Hg/Cd, Ni/V, Zn/Cr, Zn/Cu and Zn/Pb, and the general diminution of heavy metal pollutant levels independently within shelf sands and waste sediments from the Christiaensen Basin to the shore area indicate that inshore sludge patches have been derived by advective subsurface current transport from the Christiaensen Basin sludge bed. Age classes of mud-bottom species within inshore sludge patches suggest that they have formed during the past three years, a period of time when the offshore sludge bed spread 4.5 miles northward to within three nautical miles of the shoreline.

Some of the sewage sludge transported in the Apex gyre appears to complete the circuit and is deposited in the shelf channel as far as 15 to 20 km south of the disposal sites. In contrast, dredge spoils are largely confined to their disposal area where they have formed a relative topographic high in the Bight Apex.

614 - Harris, William H. 1976. Spatial and temporal variation in sedimentary grain-size facies and sediment heavy metal flow ratios in the New York Bight Apex. In: M. G. Grant (editor), Amer. Soc. Limnol. Oceanogr. Spec. Symp. 2:102-123.

Abstract: In northwest New York Bight a pronounced spatial and temporal variation occurs in sediment grain-size, heavy metal ratios, and concentrations. Substrate mobility is most pronounced near Long Island and in northern Christiaensen Basin. Long Island mud patches are most extensive during late spring through summer and may coalesce across intervening sand-wave crests obliterating the shore-zone sand-wave topography. Mud patches are either restricted to sand-wave trough axes or are absent or covered during early fall through early spring, suggesting that muddy sediments are either flushed from the nearshore sand-wave system by bottom currents or undergo in situ microbial degradation.

The northwest periphery of the Christiaensen Basin mud-sandy mud facies apparently oscillates in position. Mud patches in the peripheral zone expanded shoreward from 6.9 km to within 5.0 km of the Long Island shoreline between winter and summer 1974 and contracted seaward to within 9.3 km of the shoreline by early spring 1975; isolated mud patches remained within 6.5 km of the shoreline. Shoreward expansion and seaward contraction of the mud facies of the northwest Christiaensen Basin occurred during the same months as did nearshore mud patches growth and size reduction or disappearance; probably both instances were the result of the processes.

Individual mud patches near Long Island have not existed for a long time. Mud patches off Atlantic Beach and Lido Beach, Long Island, began to develop as early as summer 1972; others developed later.

The Cr:Zn ratio varies in mud deposits at the sludge disposal area, northern Christiaensen Basin, and Hempstead Bay and in mud patches near Long Island. Maximum Cr:Zn values occur during late spring through early fall, minimum values during early winter.

Zinc was released to Bight waters preferentially relative to Cr for which carbon-rich muds apparently served as sink during spring-summer 1974 expansion of northwest Christiaensen Basin mud facies and growth of mud patches near Long Island. During later contractions of the northwest Christaensen Basin mud facies and attendant decrease in size or disappearance of nearshore mud patches from fall 1974 through early spring

1975, Zn and Cr both were released at nearly equal rates or Zn was preferentially retained in the sediment. The Cr:Zn annual maximum may be explained by preferential zinc desorption from sludge or mud solids by cation exchange, or preferential adsorption of organic chromium chelates from solution by organic mud substrates, or zinc uptake by phytoplankton or bacteria. The Cr:Zn annual minimum may be explained by dissolution of Mn and Fe hydroxides releasing absorbed Zn and Cr or oxidation of the organosulfur compounds of the two metals at nearly equal rates.

615 - Harris, R., R. Jolly, R. Huggett, and G. Grant. 1977. Trace metals. In: Middle Atlantic Outer Continental Shelf Environmental Studies, VII-B. Chemical and Biological Benchmark Studies. U.S. Department of the Interior and Virginia Institute of Marine Science. Gloucester Point, VA. 8:1-60.

616 - Harrison, A. J. 1973. Environmental surveys in connection with North West Acid waste disposal, Pre-dumping Survey, 1970 and Post-dumping Survey, 1971. Tasmanian Fish. Res. Vol. 7. No 3.

617 - Hartenstein, Roy. 1981. Sludge decomposition and stabilization. Science. 212(4496):743-749.

Summary: Sludges are labile putrescible materials which undergo a heterotrophic decomposition process and be subject to humification before becoming stabilized substances. An understanding of the principles underlying these processes may give rise to better management plans for the application of sludge to land. Procedures are suggested for the evaluation of sludge stabilization rates.

618 - Harwood, F. L., and K. C. Wilson. 1957. An investigation into a proposal to dispose of power-station ash by discharging it into the sea at low water. Proc. Institution of Civil Engineers. Paper no. 62222. 8:53-70.

619 - Haskin, H. H., and A. S. Merrill. 1973. Preliminary report under cooperative agreement dated June 15, 1972, between National Marine Fisheries Service and Rutgers the State University to conduct an inshore inventory of surf clams along the New Jersey coast. 11p. (mimeo).

620 - Hatcher, Patrick G. 1976. Steroids as sewage specific indicators in New York Bight sediments. NOAA/AOML. Marine Geology & Geophysics Lab. Miami, FL. 13p.

Abstract: Sediments of the New York Bight were examined for steroidal compounds and found to contain relatively large amounts of coprostanol and 24 B-ethyl coprostanol. These steroids were found to be derived exclusively from sewage, and it is suggested that they be used as sewage tracers.

621 - Hatcher, P. G., and L. E. Keister. 1976a. Sediments of the New York Bight: Their bulk organic chemical properties. NOAA DR ERL MESA-21. Boulder, CO. 24p. NTIS No. PB 271399/AS.

622 - Hatcher, Patrick G., and Larry E. Keister. 1976b. Carbohydrates and organic carbon in New York Bight sediments as possible indicators of sewage contamination. In: M. G. Gross (editor), Amer. Soc. Limnol. Oceanogr. Spec. Symp. 2:240-248.

Abstract: Sediments of the New York Bight were analyzed for total organic carbon (TOC) and total carbohydrates (TCH). The TCH:TOC ratio was significantly more elevated than comparable sediments from other areas. High TCH:TOC values (>30) may be attributed to sewage-derived materials which contribute significant quantities of refractory organic matter to the Bight annually. The observed TCH:TOC ratios suggest that organic material being deposited in the Christiaensen Basin and mud patches near Long Island is predominantly of sewage origin, and that seaward of the Apex, the sedimentary organic matter becomes less influenced by sewage-derived organic matter and oceanic organic matter becomes a more significant fraction.

623 - Hatcher, P. G., L. E. Keister, and P. A. McGillivary. 1977. Steroids as sewage specific indicators in New York Bight Sediments. Bull. Environ. Contam. Toxicol. 17:491-498.

Abstract: Sediments of the New York Bight were examined for steroidal compounds and found to contain relatively large amounts of coprostanol and 24 B-ethyl corpostanol. These steroids were found to be derived from sewage, and it is suggested that they be used as sewage tracers in marine sediments.

624 - Hatcher, Patrick G., and Philip A. McGillivary. 1979. Sewage contamination in the New York Bight. Coprostanol as an indicator. Envir. Sci. and Tech. 13(10):1225-1229.

Abstract: Sediments of the New York Bight are analyzed for coprostanol, a fecal steroid, to determine the degree of sewage contamination. Coprostanol, when reported as a percentage of total steroids (% coprostanol), can be quantitatively related to the amount of sewage-derived organic matter. Furthermore, coprostanol is quite persistent in anoxic silts of the Bight and, thus, can be used to delineate historical contamination in these silts. Based on the sediments analyzed, the New York Bight is shown to be highly contaminated with sewage (most likely ocean-dumped sewage sludge), especially in the topgraphically low areas near the dump site, where black silts have been known to accumulate.

625 - Hatcher, Patrick G., George A. Berberian, Adriana Y. Cantillo, Philip A. McGillivary, Philip Hanson, and Richard H. West. 1981. Chemical and physical processes in a dispersing sewage sludge plume. In: B. H. Ketchum, D. R. Kester, and P. K. Park (editors), Ocean Dumping of Industrial Wastes. Plenum Press. New York, NY. pp. 347-378.

Abstract: In July 1976, an experiment was conducted to determine the physical and chemical changes occurring during a sewage sludge dump in a thermally stratified water column in the New York Bight with a line dump, a modified line dump, and one spot dump. The results indicate that a rapid dilution occurs in the line dump such that dissolved and particulate trace metals, nutrients, and bulk organic compounds are diluted to near background levels. Only the trace organic compounds, ammonia, and dissolved Cu provide any indication for presence of the plume for a period of 5 hours after dumping. The fecal steroid coprostanol is used as a quantitative sludge tracer to determine the initial 10^4 dilution of the plume, to measure the amount of mixing as the sludge disperses in the stratified water column, and to determine if any other chemical compounds deviate from conservative mixing. There is physical and chemical fractination as the sludge settles. Initially, a large portion rapidly settles.

626 - Hausknecht, Keith A. 1977. Results of studies on the distribution of some transition and heavy metals at Deepwater Dumpsite 106. In: Baseline Report of Environmental Conditions in Deepwater Dumpsite 106. NOAA/ODP Dumpsite Evaluation Report 77-1. Rockville, MD. Vol. III. pp. 499-541.

627 - Hausknecht, Keith A., and Dana R. Kester. 1981. 106-Mile ocean waste disposal site chemical data report from R/V KNORR cruise 58, August 27-September 7, 1976. In: Assessment Report on the Effects of Waste Dumping in 106-Mile Ocean Waste Disposal Site. NOAA/ODP Dumpsite Evaluation Report 81-1. Rockville, MD. pp. 70-79.

628 - Haverinen, A., and H. Vuoristo. 1978. Direct wastewater load into the Gulf of Bothnia from the Finnish coast. Finn. Mar. Res. No. 244. pp. 165-172.

629 - Hawkes, H. A. 1971. Disposal by dilution? - an ecologist's viewpoint. In: C. Sykes and F. A. Skinner (editors). Microbial aspects of pollution. Academic Press. New York, NY. pp. 149-179.

630 - Hayashi, T. 1978. Control of toxic effluents and management of toxic bottom sediments. In: S. A. Peterson and K. K. Randolph (editors), Management of Bottom Sediments Containing Toxic Substances. Proceedings of the Third U.S.-Japan Experts' Meeting November 1977, Easton, Maryland. U.S. Environmental Protection Agency. Corvallis, OR. (EPA-600/3-78-084). pp. 1-9.

631 - Hays, H., and G. Cormons. 1974. Plastic particles found in tern pellets, on coastal beaches and at factory sites. Mar. Pollut. Bull. 5(3):44-46.

Abstract: Small polystyrene particles, evidently of industrial origin, now appear as a contaminant of the sea in several parts of the world. They have been discovered in pellets of indigestible food regurgitated by gulls and terns, so are clearly entering the food chain at some point. So far as is known at present, they are harmless, but it would be as well to exercise caution in releasing plastic to the environment.

632 - Hazelworth, J. B., B. L. Kolitz, R. B. Starr, R. L. Charnell, and G. A. Berberian. 1974. New York Bight Project Water Column Sampling Cruises #1-5 of the NOAA Ship FERRELL, August-November 1973. NOAA DR ERL MESA-74-2. Boulder, CO. 191p. NTIS No. PB 247921/AS.

633 - Hazelworth, J. B., B. Kolitz, R. Starr, R. Charnell, G. Berberian, and M. Weiselberg. 1975a. New York Bight Project, Water Column Sampling Cruises #6-8 of the NOAA Ship FERRELL, April-June 1974. NOAA DR ERL MESA-1. Boulder, CO. 177p. NTIS No. COM 75-10726.

634 - Hazelworth, J., B. Kolitz, R. Starr, R. Charnell, G. Berberian, and M. Weiselberg. 1975b. New York Bight Project, Water Column Sampling Cruises #9-12 of the NOAA Ship FERRELL, July-November 1974. NOAA DR ERL MESA-3. Boulder, CO. 231p. NTIS No. PB 271340/AS.

Abstract: During the period July-November 1974, four oceanographic cruises, denoted 9, 10, 11, and 12, were made by the NOAA Ship FERREL in the New York Bight. The objective of the cruises was to supply data to provide a base for analysis of the water movements on the highly impacted ecosystem. This report presents the corrected physical and chemical data from these cruises and describes the parameters measured, the measurement methods, and the corrections applied to the data.

635 - Hazelworth, J. B., and M. E. Darnell. 1976. MESA New York Bight Project, Expanded Water Column Characterization Cruises (XWCC - 2 and 3), NOAA Ship RESEARCHER, 22 February - 5 March, 9-12 April, 1975. NOAA DR ERL MESA-23. Boulder, CO. 237p. NTIS No. PB 271398/AS.

Abstract: During spring 1975, the NOAA Ship RESEARCHER made two expanded water column characterization cruises (XWCC-2 and 3) in the central continental shelf of the New York Bight. The objective of these cruises was to collect data to provide a base for analysis of the water movement. Current meter arrays and bottom-mounted pressure gauges were deployed and recovered for 76 stations during XWCC-2. Geological oceanographic data were collected on 28 stations during XWCC-3. In addition, physical and chemical oceanographic data were collected on both cruises. This report presents the corrected water column data from these two cruises and a preliminary descriptive analysis.

636 - Hazelworth, John B., Shailer R. Cummings, Robert B. Starr, and George A. Berberian. 1977a. MESA New York Bight Project - Expanded Water Column Characterization Cruise (XWCC-8) - NOAA Ship GEORGE B. KELEZ, April, 1976. NOAA DR ERL MESA-27. Boulder, CO. 112p.

Abstract: During the period 12-16 April 1976, an oceanographic cruise was made by the NOAA Ship KELEZ in the New York Bight. The objective of the cruise was to supply data to provide a base for analysis of the water movements in the highly impacted ecosystem. This report presents the corrected physical and chemical (nutrient) data from this cruise.

637 - Hazelworth, J. B., R. B. Star, S.R. Cummings, and G. A. Berberian. 1977b. MESA New York Bight Project, Expanded Water Column Characterization Cruise (XWCC-9), NOAA Ship GEORGE B KELEZ, May 1976. NOAA DR ERL MESA-31. Boulder, CO. 184p. NTIS No. PB 283574/AS.

Abstract: During the period 17-24 May 1976, an oceanographic cruise was made on NOAA Ship GEORGE B. KELEZ in the New York Bight. The objective of the cruise was to supply data to provide a base for analysis of the water characteristics and movements in the highly impacted ecosystem. This report presents the corrected physical and chemical (nutrient) data from this cruise.

638 - Hazelworth, John B., Shailer R. Cummings, S. Michael Minton, George A. Berberian. 1977c. MESA New York Bight Project, Expanded Water Column Characterization Cruise (XWCC-11), NOAA Ship RESEARCHER, Sept. 1976. NOAA DR ERL MESA-29. Boulder, CO. 192p.

Abstract: During the period 8-27 September 1976, an oceanographic cruise, denoted XWCC-11, was made by the NOAA Ship RESEARCHER in the New York Bight. Bottom mounted pressure gauges were recovered. Physical, chemical, and biological oceanographic data were collected. The objective of the cruise was to supply data to provide a base for analysis of the water movement in the highly impacted ecosystem. This report presents the corrected physical and chemical (nutrient) data, and a preliminary descriptive analysis of the physical data.

639 - Hazelworth, J. B., A. Herman, T. M. Niedrauer, and P. G. Hanson. 1978a. MESA New York Bight Project, Expanded Water Column Characterization Cruise (XWCC-12), NOAA Ship GEORGE B. KELEZ, April-May 1977. NOAA DR ERL MESA-33. Boulder, CO. 176p. NTIS No. PB 283861/AS.

Abstract: During the period 28 April-6 May 1977, an oceanographic cruise was made on the NOAA Ship GEORGE B. KELEZ in the New York Bight. The objective of the cruise was to supply data to provide a base for analysis of the water characteristics and movements in the highly impacted ecosystem. This report presents the corrected physical and chemical (nutrient) data from this cruise.

640 - Hazelworth, J. B., T. M. Niedrauer, P. G. Hanson. 1978b. MESA New York Bight Project Expanded Water Column Characterization Cruise (XWCC-13), NOAA Ship GEORGE B. KELEZ, May-June 1977. NOAA DR ERL MESA-34. Boulder, CO. 215p. NTIS No. PB 284525.

Abstract: During the period 31 May-7 June 1977, an oceanographic cruise was made on the NOAA Ship GEORGE B. KELEZ in the New York Bight. The objective of the cruise was to supply data to provide a base for analysis of the water characteristics and movements in the highly impacted ecosystem. This report presents the corrected physical and chemical (nutrient) data from this cruise.

641 - Hazelworth, J. B., T. M. Niedrauer, and P. G. Hanson. 1978c. MESA New York Bight Project Expanded Water Column Characterization Cruise (XWCC-14), NOAA Ship GEORGE B. KELEZ, June-July 1977. NOAA DR ERL MESA-37. Boulder, CO. 117p. NTIS No. PB 285695/AS.

Abstract: During the period 27 June-1 July 1977, an oceanographic cruise was made on the NOAA Ship GEORGE B. KELEZ in the New York Bight. The objective of the cruise was to supply data to provide a base for analysis of the water characteristics and movements in the highly impacted ecosystem. This report presents the corrected physical and chemical (nutrient) data from this cruise.

642 - Hazelworth, J. B., T. M. Niedrauer, and P. G. Hanson. 1978d. MESA New York Bight Project Expanded Water Column Characterization Cruise (XWCC-15), NOAA Ship GEORGE B. KELEZ, August 1977. NOAA DR ERL MESA-38. Boulder, CO. 192p. NTIS No. PB 285696/AS.

Abstract: During the period 1-9 August 1977, an oceanographic cruise was made on the NOAA Ship GEORGE B. KELEZ in the New York Bight. The objective of the cruise was to supply data to provide a base for analysis of the water characteristics and movements in the highly impacted ecosystem. This report presents the corrected physical and chemical (nutrient) data from this cruise.

643 - Hazelworth, J. B., T. M. Niedrauer, and P. G. Hanson. 1978e. MESA New York Bight Project Expanded Water Column Characterization Cruise (XWCC-16), NOAA Ship GEORGE B. KELEZ, October 1977. NOAA DR ERL MESA-39. Boulder, CO. 122p. NTIS No. PB 286047/AS.

Abstract: During the period 11-19 October 1977, an oceanographic cruise was made on the NOAA Ship GEORGE B. KELEZ in the New York Bight. The objective of the cruise was to supply data to provide a base for analysis of the water characteristics and movements in the highly impacted ecosystem. This report presents the corrected physical and chemical (nutrient) data from this cruise.

644 - Hazelworth, J. B., and Shailer R. Cummings. 1979. Physical conditions compared with previous years. In: R. L. Swanson and C. J. Sindermann (editors), Oxygen Depletion and Associated Benthic Mortalities in New York Bight, 1976. NOAA Professional Paper No. 11. U.S. Dept. of Commerce. Rockville, MD. pp. 125-135.

Abstract: Seasonal density stratification began about 1 month earlier in 1976 than in 1975 (April vs. May) because of early seasonal heating (thermal stratification) and, more importantly, strong and sustained river discharge (salinity stratification). Subsequently, the rate of bottom replenishment of dissolved oxygen from the sea surface was less than that of 1975.

645 - Hazelworth, J. B., and G. A. Berberian. 1981a. MESA New York Bight Project Expanded Water Column Characterization Cruise (XWCC-17), NOAA ship GEORGE B. KELEZ, April 1978. NOAA DR ERL OMPA-3. Boulder, CO. 200p.

646 - Hazelworth, J. B., T. M. Niedrauer, G. A. Berberian. 1981b. MESA New York Bight Project Expanded Water Column Characterization Cruise (XWCC-18), NOAA Ship GEORGE B. KELEZ, June 1978. NOAA DR ERL OMPA-4. Boulder, CO. 22p.

647 - Hazelworth, J. B., and G. A. Berberian. 1981c. MESA New York Bight Project Expanded Water Column Characterization Cruise (XWCC-19), NOAA Ship GEORGE B. KELEZ, July 1978. NOAA DR ERL OMPA-5. Boulder, CO. 240p.

648 - Hazelworth, J. B., and G. A. Berberian. 1981d. MESA New York Bight Project Expanded Water Column Characterization Cruise (XWCC-20), NOAA Ship GEORGE B. KELEZ, July-August 1978. NOAA DR ERL OMPA-6. Boulder, CO. 240p.

649 - Hazelworth, J. B., and G. A. Berberian. 1981e. MESA New York Bight Project Expanded Water Column Characterization Cruise (XWCC-21), NOAA Ship GEORGE B. KELEZ, April 1979. NOAA DR ERL OMPA-7. Boulder, CO. 221p.

650 - Hazelworth, J. B., and G. A. Berberian. 1981f. MESA New York Bight Project Expanded Water Column Characterization Cruise (XWCC-22), NOAA Ship GEORGE B. KELEZ, May-June 1979. NOAA DR ERL OMPA-8. Boulder, CO. 219p.

651 - Hazelworth, J. B., and G. A. Berberian. 1981g. MESA New York Bight Project Expanded Water Column Characterization Cruise (XWCC-23), NOAA ship GEORGE B. KELEZ, July 1979. NOAA DR ERL OMPA-9. Boulder, CO. 271p.

652 - Hazelworth, J. B., and G. A. Berberian. 1981h. MESA New York Bight Project Expanded Water Column Characterization Cruise (XWCC-24), NOAA ship GEORGE B. KELEZ, August 1979. NOAA DR ERL OMPA-10. Boulder, CO. 285p.

653 - Heaney, F. L. 1977. Disposal at sea. In: D. G. Wilson (editor), Handbook of Solid Waste Management. Van Nostrand Reinhold Co., New York, NY. pp. 264-274.

654 - Hebard J. Frank, and Michael A. Champ. 1981. Legislative history and philosophy of the development of the U.S. marine pollution strategy. In: Towards a National Marine Pollution Policy. University of Rhode Island. Center for Ocean Management Studies. Kingston, RI. pp. 11-34.

655 - Hedgpeth, J. W. 1970. The Oceans: World symp. environment. 12(3):40-47.

656 - Hedgpeth, Joel W. 1972. Atomic waste disposal in the sea: An ecological dilemma? In: F. M. Taghi and J. P. Milton (editors), The Careless Technology: Ecology and International Develoment. Conference on the Ecological Aspects of International Development. (Held in Warrenton, VA, 8-11 Dec. 1968). Jointly sponsored by Conservation Foundation. Washington, D.C. and Washington University, Center for the Biology of Natural Systems, St. Louis, Missouri. Doubleday and Company, Inc. Natural History Press, Garden City, NY. pp. 812-828.

657 - Heemstra, F. W. 1971. Second post-dump survey of the chase x disposal site. U.S. Naval Research Laboratory. Washington, DC. Memorandum Report No. 2319. 24p.

658 - Heezen, Bruce C. 1975. Photographic reconnaissace of continental slope and upper continental rise. In: May 1974 Baseline Investigation of Deepwater Dumpsite 106. NOAA/ODP Dumpsite Evaluation Report 75-1. Rockville, MD. pp. 27-103.

659 - Heezen, Bruce C. 1977. Six dives to the lower continental slope and upper continental rise southwest of Hudson Canyon, geological aspects. In: Baseline Report of Environmental Conditions in Deepwater Dumpsite 106. NOAA/ODP Dumpsite Evaluation Report 77-1. Rockville, MD. Vol. 1:9-27.

660 - Henry, R. 1976. Studies of current circulation at ocean waste disposal sites. Scientific and Technical Aerospace Reports. 14(3):319.

661 - Henry, W. 1961. Sludge and effluent disposal by dispersal. Wastes Engr. pp. 677-698.

662 - Herkimer, K. W. 1974. Navigation aids for ocean waste disposal control. Government Reports Announcements. 74(12):131.

663 - Hetling, L. J. 1974. An analysis of past, present, and future Hudson River wastewater loadings: New York State Dept. Environmental Conserv. Tech. Paper 37. 35p.

Abstract: An inventory of historical, present, and predicted 1980 wastewater loadings to the Hudson River basin is presented in tabular form. The historic analysis is limited to municipal wastewater discharges in the lower Hudson River basin. Conclusions indicate a decreasing trend in organic wastewater loadings since 1964 that should continue through 1980.

664 - Hickel, W. 1969. Sedimentbeschaffenheit und bakteriengehalt im sediment eines zukunftigen verklappungsgebietes von industrieabwassern nordwest lich Helgolands. (Sediment properties and bacterial numbers in the sediment of a sea area selected for industrial waste disposal north west of Helgoland.) Helgoland wiss. Meeresunters. Vol 19. Part 1:1-20.

665 - Hickman, Robert E. 1970. Ocean disposal vs. inland disposal of treated wastewater for Falmouth, Massachusetts. New England Water Poll. Control Assoc. 4(12):140-149.

666 - Hill, H. W., and J. Pawlak. 1977. Site selection criteria for dumping grounds. ICES. C.M. 77/E:28.

667 - Hill, M. D., and G. Lawson. 1980. An assessment of the radiological consequences of disposal of high-level waste in coastal geologic formations. National Radiological Protection Board. Harwell. 65p. NRPB-R108.

668 - Hinga, K. R., G. Ross Heath, D. Richard Anderson, and Charles D. Hollister. 1982. Disposal of high-level radioactive wastes by burial in the sea floor. Environ. Sci. Technol. 16(1):28-37.

Conclusions: The nuclear age has been with us for more than a third of a century. Both the fabrication of nuclear weapons and commercial power production have produced large quantities of radioactive waste that, if handled incorrectly, will cause serious health and safety problems not only for us today, but also for future generations. Nuclear waste inventories in 1980 included 75 million gallons of high-level waste from weapons production and 12000 tons of spent reactor fuel. The fundamental problem at hand is to find a way to safely dispose of these two types of high-level waste.

The disposal issue is largely independent of the future of the nuclear industry. An immediate curtailment of nuclear power production would reduce the expected volume of waste, but not the necessity for a disposal plan. Nor would a reduction in the volume of waste change the chemical and physical properties of the nuclides, or the time and barrier requirements of a repository. It must also be remembered that the U.S. is not alone in this situation. A number of countries use nuclear materials and have similar disposal problems.

The subseabed concept has been under consideration in the U.S. for seven years. The Subseabed Disposal Program is now formulating the necessary models and progressing to verify them. Results to date indicate that this method of high-level waste disposal remains promising. In the U.S. the concept is viewed as a longer-term alternative to land-based disposal, and a seabed disposal area could possibly become a regional repository for the eastern U.S. Many foreign nations, however, may not have land areas with suitable geology for high-level waste disposal, or have the desire to develop repositories on land, and may turn to the ocean for disposal. As more nations consider the concept, the focus of research on subseabed disposal is rapidly leaving our shores for foreign ports.

669 - Hirata, S., O. Takimura, and T. Shiozawa. 1978. Behavior of manganese in aqua-benthic environment in the Hiro Bay. Jour. Oceanogr. Soc. Japan. 34(5):237-241.

670 - Hirsch, Allan. 1974. NOAA's New York Bight Marine Ecosystems Analysis Project: An interdisciplinary study of the marine environment. Mar. Techn. Soc. Jour. 8(9):29-34.

Abstract: NOAA initiated a new program in 1972-the Marine Ecosystems Analysis Program (MESA). The objective of the MESA Program is to develop information on the biological, physical, and chemical processes of selected coastal areas, which will improve our ability to assess and predict the impact of man-made alterations or natural phenomena. The first area selected for study is the New York Bight, one of the Nation's most complex and heavily used coastal areas. The New York Bight MESA Project involves a number of NOAA organizational elements, outside cooperators, and contractors in an integrated study approach.

671 - H. M. S. O. 1972. Final Act of the Iner-governmental Conference on the Convention of the Dumping of Wastes at Sea. Cmnd. 5169.

672 - H. M. S. O. 1972. Convention for the Prevention of Marine Pollution by Dumping from Ships and Aircraft. Cmnd. 4984.

673 - H. M. S. O. 1974. Control of Pollution Act. Chapter 40. ISBN 0 10 5440744.

674 - H. M. S. O. 1975. Convention for the Prevention of Marine Pollution from Land-based Sources. Cmnd. 5803.

675 - H. M. S. O. U. K. Dept. of the Environment. 1977. Sewage sludge disposal data and technical notes. Water Engineering 1. Research and Development Division Tech. Note. No. 6. 44p.

676 - Hodges, J. W. 1974. International law and radioactive pollution by ocean dumping: "With all their genius and with all their skill...." San Diego Law Review. 11(3):757-775.

677 - Holden, A. V., and A. A. Cowan. 1976. Organochlorine residues in commercial species of fish from Scottish inshore waters. ICES. C.M. 76/E:45.

678 - Holden, A. V., and A. A. Cowan. 1976. Input of organochlorines to the North Sea via atmospheric deposition. ICES. C.M. 76/E:46.

679 - Hollman, Rudolph. 1970. The physical oceanography of the New York Bight. In: Alan A. Johnson (editor), Water Pollution in the Greater New York Area. Gordon and Breach Science Publishers. New York, NY. p. 2-12.

680 - Holof, A. H. 1970. Wastewater treatment and its effects on receiving waters. In: A. A. Johnson (editor), Water Pollution in the Greater New York Area. Proceedings of a Symposium held on 13 December 1959 at the City University of New York. Gordon & Breach Science Publishers. New York, NY. pp. 145-146.

681 - Hood, Donald W. 1957. Waste dispersal in marine waters. Coastal Engineering. pp. 607-624.

682 - Hood, D. W., B. Stevenson, and L. M. Jeffrey. 1958. Deep sea disposal of industrial wastes. Indust. and Engin. Chem. 50(6):885-888.

683 - Hood, D. W. 1958. Waste disposal in marine waters. Proc. Sixth Conf. Coastal Engineering. Florida. pp. 607-624.

684 - Hood, D. W., T. Ward Duke, and B. Stevenson. 1960. Measurement of toxicity of organic waste of marine organisms. Jour. Water Poll. Control Fed. 32(9):982-993.

685 - Hood, D. W., and E. Kelley. 1972. Contamination and coastal pollution through waste disposal practices. In: B. H. Ketchum, (editor). The Water's Edge: Critical Problems of the Coastal Zone. The Coastal Zone Workshop, held in Woods Hole, Massachusetts. 22 May-3 June, 1972. Cosponsored by the (U.S.) Institute of Ecology and Woods Hole Oceanographic Institution. MIT Press. Cambridge, MA. pp. 146-186.

686 - Horne, R. A., A. J. Mahler, and R. C. Rosells. 1971. The marine disposal of sewage sludge and dredge spoil in the waters of the N.Y. Bight. Woods Hole Oceanographic Institution TM 1-71. Woods Hole, MA. NTIS No. AD 722791.

687 - Hovis, W. A., Jr. 1977. Remote sensing of water pollution. Proceedings of the Eleventh International Symposium on Remote Sensing of Environment. Ann Arbor, MI. 25-29 Apr. 1977. pp. 361-362.

688 - Howells, Gyneth Parry. 1968. The present biological status of the Hudson River. In: National Estuarine Pollution Study Proceedings. pp. 333-355.

689 - Hu, V. J. H. 1981. Ingestion of deep-sea mining discharge by five species of tropical copepods. Wat. Air Soil Pollut. 15:433-440.

690 - Hulburt, E. M., and C. M. Jones. 1977. Phytoplankton in the vicinity of Deepwater Dumpsite 106. In: Baseline Report of Environmental Conditions in Deepwater Dumpsite 106. NOAA/ODP Dumpsite Evaluation Report 77-1. Rockville, MD. Vol. II. pp. 219-231.

691 - Hume, N. B., and W. F. Garber. 1966. Marine disposal of digested screened wastewater solids. Third international conference on water pollution research. Munich, Germany. Section III. paper No. 12.

692 - Hundemann, A. S. 1979. Remote sensing applied to environmental pollution detection and management. A bibliography with abstracts. Progress Report, 1964 - July 1979. NTIS/PS-79/0842/9. 176p.

Abstract: The application of remote sensing methods to water pollution problems is discussed. Topic areas cover characteristics of dispersion and diffusion by which pollutants are transported, discharges from outfalls of industrial plants, and application of remote sensing to estuarian problems. This updated bibliography contains 167 abstracts, 11 of which are new entries to the previous edition.

693 - Hunter, J. V., and H. Heukelekian. 1965. The composition of domestic sewage fractions. Jour. Water Poll. Control Fed. 37:1142-1163.

694 - Hutzinger, O., I. E. van Lelyveld, and B. C. J. Zoeteman. 1978. Aquatic Pollutants: Transformation and biological effects. Proc. of the Second International Symposium on aquatic pollution. Noordwigkerhout (Amsterdam), The Netherlands. 26-28 Sept. 1977. Pergamon Press, Oxford. 519p.

695 - Hydroscience, Inc. 1974. Water quality evaluation for ocean disposal system, Suffolk County, N.Y. Hydroscience, Inc. Westwood, NJ.

Abstract: Disinfected secondary municipal effluent is to be conveyed from the Suffolk County Southwest Sewer District No. 3 water pollution control plant at Fleet Point across Great South Bay and the barrier beach to the Atlantic Ocean in the vicinity of Fire Island Inlet. The principal purpose of this study is to provide quantitative analytical procedures with which to determine the effects of wastewater disposal at various alternative locations on water quality in the Atlantic Ocean and Great South Bay.

In this regard, various types of mathematical water quality modeling studies were conducted, together with a field and laboratory data collection program to support the model verification and application. By these means, the location of the terminal site for the marine outfall structure is recommended to be 3.5 miles offshore and south of Cedar Island Beach. This site has been selected because it assures, to the maximum practical extent, the maintenance of public health and recreational values in the study area, and minimizes adverse effects on general water quality and the marine ecosystem.

696 - Hydroscience, Inc. 1978. Estimation of PCB reduction by remedial action on the Hudson River ecosystem. Report for NYS-DEC. Hydroscience, Inc. Westwood, NJ.

Abstract: The report explains the possible effects of remedial action to reduce the PCB sources in the Upper Hudson, on the Hudson River Ecosystem, with particular emphasis on the Lower Hudson Fishery. Limited dredging shows a lower PCB concentrion by 50% to about 20-40 ppm with a lower fish level by 20% to 8-12 ppm. Extensive dredging upstream may lower the levels by 30-50% to the action level. The no action response will take longer than a decade for flushing out of PCB.

697 - Hypes, W. D., and C. W. Ohlhorst. 1977. A summary of the test procedures and operational details of a Delaware River and an ocean dumping pollution monitoring experiment conducted 28 August 1975. NASA-TM-X-74405. 42p.

698 - ICES. 1969. Report of the ICES Working Group on pollution of the North Sea. ICES Co-operative Research Report. Series A. No. 13. pp. 11-12. [For additional ICES, see Ref. # 712].

699 - ICES. 1974. Report of the Working Group for the international study of the pollution of the North Sea and its effects on living resources and their exploitation. Co-operative Research Report No. 39.

700 - Ichiye, T., and M. Carnes. 1981. Application of aerial photography to the study of small-scale upper ocean phenomena. Pure and Applied Geophys. (PAGEOPH) 119. Basel. 1-15.

701 - Ichiye, T., Masamichi Inoue, and Michael Carnes. 1981. Horizontal diffusion in ocean dumping experiments. In: B. H. Ketchum, D. R. Kester, and P. K. Park (editors), Ocean Dumping of Industrial Wastes. Plenum Press. New York, NY. pp. 131-159.

Abstract: During two experiments monitoring ocean dumping operations in the Gulf of Mexico and off Arecibo, Puerto Rico, a number of drifters were tracked with a ship-borne radar and aerial photographs of the plumes were analyzed by use of the results of drifter experiments in the Gulf of Mexico. Horizontal eddy diffusivity was determined from increases of the variances of the drifter positions with time and were calculated with Okubo's method (1976). The variances of drifter positions did not increase with time when the wind speed exceeded critical values off Puerto Rico. It was speculated that the wind rows caused the drifter convergence during the Pierto Rico experiments. The speculation is partially confirmed by analyzing the aerial photos of the plumes which show striations along the direction of the prevailing winds.

702 - Idler, D. R. Effects of pollants on quality of marine products and effects on fishing. In: M. Ruivo (editor), Marine Pollution and Sea Life. Fishing News (Books) Ltd. London. pp. 535-537.

703 - IMCO - The Inter-Governmental Maritime Consultative Organization. 1976. Convention on the Prevention of Marine Pollution by Dumping of Wastes and Other Matter. 1972. London. 36 pp + 1979 Amendments (Pub. No. 76. 14E).

704 - IMCO - The Inter-Governmental Maritime Consultative Organization. 1979a. Report of permits issued for dumping in 1976. LDC 2/Circ. 31. 25p.

705 - IMCO - The Inter-Governmental Maritime Consultative Organization. 1979b. Report of permits issued for dumping in 1977. LDC 2/Circ. 33. 23p.

706 - IMCO - The Inter-Governmental Maritime Consultative Organization. 1980. Report of permits issued for dumping in 1979. LDC 2/Circ. 47. London. 24p.

707 - Inaba, D., et al. (13 authors). 1974. Some observations on behavior of the treated sewage disposed in the sea. Bull. Coll. Agr. & Vet. Med. Nihon Univ. No. 31:54-78.

708 - Ingram, W. T., and H. Mitwally. 1966. Paths of pollution in New York harbor: A model study. Jour. Water Poll. Control Fed. 38(10):1563-1581.

709 - Ingham, Merton C., James J. Bisagni, and David Mizenko. 1977. The general physical oceanography of Deepwater Dumpsite 106. In: Baseline Report of Environmental Conditions in Deepwater Dumpsite 106. NOAA/ODP Dumpsite Evaluation Report 77-1. Rockville, MD. Vol. 1. pp. 29-54.

710 - Inoue, N. 1971. Safety of disposing sludge containing heavy metals to the ocean. Mizu Shori Gijutsu (Jap.). 12(11):25-32. (In Japanese).

711 - Institute of Oceanographic Science. 1978. Oceanography related to deep sea waste disposal. (A survey commisssioned by the Department of the Environment). Rept. Inst. Oceanogr. Sci. No. 77.

712 - International Commission for the Exploration of the Sea (ICES). 1978. Sea bed charting. ICES. C.M. 1978/E:12. 53p. [See Ref. #'s 698, 699].

Abstract: Includes positions of dumpsites, mining operations, and dredging sites: charts submitted by delegates from Belgium, Finland, Norway, UK, and USA.

713 - International Hydronics Corporation. 1974. Report to Allied Chemical Corporation on dispersion of waste HCI. Elizabeth Works, NJ.

714 - Interstate Electronics Corporation. 1973a. Ocean waste disposal in selected geographic areas. U.S. Environmental Protection Agency, Ocean Disposal Program. Washington, DC. 374p. Report IEC-446 OC 1541. NTIS No. PB-224 793/HP.

715 - Interstate Electronics Corporation. 1973b. A bibliography on ocean waste disposal. Anaheim, CA. 105p.

716 - Interstate Electronics Corporation. 1973c. An atlas of ocean waste disposal sites. Interstate Electronics Corporation, Environmental Engineering Division. Anaheim, CA. Report No. 4460C1545.

717 - Interstate Electronics Corporation. 1973d. Directory of managers, engineers and scientists in ocean waste disposal and related environmental science fields. Interstate Electronics Corporation, Environmental Engineering Division. Anaheim, CA. Report No. 4460C1543.

718 - Interstate Electronics Corporation. 1973e. Ocean waste disposal in the New York Bight. Interstate Electronics Corporation. Environmental Engineering Division. Anaheim, CA. Report No. 4460C1559.

719 - Interstate Electronics Corporation. 1973f. Ocean waste disposal in selected geographic areas. Interstate Electronics Corporation. Environmental Engineering Division. Anaheim, CA. Report No. 4460C1541.

720 - Interstate Electronics Corporation. 1974a. Guidelines for development of criteria for control of ocean waste disposal. Interstate Electronics Corporation. Anaheim, CA. 76p.

721 - Interstate Electronics Corporation. 1974b. Development of criteria to control ocean dumping. Interstate Electronics Corporation. Environmental Engineering Division. Anaheim, CA. NTIS No. PB-233 018/1. 10p.

722 - Interstate Electronics Corporation. 1974c. Ocean waste diposal in metropolitan areas of California. Interstate Electronics Corporation. Environmental Engineering Division. Anaheim, CA. Report No. 4460C1635.

723 - Interstate Electronics Corporation. 1974d. Guidelines for development of criteria for control of ocean waste disposal. Interstate Electronics Corporation. Environmental Engineering Division. Anaheim, CA. NTIS No. PB-233 019/9. 99p.

724 - Interstate Electronics Corporation. 1974e. Navigation aids for ocean waste disposal control. Interstate Electronics Corporation. Environmental Engineering Division. Anaheim, CA. NTIS No. PE-230 945/8. 33p.

725 - Interstate Electronics Corporation. 1974f. Oceanographic surveys report. Interstate Electronics Corporation. Environmental Engineering Division. Anaheim, CA. Report No. 44 OC1647.

726 - Interstate Electronics Corporation. 1974g. A technical review of ocean waste disposal at a site in the Gulf of Mexico. Interstate Electronics Corporation. Environmental Engineering Division. Anaheim, CA.

727 - Interstate Electronics Corporation. 1974h. Radioactive waste disposal activities in the vicinity of the Farallon Islands. Interstate Electronics Corporation. Environmental Engineering Division. Anaheim, CA. NTIS No. PB 241 300/3ST. 23p.

728 - Interstate Electronics Corporation. 1976. A bibliography on ocean waste disposal. Second Edition. Interstate Electronics Corporation. Environmental Engineering Division. Anaheim, CA. Report No. 4460C0417.

729 - Interstate Sanitation Commission: New York, New Jersey, Connecticut. 1975. 1974 Report of the Interstate Sanitation Commission on the Water Pollution Control Activities and the Interstate Air Pollution Program. 63p.

Abstract: Pretreatment of industrial wastes, removal of oils from District Waters, thermal pollution, enforcement and combined sewers water quality monitoring.

730 - Interstate Sanitation Commission: New York, New Jersey, Connecticut. 1976. New York - New Jersey metropolitan area sewage sludge disposal management problem. 68p.

Abstract: Report on Sludge Management. To discontinue ocean sludge disposal. Feasible methods are composting, land spreading, and pyrolysis. Composting may produce a soil conditioner. Pretreatment and landfills are being considered, but may involve leaching problems.

731 - Interstate Sanitation Commission: New York, New Jersey, Connecticut. 1976. 1975 Report of the Interstate Sanitation Commission on the Water Pollution Control Activities and the Interstate Air Pollution Program. 108p.

Abstract: Water and air pollution activities of Commission on technical assistance, planning, laboratory analysis, monitoring, and coordination of interstate problems. Contract to model entire N.Y. Harbor area.

732 - Isaacs, J. D. 1975. Assessment of man's impact on marine biological resources. In: E. A. Pearson and E. de Fraja Frangipane (editors), Marine Pollution and Marine Waste Disposal. (Supplement to Progress in Water Technology.) Proceedings of the 2nd International Congress, San Remo, 17-21 December, 1973. Pergamon Press. pp. 329-340.

733 - Issacs, J. D., et al. [U.S. National Academy of Sciences, Committee on Ocenography]. 1962. Disposal of low level radioactive wastes in Pacific coastal waters. NAS-NRS Report No. 985. 87p. National Research Council, Washington, DC.

734 - Isotalo, I., and K. Hakkila. 1978. The load and quality in the coastal waters of the Archipelago Sea and the southern Bothnian Sea. Finn. Mar. Res. No. 244. pp. 198-214.

735 - Italiano, Michael. 1976. Marine waste disposal in the New York Bight - Public policy environmental impacts, and alternative futures. Masters Thesis. State Univ. of New York. College of Environmental Science and Forestry. Syracuse, NY. 288p.

Abstract: The objective of this program and policy review is to examine the range of factors contributing to the problems of ocean dumping in the New York Bight, and the range of resulting impacts. A program review approach is appropriate to examine the key issues of environmental impacts, public policy, and alternative futures because the problem of marine pollution is dependent upon and affected by many forces.

Specificially, ocean dumping in the New York Bight is a broad and complex marine pollution problem eliciting a great deal of fragmented, and sometimes contradicting research projects. Consequently, there is a need to synthesize and coordinate this research, and to present an accurate assessment of the basic state of knowledge on the Bight. An attempt is therefore made to describe the environmental effects from waste disposal, and public legislative and regulatory policies, realizing however, that this state of knowledge is continually changing.

Hopefully, this compilation and synthesis will serve as a base upon which further research may be facilitated; and it may serve as a reference tool for those who are interested in ocean waste disposal.

736 - Izzo, Leonard, Michael A. Champ, and Donald W. Lear. 1974. Dissolved and particulate biological iron. Supplemental Report -- Environmental Survey of Two Interim Dump Sites -- Middle Atlantic Bight. EPA Region III, Philadelphia, PA. pp. 9-12.

737 - James, W. P., F. J. Burgess, and D. Baumgartner. 1971. An aerial photographic study of waste field from three ocean outfalls. Ann. Offshore Technology Conf. Vol. 1.

738 - Jay, David A. 1974. The hydrography and circulation of New York Harbor and Western Long Island Sound: Implications for water quality. Masters Thesis. State Univ. of New York. Stony Brook, NY. 71p.

Abstract: The literature on the tides, currents, hydrography, and hydrology of New York Harbor from 1848 to date is reviewed. Particular emphasis is placed on the relationship between water, salt, and pollutant transport mechanisms in the East River, the Harbor, and western Long Island Sound.

The East and Harlem Rivers are both hydraulic tidal straits in which the tides and currents are governed to a first approximation by the tidal elevation at each end. An estuarine circulation in the Upper East River modifies the hydraulic regime and transports Upper East River surface water toward Long Island Sound.

All available measurements of water and salt transport taken over the last 105 years in the East and Harlem Rivers are compiled. The variabilities are high and the data insufficient to directly determine seasonal and yearly means of nontidal transports. Theoretical calculations based on a detailed analysis of the nontidal transport of 10 to 15×10^6 m^3 tide $^{-1}$ toward New York Harbor. However, the critical assumption of whether tidal friction is the same for both ebb and flood flows has not been verified.

The observed distributions of nutrients and sewage effluents suggest that the estuarine and dispersive mechanisms in the East River transport a considerable fraction of the $35\ m^3\ sec^{-1}$ of sewage emptied into the East River to the Sound, while the advective component is expected, on the average, to transport effluents into New York Harbor.

Transport of sewage pollutants from the East River is probably a major factor causing the observed violations of state and federal water quality standards in western Long Island Sound. It is concluded that the East River, particuarly the upper section, is a very poor location to release sewage effluents.

739 - Jay, David A., and Malcolm J. Bowman. 1975. The physical oceanography and water quality of New York Harbor and western Long Island Sound. Marine Sciences Research Center. State Univ. of New York. Stony Brook, NY. Tech. Report 23. Ref. 75-7. 71p.

Abstract: The literature on the tides, currents, hydrography, and hydrology of New York Harbor from 1848 to date is reviewed. Particular emphasis is placed on the relationship between water, salt, and pollutant transport mechanisms in the East River, the Harbor, and western Long Island Sound.

The East and Harlem Rivers are both hydraulic tidal straits in which the tides and currents are governed to a first approximation by the tidal elevation at each end. An estuarine circulation in the upper East River modifies the hydraulic regime and transports upper East River surface water toward Long Island Sound.

All available measurements of water and salt transport taken over the last 105 years in the East and Harlem Rivers are compiled. The variabilities are high and the data insufficient to directly determine seasonal and yearly means of nontidal transports. Theoretical calculations based on a detailed analysis of the nontidal transport of 10 to 15 x 10^6 m^3 $tide^{-1}$ toward New York Harbor. However, the critical assumption of whether tidal friction is the same for both ebb and flood flows has not been verified.

The observed distributions of nutrients and sewage effluents suggest that the estuarine and dispersive mechanisms in the East River transport a considerable fraction of the 35 m^3 sec^{-1} of sewage emptied into the East River to the Sound, while the advective component is expected, on the average, to transport effluents into New York Harbor.

Transport of sewage pollutants from the East River is probably a major factor causing the observed violations of state and federal water quality standards in western Long Island Sound. It is concluded that the East River, particularly the upper section, is a very poor location to release sewage effluents.

740 - Jeffress, W. S. 1977. Geologic effects of ocean dumping on New York Bight inner shelf. Amer. Assoc. of Petroleum Geologists Bull. 61(5):841.

Abstract: A detailed study intended to inventory marine sand and gravel resources was conducted using 824 km of high-resolution seismic-reflection subbottom profiles and 61 vibratory sediment cores. Analyses of these data together with bathymetric maps from 1845 to 1973 and ancillary deep-boring logs have revealed that many of the natural bathymetry and geologic conditions on the inner New York Bight shelf have been modified significantly by ocean disposal of waste solids since about 1888. Innermost parts of the Hudson Shelf channel have aggraded by approximately 15 m and broad hills mask the former natural-channel morphology. Composition of the waste solids varies considerably; much of the early material was natural soil and excavation debris from construction of buildings, subway tunnels, and navigation channels. More recent wastes consist primarily of fine-grained dredge spoil, treated sewage sludge, and industrial acids. The natural geologic stratigraphy shows considerable variation also, making the distinction between natural and anthropogenic sediment difficult. Upper Cretaceous clastic strata crop out at Shrewsbury Rocks and variable thicknesses of stratified Quaternary sand and gravel predominate over the shelf. Cores from the Hudson Channel depression contain relict-estuarine deposits. Deposition rates of solid wastes have been far greater than for natural sediment; most of the waste solids have remained in the original dump sites and are modified only slightly by present shelf processes.

741 - Jeffries, Harry P. 1962. Environmental characteristics of Raritan Bay, a polluted estuary. Limnol. and Oceanogr. 7(1):21-31.

742 - Jenkinson, Ian. R. 1972. Sludge dumping and benthic communities. (Southampton region). Mar. Pollut. Bull. 3(7):102-105.

743 - Jensen, L. D. 1969. Biological processes which interact with and influence the distribution of wastes introduced into the marine environment. National Academy of Sicences. Background papers on coastal wastes management. Washington, DC. p. XIV-1 to XIV-9.

744 - Jensen, S. 1976. Havet som affaldsplads. (Dumping of wastes in the sea.) [In Danish]. Naturens Verden. pp. 97-102.

745 - Johnson, J. W., and R. L. Wiegel. 1970. Oceanographic factors in the functional design of waste disposal systems. University of California. Water Resources Center Report No. 21. pp. 133-139.

746 - Johnson, P. P., and S. A. Wood. 1977. Seasonal variability of sediment texture in the Middle Atlantic region. In: D. W. Folger (editor), Geologic Studies of the Middle Atlantic Outer Continental Shelf, Vol. III Geologic Studies. U.S. Geol. Survey, Woods Hole, Mass. U.S. Dept. of Interior. Bureau of Land Management and Virginia Institute of Mar. Sci. Gloucester Point, VA.

747 - Johnson, R. J., and V. T. McCauley. 1972. Technical requirements for control of ocean dumping. 8th Ann. conf. and Exposition. Mar. Technol. Soc. Sept. 11-13, 1972. Mar. Techn. Soc. Washington, DC. pp. 743-761.

748 - Johnson, R. W., and J. B. Hall Jr. 1977a. Remote sensing operations (multispectral scanner and photographic), In the New York Bight September 22, 1975. NASA Langley Research Center. Langley, VA. NASA TM X-73993. 14p.

Abstract: Ocean dumping of waste materials is a significant environmental concern in the New York Bight. One of these waste materials, sewage sludge, was monitored in a joint National Aeronautics and Space Administration/National Oceanic and Atmospheric Administration (NASA/NOAA) experiment conducted in the New York Bight on September 22, 1975. NASA remote sensing over controlled sewage sludge dumping included an 11-band Johnson Space Center NP-3A aircraft. Concurrent in situ water samples were taken and acoustical measurements were made of the sewage sludge plumes resulting from line (moving barge) and spot (stationary barge) dumps. Multiple aircraft overpasses were made to evaluate temporal effects on the plume signature.

749 - Johnson, R. W., I. W. Duedall, R. M. Glasgow, J. R. Proni, and T. A. Nelson. 1977b. Quantitative mapping of suspended solids in wastewater sludge plumes in the New York Bight Apex. Jour. Water Poll. Control Fed. 49(10):2063-2073.

Abstract: Statistical analysis techniques were applied to develop quantitative relationships between aircraft remotely sensed data and measurements made in the plume resulting from ocean dumping of wastewater in the New York Bight. A correlations coefficient of 0.96 was obtained between suspended solids in the plume and remotely sensed data from an aircraft multispectral scanner. The standard error of estimate was 4.11 mg/l over a measured range of 1.11 to 32.20 mg/l. A calibrated regression equation from the analysis was used to develop a map of quantitative distributions of suspended solids in wastewater sludge dump plumes. Spectral signatures were obtained that may be useful for identification of wastewater sludge plumes relative to other pollution features, such as acid waste plumes.

750 - Johnson, Robert W. 1977c. Multispectral analysis of ocean dumped materials. Proceedings of the Eleventh International Symposium on Remote Sensing of Environment. April 25-29. Ann Arbor, MI.

Abstract: Experiments conducted in the Atlantic coastal zone indicate that plumes resulting from ocean dumping of acid wastes and sewage sludge have unique spectral characteristics. Remotely sensed wide area synoptic coverage provides information on these pollution features that is not readily available from other sources. Aircraft remotely sensed photographic and multispectral scanner data were interpreted by two methods. First, qualitative analyses in which pollution features are located, mapped, and identified without concurrent sea truth and, second, quantitative analyses in which concurrently collected sea truth is used to calibrate the remotely sensed data and to determine quantitative distributions of one or more parameters in a plume.

Location and mapping of pollution features in the coastal zone previously have been based on radiance differences between the plumes and receiving waters, using sea truth or other data for identification. As a result of the data analyses in these experiments, an in-scene calibration technique was developed that "normalizes" atmospheric effects, thereby potentially providing a means of plume identification that is independent of the specific scene and the multispectral scanner used. The in-scene calibration technique is based on ratio values in each band of a multispectral scanner where the ratios are radiance values in the pollutant plume divided by radiance values of ocean water in the same scene. Application of this technique to data from several experiments indicates that plumes resulting from acid wastes and sewage sludge have distinctive spectral characteristics over a range of environmental conditions and for two multispectral scanners flown at altitudes of 3.0 and 19.7 kilometers.

In addition to qualitative analyses that used the in-scene calibration, quantitative analysis techniques were applied to sewage sludge dump plumes. Results indicated that quantitative relationships exist between suspended solids in the plumes and remotely sensed radiance values. The calibrated regression equation that used Modular Multispectral Scanner bands 4 (540-580 nanometers (nm) and 9 (760-860 nm) had a correlation coefficient of 0.961 and a standard error of estimate or 4.11 mg/l for a measured range from 1.11 to 32.20 mg/l. The calibrated regression equation was applied to map quantitative distributions of suspended solids in two plumes in the remotely sensed scene.

751 - Johnson, R. W., R. M. Glasgow, I. W. Duedall, and J. R. Proni. 1978. Monitoring the temporal dispersion of a sewage sludge plume in the New York Bight by remote sensing. In: Proceedings of the American Society of Photogrammetry. 44th Annual meeting, 26 February-4 March 1978. Washington, DC. pp. 423-440.

752 - Johnson, R. W., R. M. Glasgow, I. W. Duedall, and J. R. Proni. 1979. Monitoring the temporal dispersion of a sewage sludge plume. Photogrametric Engineering and Remote Sensing. 45(6):763-768.

Abstract: Sewage sludge from the Metropolitan New York City area is dumped in the Apex of the New York Bight. Possible environmental effects due to this dumping are being studied. Results of a joint National Aeronautics and Space Administration (NASA), National Oceanic and Atmospheric Administration (NOAA), and State University of New York (SUNY) experiment on July 15, 1976, indicate remotely sensed monitoring and data analysis techniques may be used to study temporal dispersion characteristics of plumes resulting from dumping of sewage sludge. Remotely sensed data were collected by multispectral scanner radiance data to obtain a regression equation which was then used to map quantitative distributions of suspended solids in the plume. For the spot dump monitored, reflected radiances from the surface waters reached peak values about 45 minutes after the dump. After this time, spectral analysis indicated that the calibrated equation could be applied. The plume was dispersed within about five hours. Spectral characteristics of the experiment plume were similar to sewage sludge plumes that have been monitored in other experiments in the Atlantic Coastal Zone. These plumes are readily distinguished from acid wastes that are dumped in the same geographical area.

753 - Johnson, Robert W. 1980. Remote sensing and spectral analysis of plumes from ocean dumping in the New York Bight Apex. In: Remote Sensing of Environment. 9:197-209. Elsevier North Holland Inc. Amesterdam.

Abstract: Experiments conducted in the Atlantic Coastal Zone indicate that plumes resulting from ocean dumping of acid wastes and sewage sludge have distinguishable spectral characteristics. Remotely sensed wide-area synoptic coverage provides information on these pollution features that is not readily available from other sources. Photographic and multispectral scanner data remotely sensed from aircraft were interpreted by two methods. First, qualitative analyses in which pollution features are located, mapped, and identified without concurrent sea truth and, second, quantitative analyses in which concurrently collected sea truth is used to calibrate the remotely sensed data and to determine quantitative distributions of one or more parameters in a plume. For the qualitative analyses an in-scene calibration technique was developed that "normalizes" environmental effects, thereby potentially providing a means of plume identification that is independent of the specific scene and the multispectral scanner used. Application of this technique to data from several experiments indicates that plumes resulting from acid wastes and sewage sludge have distinctive spectral characeristics over a range of environmental conditions and for two multispectral scanners flown at altitudes of 3.0 and 19.7 kilometers. In addition to qualitative analyses that used the in-scene calibration, quantitative analysis techniques were applied to sewage sludge dump plumes. A calibrated regression equation that related remotely sensed radiances to suspended solids concentrations was developed in order to map synoptic suspended solids distributions in plumes.

754 - Johnson, Robert W., and Craig W. Ohlhorst. 1981. Application of remote sensing to monitoring and studying dispersion in ocean dumping. In: B. H. Ketchum, D. R. Kester, and P. K. Park (editors), Ocean Dumped Industrial Wastes. Plenum Press. New York, NY. pp. 175-191.

Abstract: Experiments conducted in the coastal waters of the United States indicate that plumes resulting from ocean dumping of sewage sludge and industrial wastes have distinguishable spectral characteristics. Remotely sensed wide-area synoptic coverage provides information on these pollution features that is not readily available from other sources. Results indicate that qualitative analysis techniques may be used for the location, identification, and mapping of plumes resulting from ocean dumping of waste materials. An in-scene background elimination technique was developed that "normalizes" atmospheric and other environmental effects, thereby potentially providing a means of plume identification that is independent of the specific scene and the multispectral scanner used. Application of this technique to data from several experiments demonstrates that plumes resulting from sewage sludge and several industrial wastes have distinctive spectral characteristics over a range of environmental conditions and for two multispectral scanners flown on aircraft at altitudes of 3.0 and 19.7 kilometers.

In addition to qualitative analyses that used the in-scene background elimination, quantitative analysis techniques were applied to remotely sensed data collected over sewage sludge and industrial waste plumes. Results indicate that quantitative relationships exist between remotely sensed data and suspended solids in sewage sludge plumes and iron concentrations in acid-iron waste plumes. Calibrated regression equations from the multiple regression analyses were applied to map quantitative distributions of those parameters in remotely sensed scenes. These quantitative values and/or radiance differences also have been used to study the temporal dispersion of plumes and surface water movement in the dump areas.

755 - Johnston, R., J. A. Adams, and H. D. Dooley. 1971. ("Unpublished Typescript"). Some observations on the hydrography, chemistry and

plankton of the Firth of Clyde in relation to nitrate-rich effluents. Dept. of Agriculture and Fisheries for Scotland. Aberdeen Marine Laboratory Internal Report No. 29. Aberdeen, Scotland. 27p.

756 - Johnston, R. 1974. ("Unpublished Typescript"). An exploratory appraisal of the inputs of domestic and industrial wastes to the North Sea with regard to Scottish Marine Fisheries. Dept. of Agriculture and Fisheries for Scotland. Aberdeen Marine Laboratory Internal Report. No. 3. Aberdeen, Scotland. 14p.

757 - Johnston, R. 1976. Marine pollution. Academic Press. London. 729p.

758 - Johnston, R. 1979. The sea and the disposal of sewage wastes. Scott. Fish. Bull. 45:32-37.

759 - Johnston, Robert. 1981. Ocean dumping-history, control and biological impact. In: B. H. Ketchum, D. R. Kester, and P. Kilho Park (editors), Ocean Dumping of Industrial Wastes. Plenum Press, New York, NY. p. 39-51.

760 - Jokinen, S. 1978. The state of the sea off Kokkola and Pietarsaari. Finn. Mar. Res. No. 244. pp. 183-190.

761 - Jones, Cynthia, and Richard Haedrich. 1977. Epibenthic invertebrates. In: Baseline Report of Environmental Conditions in Deepwater Dumpsite 106. NOAA/ODP Dumpsite Evaluation Report 77-1, Rockville, MD. Vol. II. pp. 451-458.

762 - Jones, D. S., I. Thompson, and W. Ambrose. 1978. Age and growth rate determinations for the Atlantic surf clam *Spisula solidissima* (Bivalvia: Maetracea), based on internal growth lines in shell cross sections. Marine Biology. 47:63-70.63-70.

Abstract: Marked and recovered surf clams, *Spisula solidissima* Dillwyn, from Virginia (USA) deposited one internal growth line during a period of 1 1/2 years, probably in response to spawning during late summer. We have used these annual growth lines to make growth curves for two samples of New Jersey (USA) clams, one from inshore (1.8 km from shore, 15 m deep), the other from offshore (17.5 km from shore, 28 m deep) waters. The offshore and inshore clams grow at approximately the same rate until age 3 years, after which time the growth rates differ, as does the ultimate lifespan; the offshore clams grow more rapidly and attain a greater age - up to 31 years. Comparison of our growth curves with other published curves revealed a close correspondence to a curve which was based on a study of growth over a 5-year period. Curves based on external growth lines probably underestimates growth rate in early life and overestimate it in later years.

763 - Jones, C. R., C. T. Fray, and J. R. Schubel. 1979. Textural properties of surficial sediments of lower bay of New York Harbor. Marine Sciences Research Center. State Univ. of New York. Stony Brook, NY. Special Report 21. Ref 79-4. 113p.

Abstract: This is one of a series of reports on the sand and gravel resources of the Lower Bay of New York Harbor and the environmental effects of mining in this area. The objective of this report is to describe the textural properties of the surficial sediments.

764 - Jones, P. G. W., J. L. Henry, and A. R. Folkard. 1973. The distribution of selected trace metals in the waters of the North Sea 1971-1973. ICES. C.M. 73/C:5.

765 - Josa y Castells, F. 1962. Vertides directos de aguas residuales en el mer. (The direct disposal of waste waters in the sea.) Revta Obr. publ. Deciembre. pp. 805-812.

766 - Josa, F. 1968/1969. Die Einleitung der Abwasser in Meer. (The introduction of wastewater into the sea). Abwassertechnik. 12(3) V-VII, 1968, and Abwassertechnik, 19(4), VII-VIII, 1968/69).

767 - Juda, L. 1979. The regional effort to control pollution in the Mediterranean Sea. Ocean Management 5(2):125-150.

768 - Juhat, M. and A. Toompuu. 1978. Modeling of substance transfer in the coastal regions of the Baltic Sea. Ambio Spec. Rep. 5. 1977 (1978). pp. 193-196.

769 - Kalin, R. J. 1971. Biological effects of waste disposal in western Long Island Sound: an initial survey. In: M. G. Gross, J. A. Black, R. J. Kalin, J. R. Schramel, and R. N. Smith. Survey of marine waste deposits, New York Metropolitan Region. pp. 56-72. Ch. 6. State University of New York. Stony Brook, NY. Tech. Rep. Ser. No. 8.

770 - Kamlet, Kenneth S. 1975a. It's time to stop killing the ocean. National Wildlife. April-May. pp. 19-21.

771 - Kamlet, K. S. 1975b. Deep-sixing wastes: Disaster at sea. Environ. Action. August 2. pp. 4-8.

Abstract: United States. National Wildlife Federation. 1977. Statement of Kenneth S. Kamlet on Behalf of the National Wildlife Federation at Public Hearing Before the U.S. EPA on the Desirability of Relocating Ocean Dumpsites for the Disposal of Municipal Sewage Sluge, Toms River, New Jersey, May 31-June 1, 1977. National Wildlife Federation. Washington, DC. 32p.

772 - Kamlet, Kenneth S. 1981a. The oceans as waste space: The rebuttal. Oceanus. 24(1):10-17.

Abstract: Discussion of the ocean as a common resource, as a dispersal medium, concepts of dispersal verse isolation for toxic substances with the focus on the need for multimedium management (oceans, land, or air). Monitoring and predictive abilities.

773 - Kamlet, Kenneth S. 1981b. Ocean disposal of organochlorine wastes by at-sea incineration. In: B. H. Ketchum, D. R. Kester, and P. K. Park (editors), Ocean Dumping of Industrial Wastes. Plenum Press. New York, NY. pp. 295-314.

Abstract: Among the industrial wastes with a history of ocean dumping in the U.S. and abroad, organochlorine (and other organohalogen) compounds are perhaps the most troublesome in terms of potential environmental impact. This paper describes the initial experience of the United States with at-sea incineration as a new method of ocean disposal for organochlorine wastes. It traces the development of a program for regulating such disposal under the Ocean Dumping Law, and it discusses the results of several at-sea incineration operations. The paper also briefly describes current efforts to develop international regulations and guidelines to regulate at-sea incineration throughout the world.

774 - Kamlet, K. S. 1981c. Court decision makes future of sewage sludge ocean dumping uncertain. Coastal Ocean Pollution Assessment News. State University of New York. Stony Brook, NY. pp. 46-47.

775 - Kaneta, Pamela J. 1979. Phytoplankton community structure in the New York Bight, August 1977. Master's Thesis. Marine Research Center, State Univ. of New York. Stony Brook, NY. 51p.

Abstract: The distribution and taxonomic composition of the phytoplankton in the New York Bight were investigated during August 1977. Principal component analysis permitted separation of the phytoplankton samples into three distinct groups by depth and geographical position, the onshore/offshore delineation being the 50 meter isobath. These regions were inshore New Jersey, inshore Long Island and offshore on the shelf. The first two principal components were both related to salinity and temperature. The distinct groups could be clearly seen as projections on the two-dimensional subspace defined by these axes, but were not simply related to salinity and temperature gradients, or any of the other hydrographic variables that were measured. The differences between groups could possibly be related to the past history of the water mass or some other less standard and less easily measured hydrographic variable (e.g. trace elements) which was not monitored.

776 - Kapp, Raymond M. 1981. Distribution of recent mortalities of the ocean quahog, <u>Artica islandica</u>, at two Middle Atlantic ocean dumpsites, Masters Thesis. The American University. Washington, DC. 38p.

Abstract: Spatial distribution patterns of recent mortalities of the ocean quahog, <u>Artica islandica</u>, were developed for the Middle Atlantic industrial acid waste and sewage sludge dumpsites off Delaware Bay from data collected on fourteen cruises during August 1974 through February 1979. Recent mortality is identified by the occurrence of clappers (hinged valves of a dead clam) in dredge and bottom trawl samples. Percent recent mortality, defined as the ratio of clappers to the total of live clams plus clappers, is a relative statistic because the natural rate of oean quahog mortality has not been established within the Middle Atlantic Bight.

Mortality distributions plotted against clusters of sediment heavy metals reveal a possible depositional pattern for wastes associated with relatively high recent mortality percentages. Bottom areas with highest mortality estimates are located within the surface boundaries of both dumpsites and in areas to the north and east.

777 - Kapp, R. M., and M. A. Champ. (Unpublished Manuscript). Distribution of recent mortalities of the ocean quahog, *Arctica islandica* at two Middle Atlantic Ocean Dumpsites.

778 - Kapuscinski, R. B., and R. Mitchell. 1981. Solar radiation induces sublethal injury in *Escherichia coli* in seawater. Appl. Envir. Microbiol. 41:670-674.

779 - Karbe, L., C. Schnier, and R. Niedergesass. 1978. Trace elements in mussels *Mytilus edulis* from German coastal waters. Evaluation of multielement pattern with respect to its use for monitoring programs ICES. C.M. 78/E:24.

780 - Kass, N. I. 1941. Sludge disposal at sea. Water Works and Sewage. 88:385.

781 - Kastens, K. A., C. T. Fray, and J. R. Schubel. 1978. Environmental effects of sand mining in the lower bay of New York Harbor - Phase I. Marine Sciences Research Center. State Univ. of New York. Stony Brook, NY. Special Report 15. Ref. 78-3. 139p.

Abstract: Investigation of the sand and gravel resources of Lower New York Harbor was undertaken in part to assess the assumptions for restrictions passed by the New York Department of Environmental Conservation on mining in the harbor. Formerly the harbor was a major source for construction aggregate and fill in metropolitan New York and New Jersey. As the 1st step toward developing a predictive capability for determining environmental impacts and then a management plan for sand and gravel resources, pertinent existing data (including dissertations and other unpublished reports) on the geological and physical setting,the biological resources, and the sand and gravel resources of the area were collected and interpreted. Textural data for new sediment samples collected from East Bank and adjacent areas east of Ambrose Channel and a limited number of samples from West Bank are given. Results of a preliminary geophysical survey of the harbor using a high-resolution seismic profiling system to map the distribution of sand, gravel, and other sediments are shown. Sediment characteristics are highly variable both horizontally and vertically, and sources of the sediment types are from a variety of places and geomorphic processes. More detailed mapping, possibly using an electronic navigation system, is recommended. An annotated bibliography and critical review of major literature relating to the quantity and character of the sand and gravel of Lower New York Harbor and of the processes renewing and distributing them is included.

782 - Katzenelson, E., and H. I. Shuval. 1975. Viral pollution considerations in marine waste disposal. In: E. A. Pearson, and E. de Fraja Frangipane (editors), Marine Pollution and Marine Waste Disposal. (Supplement to Progress in Water Technology.) Proceedings of the 2nd International Congress. San Remo. 17-21 December, 1973. Pergamon Press. New York, NY. pp. 125-129.

783 - Kautsky, Hans. 1972. Meeresverschmutzung. Vorkommen, verbleib und wirkung von schadstoffen im meer. Sonderdruck. UMSCHAU in Wissenschaft und Technik. Heft 18. pp. 587-589.

784 - Kayser, H. 1969. Zuchtungsexperimente an zwei marinen Flaggellaten (Dinophyta) und ihre Anwendung im toxikologischer Abwassertest. Helgolander wiss. Meeresunters. 19:21-44.

785 - Kayser, H. 1970. Experimental-ecological investigations on Phaeocystis poucheti (Haptophyceae): Cultivation and waste water test. Helgolander wiss. Meeresunters 20:195-212.

786 - Keeley, J. W., and R. M. Engler. 1974. Discussion of regulating criteria per ocean disposal of dredged materials. U.S. Army Eng. Misc. Paper. No. AD775826. Vicksburg, MS. 21p.

787 - Keller, George H., Douglas Lambert, Gilbert Rowe, and Nicholas Staresinic. 1973. Bottom currents in the Hudson Canyon. Science. 180(4082):181-183.

Abstract: In-place measurements of the bottom currents in the Hudson Canyon reveal that the current regime is characterized by a pronounced reversal of flow up and down the canyon. Velocities are commonly of the order of 8 to 15 centimeters per second, reaching 27 centimeters per second on occasion in the upper and central portion of the canyon. Although a 2.5-day recording of currents showed a net transport upcanyon, a combination of 66 current measurements from the submersible ALVIN, the analysis of sediment texture and organic carbon, and the determination of the benthic fauna-nutrient relationship indicate that over the long term there is a net transport of fine material through the canyon to the outer continental rise.

788 - Kendall, Arthur W., Jr., and John W. Reintjes. 1975. Geographic and hydrographic distribution of atlantic menhaden eggs and larvae along the Middle Atlantic coast from RV DOLPHIN Cruises, 1965-1966. Fishery Bull. 73(2):317-335.

789 - Kendall, A. W., Jr. 1975. Patterns of larval fish distributions in the Middle Atlantic Bight. In: B. Mar. (editor), Effects of Energy-Related Activities on the Atlantic Continental Shelf, Conference of Brookhaven National Laboratory. Upton, NY. pp. 126-145.

Abstract:

1. There is no place or time not used by some fish in the Middle Atlantic Bight for spawning and larval development. Eggs and larvae of fish are found from the shore zone to the edge of the continental shelf and beyond. Several species center their area of early development in the Bight, while others use the Bight as southern or northern extents of their spawning areas.

2. Most marine fish probably do not seek a specific geographic site for spawning at sea. Rather, they seem to be influenced by temperature and other factors to spawn wherever they find themselves within broad geographic areas. Spawning areas for each species generally shift along the coast seasonally.

3. The sizes of year-classes of fish are largely determined by differential mortality during the egg and larval stages. These stages are quite vulnerable and sensitive to many forms of environmental stress, even conditions lasting only a few days.

4. Enormous quantities of planktonic eggs and larvae are produced over fairly large areas and long perids of time. To maintain stable populations, most of these must die before reaching adulthood. Man-induced mortality must be considered in light of the high level of natural mortality.

5. Fishes with dissimilar habits and economic importance as adults may be competitors or predators as larvae and thus influence each others' year-class strength. Density-dependent mortality may also occur during the larval period.

790 - Kenyon, K. 1972. Test dumps and surveys at Pt. Judith Station; 15 M depth. In: M. W. First, (editor). Municipal Waste Disposal by Shipborne Incineration and Sea Disposal of Residues. Report. Harvard University School of Public Health. Boston, MA. pp. 66-101.

791 - Kester, Dana R., Keith A. Hausknecth, and Richard C. Hittinger. 1977. Recent analyses of copper, cadium and lead at Deepwater Dumpsite 106. In: Baseline Report of Environmental Conditions in Deepwater Dumpsite 106. NOAA/ODP Dumpsite Evaluation Report 77-1. Rockville, MD. Vol. III. pp. 543-546.

792 - Kester, D. R., R. C. Hittinger, Prithviraj Mukherji, and K. Hausknecht. 1978. Metal studies at DWD-106. Final Report NOAA Grant 04-7-158-44027, 04-7-158-44064 and 04-8-M01-39 NOAA Ocean Dumping Program. University of Rhode Island. Kingston, RI. 49p.

793 - Kester, D. R. 1978. The ocean as a dump site. Studies basic to a national policy. Maritimes 22(2):10-11.

794 - Kester, Dana R., and Keith A. Hausknecht. 1981. Transition and heavy metal chemistry associated with 106-Mile site. In: Assessment Report on the Effects of Waste Dumping in 106-Mile Ocean Waste Disposal Site. NOAA/ODP Dumpsite Evaluation Report 81-1. Rockville, MD. pp. 39-69.

795 - Kester, Dana R., Richard C. Hittinger, Prithviraj Mukherji, and Keith A. Hausknecht. 1981a. Metal studies at 106-Mile site. In: Assessment Report on the Effects of Waste Dumping in 106-Mile Ocean Waste Disposal Site. NOAA/ODP Dumpsite Evaluation Report 81-1. Rockville, MD. pp. 81-129.

796 - Kester, Dana R., Richard C. Hittinger, and Prithviraj Mukherji. 1981b. Transition and heavy metals associated with acid-iron waste disposal at Deep Water Dumpsite 106. In: B. H. Ketchum, D. R. Kester, and P. K. Park (editors), Ocean Dumped Industrial Wastes. Plenum Press. New York, NY. pp. 215-232.

Abstract: Liquid acid-iron waste has been discharged into the surface waters at Deep Water Dumpsite 106 in 4 x 10^6 liter quantities since 1976. Upon mixing with seawater, hydrous ferric oxide precipitates and can scavenge potentially toxic metals (Cu, Cd, and Pb) and organic substances from the waste plume and the seawater. A series of water samples were collected at various times from 0.5 to 27 hours after a dump of acid-iron waste using a noncontaminating pumping system and acoustic backscattering to locate the waste plume in the mixed layer (upper 20 m). The samples were anlayzed for pH, total suspended matter, phosphate, silicate, temperature, and salinity. Because of the high Fe concentrations in the waste (0.5 molar), the total Fe concentration in the waste plume provides a good index of the dilution of the waste. During the first 27 hours after the dump, the dilution occurred as a two-stage process. There was rapid initial dilution by 10^4 in the first 0.5 hr followed by slower dilution to 10^5 after 27 hr. Correlation analysis of the metal concentrations in the waste plume showed that Fe and Pb disperse in a closely coupled fashion related to particulate matter. The Cd and Cu disperse at a slower rate than Fe and Pb and they are less associated with the particulate phase.

797 - Kester, Dana R., Bostwick H. Ketchum, and P. Kilho Park. 1981c. Future prospects of ocean dumping. In: B. H. Ketchum, D. R. Kester, and P. K. Park (editors), Ocean Dumping of Industrial Wastes. Plenum Press. New York, NY. pp. 505-517.

Abstract: The dumping of wastes in the ocean will be an important consideration in waste disposal management in the coming years. Maintaining the quality of the marine environment will require an improved understanding of the fate and effect of wastes in the ocean. In addition to the industrial chemicals, dredged material, and the sewage sludge presently dumped in the ocean, consideration should be given to the behavior in the ocean of radioactive wastes, ocean mining wastes, and incineration residues. The concept of assimilative capacity provides a basis for identifying the physical, chemical, and biological processes in the ocean which are important in waste disposal considerations, and one can recognize the distinctive behaviors in the ocean of several types of chemicals such as biodegradable substances, toxic metals, and persistent synthetic organic chemicals. There is a need to improve the assessment of the biological effects of pollutants in the marine environment. Strategies for monitoring waste disposal in the sea should provide an early warning of environmental degradation and should enhance the understanding of marine processes in order to assure effective management of ocean dumping practices.

798 - Ketchum, B. H., and W. L. Ford. 1948. Waste disposal at sea. Preliminary report on acid-iron waste disposal. Woods Hole Oceanographic Institution Report to National Research Council. Woods Hole, MA.

799 - Ketchum, Bostwick. 1950a. Progress report on waste disposal in the New York Bight. Woods Hole Oceanogr. Inst. Ref. 50-10. Woods Hole, MA. 2p.

800 - Ketchum, Bostwick H. 1950b. Hydrographic factors involved in the dispersion of pollutants introduced into tidal waters. Jour. Boston Soc. Civ. Eng. 37:296-314.

801 - Ketchum, Bostwick H., Alfred C. Redfield, and John C. Ayers. 1951. The oceanography of the New York Bight. WHOI Papers in Physical Oceanography and Meteorology. Woods Hole, MA. 12(1):1-46.

802 - Ketchum, B. H., and W. L. Ford. 1952. Rate of dispersion in the wake of a barge at sea. Trans. of the Amer. Geophys. Union. 33(5):680-684.

803 - Ketchum, B. H. 1953. Preliminary evaluation of the coastal water off Delaware Bay for disposal of industrial wastes. Woods Hole Ref. No. 53-51. Woods Hole, MA.

804 - Ketchum, B. H., and D. J. Keen. 1955. The accumulation of river water over the continental shelf between Cape Cod and Chesapeake Bay. Pap. Mar. Biol. and Oceanogr. Deep-Sea Research. Supplement to Vol. 3:346-357.

805 - Ketchum, B. H., C. S. Yentsch, and N. Corwin. 1958. Some studies of the disposal of iron wastes at sea. Woods Hole Oceanographic Institution. Reference 58-7. Woods Hole, MA. 17p.

806 - Ketchum, B. H., C. S. Yentsch, N. and Corwin. 1958. Some Studies of the disposal of iron wastes at sea: Summer, 1958. I. The distribution of plankton in relation to the circulation and chemistry of the water. Woods Hole Oceanographic Institution. Ref. No. 58-55. Woods Hole, MA. 35p.

807 - Ketchum, Bostwick. 1970. Ecological effects of sewer sludge disposal at sea. Paper presented to Water Poll. Control Fed. Convention. Boston, MA. Session No. 16.

808 - Ketchum, B. H., (editor). 1972. The water's edge: Critical problems of the coastal zone. MIT Press. Cambridge, MA.

809 - Ketchum, B. 1973. A realistic look at ocean pollution. Jour. Mar. Tech. Society. 7(7):8-15.

810 - Ketchum, B. H. 1974. Sea disposal of acid-iron wastes resulting from production of titanium dioxide. ICES. C.M. 74/E:52.

811 - Ketchum, Bostwick H. 1976. The old and the new: New York Bight research in perspective. In: Middle Atlantic Continental Shelf and the New York Bight. In: M. G. Gross (editor), Amer. Soc. of Limnol. and Oceanogr. Spec. Symp. 2:14-19.

812 - Ketchum, Bostwick H., Dana R. Kester, and P. Kilho Park (editors). 1981. Ocean Dumping of Industrial Wastes. Plenum Press. New York, NY. 525p.

Abstract: Presents a wealth of previously unpublished material. An introductory section considers the regulatory aspects of ocean dumping, the international agreements concerning waste disposal in the ocean, and a summary of recent ocean dumping practices in the United States. Three subsequent sections contain new data and interpretations related to the physical, chemical, and biological aspects of industrial waste dumping. Specific chapters highlight hydrographic features and dispersion characteristics of wastes at U.S. dumpsites, and new techniques applied to the problems of waste disposal, laboratory bioassay studies, an examination of field specimens for biological abnormalities, and behavioral and metabolic studies of fish response to industrial wastes. The volume concludes with an inspection of the future prospects of ocean dumping.

813 - Kieler-Meeresforschungen (Kiel). 1970. Chemische, mikrobiologische und planktologische Untersuchungen in der Schlei im Hinblick auf deren Abwasserbelastung. (Chemical, microbiological and planktological investigations in the Schlei Fjord under the aspect of sewage and waste disposal.) Kieler Meeresforsch. 26:105-216.

814 - King, Kenneth, Jr. 1970. The New York Bight: A case study of ocean pollution. Special report to the Urban Ecology Program. Salk Inst. Columbia Univ. New York, NY. 93p.

815 - Kingston, J. M., I. L. Barnes, T. J. Brady, T. C. Rains, and Michael A. Champ. 1978. Separation of the transition elements from alkali and alkaline earth elements in estuarine and sea water with chelating resin and their determination by graphic furnace atomic absorption spectrometry. Analytical Chemistry. 50(14):2064-2070.

816 - Kinne, O., and H. Rosenthal. 1967. Effects of sulphuric acid on fertilization, embroyonic development and larvae of the herring, Clupea harengus. Mar. Biol. 1:65-83.

817 - Kinne, O., and K. H. Schumann. 1968. Biologische konsequenzen schwefelsaure und eisensulfathaltiger Industrieabwasser. Mortalitat junger _Gobius pictus_ und _Solea solea_ (Pisces). Helgolander wiss. Meeresunters. 17:141-155.

818 - Kinne, O. 1980. 14th European marine biology symposium: "Protection of life in the sea": Summary of symposium papers and conclusions. Helgolaender Meeresuntersuchungen 33(1-4):732-761.

819 - Klein, C. Unpublished Diplomarbeit, 1973. Populationsdynamik und produktionsleistung des benthischen rohrenbauenden Amphioden _Ampelisca brevicornis_ an zwei Stationen in der Helgolander Bucht. Ruhg-Universitat Bochum. 28p.

820 - Klein, L. A., M. Lang, N. Nash, and S. L. Kirschner. 1974. Sources of metals in New York City wastewater. Department of Water Resources. City of New York. New York Water Pollution Control Association. 18p.

821 - Klemas, V., G. Davis, and R. Henry. 1976. LANDSAT observations of ocean dump plume movement and dispersion. University of Delaware. College of Marine Studies Report No. DU-CMS-76. Newark, DE. 4p.

Abstract: Eighteen LANDSAT images were analyzed to study the dispersion and movement of ocean dump plumes thirty-eight miles southeast of Cape Henlopen, Delaware, at the disposal site for waste discharged from a plant producing titanium dioxide. The discharge, which is a greenish-brown liquid containing up to 10% acidity and 4% chloride salts, is discharged several times per month. The following preliminary results were obtained: the frequency of the dumping made it possible for the LANDSAT satellite to image the waste plume in various stages of degradation; the predominant direction of movement of the waste plumes imaged was to the southeast; the average drift velocity for surface drogues and the waste plumes was about 0.5 knots, and the waters at the test site were highly stratified and stable in the summer and nearly homogeneous in the winter. In general, it appears that rapid movement toward shore can occur primarily during storms. However, the plume is rapidly dispersed and diluted and the probability of an identifiable plume containing heavy concentration of acid waste reaching the shore is quite low.

822 - Klemas, V., G. R. Davis, and R. D. Henry. 1977. Satellite and current drogue studies of ocean-disposed waste drift. Jour. Water Poll. Control Fed. 49:757-763.

823 - Klemas, V., W. Philpot, and G. Davis. 1978. Determination of spectral substances in natural waters. Final Report. University of Delaware. Newark, DE. NASA-CR-156998. 100p.

Abstract: Optical remote sensing of water pollution offers the possibility of fast, large-scale coverage at a relatively low cost. The possibility of using the spectral characteristics of the upwelling light from water for the purpose of ocean water quality monitoring was explained. The work was broken into several broad tasks as follows: (1) definition of a remotely measured spectral signature of water, (2) collection of field data and testing of the signature analysis, and (3) the possibility of using LANDSAT data for the identification of substances in water. An attempt to extract spectral signatures of acid waste and sediment was successful.

824 - Klemas, V., D. Bartlett, and W. Philpot. 1978. Remote sensing of coastal environment and resources. Coastal Mapping Symposium. Rockville, Md. August 14-16. Proceedings. American Society of Photogrammetry. Falls Church, VA. pp. 3-16.

Abstract: The use of remote sensing analysis to identify different categories of surface features is discussed, and two remote sensing studies of Delaware's coastal environment are described. In one study, eigenvector analysis of LANDSAT/MSS data was used to distinguish clear water from water containing industrial waste.

825 - Klemas, Vytautas, and William D. Philpot. 1981. Remote sensing of ocean-dumped waste drift and dispersion. In: B. H. Ketchum, D. R. Kester, and P. K. Park (editors), Ocean Dumped Industrial Wastes. Plenum Press. New York, NY. pp. 193-211.

Abstract: The drift and dispersion of sixteen acid waste plumes 64 km off the Delaware coast were investigated using LANDSAT imagery, current drogues, and ship data. The waste plumes imaged by LANDSAT were found to be drifting at average rates from 0.59 km hr^{-1} to 3.39 km hr^{-1} into the southwest quadrant. The plumes seemed to remain above the thermocline which was observed to form from June through August at depths ranging from 13 m to 24 m. During the remainder of the year the ocean at the test site was not stratified, permitting wastes to mix throughout the water column.

The magnitudes of plume drift velocities were compatible with the drift velocities of current drogues released over a 12-month period at the surface, at middepth and near the bottom. However, during the stratified warm months, more drogues tended to move in the north-northeast direction, while during the nonstratified winter months a southwest direction was preferred.

Rapid waste movement toward shore occurs primarily during storms, particularly northeasters. During such storms, however, the plumes were rapidly dispersed and diluted. The plume width was observed to increase at a rate of about 1.5 cm sec^{-1} during calm sea conditions, yet attain spread rates in excess of 4 cm sec^{-1} on windy days. These results indicate that by the time a waste plume would reach shore, dilution would be at least one million to one.

826 - Knight, G. W. 1932. Abatement of pollution in New York harbor. Military Engineer. 24:221-224.

827 - Knutzen, J. 1972. Forurensningseffekter ved utslipp eller dumping av avfall fra titandioksydproducksjon. Unpublished report of NIVA. 47p.

828 - Knutzen, J., T. Bokn, and B. Rygg. 1974. Undesökelse av blotbunnsfauna og fastsittende alger i Hvaleromradet 18-20/9 1973. Unpublished report of NIVA. 38p.

829 - Knutzen, J., and B. Bjerkeng. 1979. Sjövannsvasking av röykgasser fra kullfyrt kraftverk. Vurdering av okologiske konsekvenser, med Oslofjorden som eksempel. (Seawater scrubber for flue gas from coal fired power plant. Evaluation of ecological consequences with the Oslofjord (S. Norway) as an example). Årbok Norsk institutt for vannforskning. 14p.

830 - Knutzen, Jon. 1981. Effects of decreased pH on marine organisms. Mar. Pollut. Bull. 12(1):25-29.

Abstract: Sea water scrubbing of flue gas may result in a significant pH decrease in the waters of coastal areas with restricted dilution capacity. A review of the literature on pH tolerance of marine organisms indicates that there is little evidence of harm caused by decrease of 0.5-1.0 pH units. However, few long term or life cycle studies have been conducted. As pH may influence seawater chemistry in several important ways, notably the carbonic acid equilibrium, chemical form of metals, and the characteristics of dissolved substances, it is concluded that pH tolerance deserves more interest from marine biologists than it has received hitherto.

831 - Koba, H. 1978. Recent progress in techniques for managing contaminated bottom sediments. In: S. A. Petersen, and K. K. Randolph (editors), Management of Bottom Sediments Containing Toxic Substances. Proceedings of the Third US-Japan Experts' Meeting November 1977, Easton, Maryland. US Environmental Protection Agency. Corvallis, OR. pp. 49-64.

832 - Koch, Pierre. 1960. Discharge of wastes into the sea in European coastal areas. In: Pearson, E. A. (editor), Proc. of the 1st International Conf. on Waste Disposal in the Marine Environment. Pergamon Press. New York, NY. pp. 122-162.

833 - Koczy, F. F. 1960. The distribution of elements in the sea. Disposal of Radioactive Wastes. Vol. 2:189-197. EAEA, Vienna.

834 - Kodischek, L. K., and P. Guyre. 1974. Antimicrobial-resistant coliforms in New York Bight. Mar. Pollut. Bull. 5(5):71-74.

Abstract: A preliminary study of sediments and overlying water samples obtained at the approximate center off the sewage sludge bed in the New York Bight revealed significant numbers of coliforms resistant to antibiotics and heavy metals. Some of the isolates were shown to transfer antibiotic resistance during conjugation with a recipient strain of _Salmonella gallinarum_. The selection of bacterial populations resistant to antimicrobials may be significant in the ecology of this impoverished area, and requires further study.

835 - Koditscheck, Leah. 1976. Antimicrobial resistant bacteria in the New York Bight. In: M. G. Gross (editor), Amer. Soc. Limnol. and Oceanogr. Spec. Symp. 2:383-393.

Abstract: Sediment samples and overlying water from stations around the sewage dump site in the New York Bight were analyzed for coliform and noncoliform bacteria resistant to mercury and antibiotics. Few or no coliforms could be found in these samples, but a pollution gradient was identified in a northeasterly direction, toward Long Island. About 1% of the sediment bacteria desorbed was consistently resistant to $HgCl_2 10^{-2M}$ and/or tetracycline 40 mcg/ml at MESA station 34. The antibiogram of a majority of the isolates from these sediments showed multiple antibiotic resistance.

Sediments from which antimicrobial resistant bacteria were isolated had high bound water content, much debris, little or no evidence of normal benthic macrofauna, and a black, gelatinous consistency. In contrast, sediment taken from the Sandy Hook transect showed no resistant bacteria, and demonstrated no other parameters related to pollution. Analysis of water and sediments containing very small numbers of coliforms from a beach at Sandy Hook State Park revealed a significant number of bacteria resistant to seven or more antibiotics. These data suggest that techniques used in these studies may detect pollution gradients which are not measurable by coliform counts.

836 - Koebel, C. T., and D. A Krueckeberg. 1975. Demographic patterns in the New York Bight region. MESA New York Bight Atlas Monograph 23. MESA New York Bight Project and New York Sea Grant Institute. Albany, NY. 43p.

Abstract: Atlas Monograph 23 summarizes what we presently know and can predict about Bight region population. The region's early and intimate ties with the port of New York brought trade, prosperity, and many people, who have been migrating into Bight region cities since the early 1800s. Koebel and Krueckeberg describe the strong urban pattern of people working in a location different from the one they live in, and the pattern of population expansion into successive outer rings. Projections suggest a stabilizing population in the Northeast as a whole, but major increases in suburban counties, 60% of them directly on the ocean, harbor, and estuaries.

837 - Koh, R. C. Y., and L. Fan. 1968. Prediction of the radioactive debris cloud distribution subsequent to a deep underwater nuclear explosion. A Final Report prepared for the Naval Radiological Defense Laboratory. Contract No. N00228-68-C-0684. 266p.

838 - Koh, R. C. Y. 1971. Ocean sludge disposal by barges. Water Resources Res. 7:1647-1651.

839 - Koh, Robert C. Y., and Y. L. Chang. 1973. Mathematical model for barged ocean disposal of wastes. Office of Research and Development, U.S. Environmental Protection Agency. U.S. Gov. Print. Office. Washington, DC. EPA No. 660 273 029. 595p.

840 - Kohn, Barbara, and Gilbert T. Rowe. 1981. Dispersion of two liquid industrial wastes dumped at 106-Mile site. In: Assessment Report on the Effects of Waste Dumping in 106-Mile Ocean Waste Disposal Site. NOAA/ODP Dumpsite Evaluation Report 81-1. Rockville, MD. pp. 133-155.

841 - Kolitz, B. L., J. B. Hazelworth, R. B. Starr, and S. R. Cummings. 1976. MESA New York Bight project, expanded water column characterization cruises (XWCC 4-5). NOAA Ship GEORGE B. KELEZ, May-June 1975. NOAA DR ERL MESA-24. Boulder, CO. 231p. NTIS No. PB 271400/AS.

Abstract: During May and June 1975, the NOAA Ship GEORGE B. KELEZ made two expanded water column characterization cruises (XWCC-4 and 5) and occupied 48 and 54 stations, respectively, in the New York Bight. The objective of these cruises was to gather data to provide a base for analysis of the water movements. This report presents the corrected physical data collected during these two cruises.

842 - Kolitz, B. L., J. B. Hazelworth, G. A., Berberian, and S. R. Cummings. 1976. MESA New York Bight Project, Expanded Water Column Characterization Cruise (XWCC-7), NOAA Ship GEORGE B. KELEZ, December 1975. NOAA DR ERL MESA-26. Boulder, CO. 122p. NTIS No. PB 271402/AS.

Abstract: During December 1975, the NOAA Ship GEORGE B. KELEZ made an expanded water column characterization cruise (XWCC-7), occupying 51 stations in the New York Bight. The objective of this cruise was to gather data to provide a base for analysis of the water movements. This report presents the corrected physical and chemical (nutrient) data from this cruise.

843 - Kolpack, Ronald L. 1979. Distribution of suspended particulate matter near sewage outfalls in Santa Monica Bay, California. In: H. D. Palmer and M. G. Gross (editors), Ocean Dumping and Marine Pollution. Dowden, Hutchinson and Ross, Inc. Stroudsburg, PA. pp. 205-239.

Abstract: Between 1961 and 1976 more than 12×10^8 1/day of primary and secondary-treated sewage effluent and 19×10^6 1/day of sludge were discharged into Santa Monica Bay in southern California. During 1971 the amount of total suspended solids discharged from the effluent and sludge outfalls was about 6.1×10^4 metric tons. No significant accumultion of particulate material around the discharge sites has been detected in precision-depth and high-resolution seismic-reflection profiles.

Distribution and transport paths of sewage particles within the water column on the Santa Monica Shelf were studied with more than 500 light-transmission profiles made on eight cruises between September, 1970, and

November, 1972. Additional measurements of temperature, salinity, and bottom currents were used to interpret the distribution of suspended matter in the water column.

The sewage plume in Santa Monica Bay rises rapidly from an initial discharge depth of about 60 meters to about 20 meters below the water surface. Wave action in the water column tends to disperse the sewage plume over a large part of the Santa Monica Shelf. The dominant direction of initial transport during most of the year is toward the shoreline southeast of the discharge site. Subsequently, at several reasonable consistent locations within the bay the net transport of the sewage material, as well as material derived form nearshore mixing, is directed toward the west and transported offshore in well-defined zones which are approximately normal to the coastline. The amount of suspended material transported to the offshore area via Santa Monica and Redondo canyons appears to be rather negligible during normal conditions. On the other hand, a significant portion of the suspended material in Santa Monica Bay is transported within the near-surface water column toward the offshore basins. This pattern of particle transport also takes place at the water surface and at the sediment-water interface, although there is somewhat more variation near these surfaces.

844 - Konop, Dane. 1978. Monitoring ocean dumping. Sea Technology. 19:26-28.

845 - Koons, C. B., and J. P. Thomas. 1979. C_{15+} Hydrocarbons in the Sediments of the New York Bight. In: Proceedings of the 6th Conference on Prevention, Behaviour, Control and Clean-up of Oil Pollution, pp. 625-628. American Petroleum Institute. Washington, DC. pp. 625-628.

Abstract: The purpose of this study was to document the distribution and abundance of C_{15} hydrocarbons in sediment samples taken from the Hudson River, the New York Harbor, and across the continental shelf to the continental rise. Collection of 35 of these samples took place in 1975-76 using a multiple corer, a bottom grab, or the submersible ALVIN. Materials obtained from these areas were considered representative and included dredge spoils, sewage sludge, and sediment from both the Deepwater Dumpsite 106 on the continental rise and the comparatively cleaner sea floor of the continental shelf beyond the Apex of the New York Bight.

Total C_{15} hydrocarbons are most abundant (3000-6000 parts per million) in areas highly impacted by man-harbor sediments and dredge spoil and sewage sludge disposal areas. Values from the continental shelf are lower (80 ppm) and values from the continental rise are the lowest (40 ppm). Gas chromatographic traces clearly distinguish the hydrocarbons in the dredge spoil and sewage sludge sediment samples from the hydrocarbons found in sediment samples relatively free of sludge or spoil materials.

846 - Kott, Y. 1975. Effluent quality of chlorinated sewage discharged from sea outfalls. Progr. Water Technol. Supp. 1975. Pap. No. 16. pp. 155-163.

847 - Krishnappan, B. G. 1977. Dispersion of granular material dumped in deep water. Unpublished report for the Canada Center for Inland Waters.

848 - Krout, J. E. 1973. Biological effects of ocean disposal of solid waste. University of Rhode Island. Marine Technical Report No. 9. Kingston, RI. 57p.

849 - Krueger, William H., Robert H. Gibbs, Jr., Robert C. Kleckner, Aimee A. Keller, and Michael J. Keene. 1977. Distribution and abundance of mesopelagic fishes on cruises 2 and 3 at Deepwater Dumpsite 106. In: Baseline Report of Environmental Conditions in Deepwater Dumpsite 106. NOAA/ODP Dumpsite Evaluation Report 77-1. Vol. II. Rockville, MD. pp. 377-422.

850 - Kullenberg, G. 1974. Some aspects of the dispersion problems in connection with marine dumping. ICES. C.M. 74/E:28.

851 - Kullenberg, G. E. B. 1974-1975. Ocean dumping sites. Ocean Management 2:189-209.

852 - Kullenberg, G. E. B. 1975a. Effects of environmental factors, Part I. Discharge of sewage from sea outfalls. Pergamon Press. New York, NY. No. 33:333-342.

853 - Kullenberg, G. E. B. 1975b. Ocean dumping sites. Ocean Mgmt. 2(3):189-209.

854 - Kullenberg, Gunnar E. B. (In Press). Physical oceanography studies related to waste disposal in the sea. In: I. W. Duedall, B. H. Ketchum, P. K. Park, and D. R. Kester (editors), Wastes in the Ocean, Vol. I: Industrial and Sewage Wastes in the Ocean. John Wiley and Sons, Inc. New York, NY.

Abstract: Theoretical models of horizontal and vertical mixing of ocean-dumped wastes on small scale to mesoscale are presented. Observations from ocean dumping and dye tracer experiments are used to test the models and to calculate mixing parameters. These are shown to be highly variable, depending upon environmental conditions such as wind, current distributions, and density stratification. The experiments show that the waste material in stably stratified conditions can become distributed in well-defined layers related to the current and density distributions. Such layers can be very persistent.

The mixing in the sea is often weak, also in the surface layer. The initial dilution achieved during a disposal operation is therefore an important factor in assessing potential effects of the waste disposal. The physical dispersion characteristics of a potential waste disposal site are often decisive site selection criteria, whereas the biological characteristics often are decisive for not using an area as waste recipient.

For small-scale mixing in the range 1 m to about 10^4 m, parameterization by means of mixing coefficients or diffusion velocities can be useful in describing the combined effects of horizontal and vertical random motion on the dispersion. For the mesoscale range, the parameterization of the mixing by means of mixing coefficients does not appear meaningful.

855 - Kunig, P. 1980. Pollution from sources other than oil: dumping at sea. In: D. J. Ousine and J. P. Grant (editors), The Impact of Marine Pollution. Crom Helm. London. pp. 181-202.

856 - Kuo, C. Y., and T. D. Modena. 1979. Mathematical model investigation of long-term transport of ocean-dumped sewage sludge related to remote sensing. Final Report, 18 Jul. 1977 - 17 Feb. 1979. Old Dominion Univ. Research Foundation. Norfolk, VA. NASA-CR-158159. 151p.

Abstract: An existing, three-dimensional, Eulerian-Lagrangian finite-difference model was modified and used to examine the transport of dumped sewage sludge in the New York Bight. Both in situ and laboratory data were utilized in an attempt to approximate model inputs such as mean current speed, horizontal diffusion coefficients, particle size distributions, and specific gravities. The results presented are a quantitative description of the fate of a negatively buoyant sewage sludge plume resulting from continuous and instantaneous barge releases. Concentrations of the sludge near the surface were compared qualitatively with those remotely sensed. Laboratory study was peformed to investigate the behavior of sewage sludge dumping in various ambient density conditions.

857 - Lacy, W. J., and G. Rey. 1975. Technology of marine disposal in USA of industry wastes. In: E. A. Pearson, and E. de Fraja Frangipane (editors), Marine Pollution and Marine Waste Disposal. (Supplement to Progress in Water Technology.) Proceedings of the 2nd International Congress. San Remo. 17-21 December, 1973. Pergamon Press. New York, NY. pp. 369-375.

858 - Laevatsu, T., and R. J. Callaway. 1974. Computation of tides, currents, and dispersal of pollutants in the New York Bight from Block Island to Atlantic City with large grid size, single- and two-layer hydrodynamical-numerical models. Environmental Prediction Research Facility. Naval Postgraduate School. Monterey, CA. 79p.

Abstract: The application of a large grid HN model to the New York Bight is to large extent a problem of treating two long open boundaries. The input at these open boundaries serves also for tuning of the results inside the computational area.

A few earlier current measurements by two lightships in the Bight as reported by Haight (1942) have been used for verification. The tidal currents predominate in the New York Bight proper, with a superimposed slow net flow toward the south. This net flow can be simulated in the models and tuned with the prescription of a proper slope to the two open boundaries.

The diffusion computations over a short period of time in the large grid models do not give fully satisfactory results with the diffusion model used at present in the HN models; instead, the transport of the centers of release points have been computed and presented. Furthermore, the net transport through various sections in the area has also been computed. The single-layer model is applicable for the winter season when the area is fully mixed from surface to the bottom.

In the two-layer HN model, which is applicable for summer conditions when stratification of the waters occurs, some inputs, such as internal friction, need some additional numerical experimentation and adjustment and, above all, proper field measurements as nearly not enough data are available for the selection of a proper internal friction coefficient. Furthermore, internal waves (internal fluctuations) occur in the interface after introduction of the wind into the model run. These fluctuations diminish with time, and subsequently require a longer real-time computation period and, as a consequence, more computer time. Such fluctuations are known to occur in the sea as recordings of thermal structure changes with depth and time over the continental shelf, as well as sea-level recordings near the coasts have shown.

Of practical importance is the knowledge of differences in the currents in upper and lower layers and the time and space variation of these differences. The results of the model show that in many areas a different net flow occurs in the lower layer than in the surface layer. This is partly demonstrated in the report with some of the output from special selected points and agrees with the results from sea-bed drifter experiments.

There is a coastward movement of the deeper layers off Long Island Sound, making this area less desirable for sludge disposal. The Hudson Submarine Canyon divides the hydrographic regime in New York Bight. The flow south of the canyon is stronger and toward the south. A potential use of this type of study is in the selection of general areas where sludge dumping may be less troublesome than in others.

859 - Laevastu, T., J. Harding, K. Rabe, and S. Larson. Ocean-atmosphere interactions off the northeast coast of North America. In: M. G. Gross (editor), Amer. Soc. Limnol. and Oceanogr. Spec. Symp. 2:35-43.

Abstract: The New York Bight-Middle Atlantic continental shelf region is typical of midlatitude continental east coast areas in the intense sea-air interaction occurring during late autumn and winter. During these seasons dry but cool air moving over warm coastal water takes up a considerable amount of heat and moisture (latent heat). The daily uptake near the coast can exceed 0.5 g of water per cm^2 on a monthly mean basis.

The response of the surface layers of the atmosphere to the properties of the sea surface is relatively rapid so that quasi-equilibrium conditions are established after surface air has traveled about 6 h over water.

The main consequences of this large heat and moisture uptake on the atmosphere are deepening of lows that pass the coastline and frequent cyclogenesis off the coast. As the heat and moisture are transported upward, a trough forms at the 850-mb level. A thermally driven cyclonic surface wind component is created along the coast, but it often escapes the attention of synoptic weather analysts.

The main effect of the atmosphere on the ocean is rapid cooling of already cool coastal waters during autumn and winter, resulting in increases in thermocline depth. The coaf lows that pass the coastline and frequent cyclogenesis off the coast. As the heat and moisture are transported upward, a trough forms at the 850-mb level. A thermally driven cyclonic surface wind component is created along the coast, but it often escapes the attention of synoptic weather analysts.

The main effect of the atmosphere on the ocean is rapid cooling of already cool coastal waters during autumn and winter, resulting in increases in thermocline depth. The coastal southerly current is influenced by the surface cyclonic wind component.

860 - Laevastu, T., M. Clancy, and A. Stroud. 1974. Computation of tides, currents, and dispersal of pollutants in Lower Bay and approaches to New York with fine and medium grid size hydrodynamical-numerical models. Environmental Prediction Research Facility, Naval Post-graduate School. Monterey, CA. Prepared for the Environmental Protection Agency. Pacific Northwest Environmental Research Laboratory. Corvallis, OR.

Abstract: The report summarizes the results of two different HN model applications with different grid sizes: one with a small grid size for the Lower Bay of New York; and the second with a larger grid size for the Approaches to New York, which includes part of the New York Bight outside the Ambrose Channel.

(1) The numerical model reproduces well the currents as known from earlier empirical studies, but presents many more details and makes it possible to compute other current-dependent processes, such as transport and diffusion.

(2) The flushing of the Lower Bay NE of the line between The Narrows and Sandy Hook is considerably more rapid than in the area SW of this line toward Raritan Bay.

(3) There is a weaker outflow from the Lower Bay between the Ambrose Channel and Coney Island (off the Long Island coast), and the main outflow is between Ambrose Channel and Sandy Hook, and turns toward the south along the New Jersey coast.

861 - Landner, L. 1975a. Criteria for marine waste disposal in Sweden. In: E. A. Pearson, and E. de Fraja Frangipane, (editors), Marine Pollution and Marine Waste Disposal. (Supplement to Progress in Water Technology.) Proceedings of the 2nd International Congress. San Remo. 17-21 December 1973. Pergamon Press. New York, NY. pp. 89-93.

862 - Landner, L. 1975b. Research on marine pollution in Sweden. In: E. A. Pearson, and E. de Fraja Frangipane (editors), (Supplement to Progress in Water Technology.) Proceedings of the 2nd International Congress. San Remo. 17-21 December, 1973. Pergamon Press. New York, NY. pp. 459-462.

863 - Lang, M., C. Samowitz, and M. Jethwani. 1975. Control of water pollution in New York City. Water Research. Vol. 9. pp. 981-991.

Abstract: Water pollution control program must 1) convince public and government of the need for a program, 2) obtain money for program, 3) design and construct program. Since it may be 20-50 years before this process results in a working program, need to consider the needs and requirements when the program begins to function, technology, facilities needed at that time and need to know concentration and location of industries and population at that time. Paper discusses these processes and problems for New York City.

864 - Lang, M. 1974. Problems of Pollution Water Resources in the New York City Metropolitan Area. In: O. A. Koels (editor), New York Academy of Science Ann. (Hudson River Colloquium) Vol. 250. pp. 178-181.

Abstract: High-degree secondary treatment for dry weather wastewater. Construction of storm water treatment to be completed by 1980. Description of treatment plants, improvements, and problems.

865 - Lassus, P. and P. Maggi. 1973. Etude de la toxicite de l'effluent Mantedisan vis-a-vis d'organismes marins. Institute Scientifique et Technique des Peches Maritimes Report. 56p.

866 - Larsson, Ake, Karl-Johan Lehtinen, and Carl Haux. 1980. Biochemical and hematological effects of a titanium dioxide industrial effluent on fish. Bull. Environm. Cont. Toxicol. 25:427-435.

867 - Lavelle, J. W., G. H. Keller, and T. L. Clarke. 1975. Possible Bottom Current Response to Surface Winds in the Hudson Shelf Channel. Jour. of Geophys. Res. 80(151):1953-1956.

Abstract: Current measurements made in the Hudson Shelf Channel during the summer of 1973 show essentially channel axial bottom current even though the channel aspect ratio is small in the area of measurement. Although the current record is of short duration, correlation of water movement with surface winds is suggested by the data. The sense of summertime nontidal bottomflow in the channel (up or down channel) would appear to be controlled by the surface wind direction (offshore or onshore). These results would suggest the likelihood of net down-channel flow during the summer months.

868 - Law, R. J. 1980. Hydrocarbon concentrations in water and sediments from the Irish Sea, Liverpool Bay and the English Channel. ICES. C.M. 80/E:26.

869 - Laycock, E. E. 1976. Ocean dumping surveillance system. In 2nd Annual Combined Conf. Mar. Tech. Soc. and I.E.E.E. Oceans 1976. Washington, DC. Paper 25D1-3.

870 - Lear, D. W. 1974a. Environmental survey of two interim dumpsites Middle Atlantic Bight, Supplemental Report. U. S. Environmental Protection Agency, Region III. Philadelphia, PA. EPA-903/9-74-010B.

871 - Lear, D. W., and H. D. Palmer. 1974b. Ocean disposal: Middle Atlantic Bight. [Sewage sludge; acid wastes]. Proc. 10th Annual Meeting Mar. Tech. Soc. pp. 115-126.

872 - Lear, D. W., S. K. Smith and M. L. O'Malley (editors). 1974c. Evironmental survey of two interim dumpsites, Middle Atlantic Bight. U.S. Environmental Protection Agency. Region III. Philadelphia, PA. EPA-903/9-74-010A.

873 - Lear, D. W., and G. B. Pesch. 1975. Effects of ocean disposal activities on mid-continental shelf environment off Delaware and Maryland. U.S. Environmental Protection Agency, Region III. Philadelphia, PA. EPA 903/9-75-015. 214p.

874 - Lear, D. W., M. L. O'Malley, S. K. Smith. 1977. Effects of ocean dumping activity Mid-Atlantic Bight 1976. U.S. Environmental Protection Agency, Region III, Philadelphia, PA. Interim Rep. EPA 903/9-77-029. 168p.

875 - Lear, D. W. 1978. Testimony at public hearing, City of Philadelphia ocean dumping permit. Georgetown, Delaware. May.

876 - Lear, D. W., M. L. O'Malley, and S. K. Smith. 1981. Effects of ocean dumping on a temperate midshelf environment. In: B. H. Ketchum, D. R. Kester, and P. K. Park (editors), Ocean Dumping of Industrial Wastes. Plenum Press. New York, NY. pp. 485-502.

Abstract: The fate and effects of ocean-dumped sewage sludge and acid iron waste were studied at two midcontinental shelf sites off the Delaware-Maryland coast. Materials accumulated in sediment, especially in winter, and tended to persist in topographic low areas. Ambient levels in sediments, of metals, organic carbon, and organohalogens were determined, and statistically significant areas of impaction delineated. Uptake of certain metals in scallops and clams was indicated. The mahogany clam, *Arctica islandica*, may have suffered mortalities due to dumping activities. Contamination of bottom sediments by sewage indicator bacteria was demonstrated. Studies of impact on bottom communities are continuing.

877 - Lear, Donald W., and Marria L. O'Malley. (In Press). Effects of sewage sludge dumping on continental shelf benthos. In: I. W. Duedall, B. H. Ketchum, P. K. Park, and D. R. Kester (editors), Wastes in the Ocean, Vol. I: Industrial and Sewage Wastes in the Ocean. John Wiley and Sons, Inc. New York, NY.

Abstract: The fate and effects of the ocean dumping of sewage sludge at a midcontinental shelf dumpsite were studied by comparison of the defined area of deposition with a reference area away from prevailing flows. Concentrations of lead, copper, and organic carbon, and the percent fine sediment fraction (silts + clays) were significantly greater in the dumpsite grid compared with the reference grid. Swales in the dumpsite grid were areas of highest concentrations of contaminants, and the pollution-sensitive ampeliscid amphipods were atypically absent from these sites. The pollution-tolerant polychaete <u>Capitella capitata</u> was found only in contaminated swales.

This study identified at least one locus of pollution due to ocean dumping, and the biological response characteristic of such contamination.

878 - Lear, Donald W., Marria L. O'Malley, William C. Muir, and George Pence. (In Press). Environmental effects of sewage sludge at the Philadelphia dumpsite. In: G. F. Mayer, A. Calabrese, F. A. Cross, D. C. Malins, E. W. Menzel, J. S. O'Connor, and F. J. Vernberg (editors), Ecological Stress in the New York Bight: Science and Management. Estuarine Research Federation, Columbia, SC.

Abstract: The Philadelphia sewage sludge disposal site is located about 70 km east of Ocean City, Maryland. Sludge has been dumped at this site since 1973, and the environmental effects of the practice have been under detailed observation since that time. The site showed no evidence of departures from typical, temperate midshelf conditions prior to the onset of sludge dumping. Since then, trends of environmental modification and degradation have appeared in areas contaminated by sludge-derived materials. The changes seem limited to the ocean bottom environment and include increased concentrations of metals in organisms, elevated metals and total organic carbon levels in sediments, changes in abundance of some species, apparent mortalities of molluscan shellfishes, unique occurrences of sewage bacteria, and appearance of pathological conditions in endemic crustaceans.

879 - Lee, G. F., M. D. Piwoni, J. M. Lopez, G. M. Mariani, J. S. Richardson, D. H. Homer, and F. Saleh. 1975. Research study for the development of dredged material disposal criteria. U.S. Army. Contract Report WES. -D-75-4. Vicksburg, MS. 381p.

880 - Lee, James A., William D. Stockton, and John C. Ayers. 1951. Wind, sea and swell conditions in the New York area. In: John C. Ayers, et al. The Hydrography of the New York Area. Status Report No. 5. (Contract N6 ONR 264, task 15). Cornell Univ. Ithaca, NY.

881 - Lee, John J. 1975. The distribution of forominifera of the New York Metropolitan region. City College of New York, NY. 17p.

Abstract: Analysis of 100 benthic samples and 20 plankton tows. Conclude that there are comparatively large numbers of Forominifera (environmental indicator species) in the benthic communities of the lower N.Y. Harbor and

Bight. Forominifera population here is similar to assemblages found in cores after Pleistocene glacier retreat. Superficial analysis of data suggests 90% reduction in Forominifera in northern parts of the dump site.

882 - Lee, W. Y., and J. A. C. Nicol. 1981. Toxicity of biosludge and pharmacectical wastes to marine invertebrates. In: B. H. Ketchum, D. R. Kester, and P. K. Park (editors), Ocean Dumping of Industrial Wastes. Plenum Press. New York, NY. pp. 439-454.

Abstract: The toxicity of two industrial wastes to marine invertebrates was assessed in the laboratory. The test animals were benthic and planktonic, and included a sea anemone, a hydromedusa, a polychaete, amphipods, an isopod, a crab, and a shrimp. Owing to the difference in their chemical composition, pharmaceutical waste was more toxic then the biosludge. Based on the survival data, biosludge was acutely toxic at about 10% (by volume) and pharmaceutical waste at a level $\leq$ 1 %. Animals differed in their sensitivity to the toxic material, juvenile being more sensitive than the adult. During chronic exposure, the reproduction of amphipods in diluted mixtures of biosludge was affected at a lower concentration ($<$ 10%) than that which induced acute mortality. Besides survival and reproduction, a feasible method for assaying toxicity at the cellular level was also suggested, involving cell aggregation in the sponge *Microciona*.

883 - Lee, Wen Yuh. (In Press). Copepods and ichthyoplankton: Laboratory studies of pharmaceutical waste toxicity. In: I. W. Duedall, B. H. Ketchum, P. K. Park, and D. R. Kester (editors), Wastes in the Ocean, Vol. I: Industrial and Sewage Wastes in the Ocean. John Wiley and Sons, Inc., New York, NY.

Abstract: Studies were carried out to determine the toxicity of six ocean-dumped pharmaceutical wastes to marine copepods and redfish (*Sciaenops ocellata*) eggs and larvae. To simulate waste concentrations at the dumpsite, copepods were consecutively immersed in a series of waste dilutions for intervals of 2 min to 2 h: 2 min in 10% waste solution, 5 min in 1%, 1 h in 0.1%, and 2 h in 0.01%. Animals were then transferred to untreated seawater to determine the delayed mortality. Samples from Merck and Pfizer wastes were acutely toxic to marine copepods; mortality was $>$ 35% at the end of 3 h exposure. Capri, Squibb, and Upjohn wastes produced low initial mortalities but high delayed mortalities ($>$ 60%).

The ichthyoplankton were treated with the Capri, Squibb, and Bristol wastes. During the exposure, observations were made on hatching success of embryos and on morphological deformity, behavioral abnormality, and survival of larvae. Merck and Squibb wastes were acutely toxic to eggs and newly hatched larvae at concentrations of 0.5% to 1%. The corresponding toxic levels for Bristol waste were 0.07% for eggs and 0.045% for larvae. The higher toxicity of Bristol waste may have been caused by its major component, N, N-dimethylaniline, which is more persistent in the marine environment than other components in the wastes.

884 - Lehmann, E. J. 1974. Ocean waste disposal. A bibliography with abstracts. National Technical Information Service, Springfield, VA. (NTIS-WIN-74-062). 146p.

885 - Lehtonen, H., and P. Tuunainen. 1974. On the effects of effluents from the titanium dioxide industry on the biota and fisheries of the coastal area of the Gulf of Bothnia near Pori, Finland. ICES. C.M. 1974/E:45.

886 - Lehtonen, H. 1975. The biological effects of the titanium dioxide industry in Finland. ICES. C.M. 75/E:26.

887 - Leighton, M. D. 1902. Sewage pollution in the metropolitan area near New York City and its effect on inland water resources. U.S. Geol. Survey Water-Supply Paper No. 72. 75p.

888 - Leiser, Arthur (Jr.). 1974. Organizational Evaluation of Industrial Water Pollution Control in the New York Region by Rae Zimmerman. Water Resources Bull. 10(6):1298-1299.

889 - Leitzell, T. L. 1973. The ocean dumping convention--a hopeful beginning. San Diego Law Review. 10(3):501-513.

890 - Leonhardt, H. W. 1968. Ablagerung von festen und flassigen Abfallstoffen auf hoher See. (Discharge of solid and liquid waste at sea.) Foderation Europaischer Gewasserschutz (FEG). Schutz der Meeresksten gegen Verunreinigung; symposium, 5-7 Oktober 1967. Hamburg, FEB, Zurich. pp. 71.

891 - Lesht, B. M. 1976. Simultaneous observations of near bottom flow and turbidity on the Long Island inner shelf. Trans. Amer. Geophy. Union, Abstr. Publ. EOS. 57(12):938.

892 - Lesht, B. M. 1977. Frequency distribution of bottom shear stresses within a tidal cycle on the inner continental shelf. Trans. Amer. Geophy. Union, Abstr. Publ. EOS. 58(6):409.

893 - Lesht, B. M. 1978. Field measurements of the bottom frictional boundary layer in the New York Bight. NOAA TM ERL MESA-28. 168p. NTIS No. PB 288194/AS.

894 - Lesht, B. M. 1979. Relationship between sediment resuspension and the statistical frequency distribution of bottom shear stress. Mar. Geol. 32:M19-M27.

Abstract: Detailed observations of near-bottom turbidity and currents are used to examine the relationship between bottom shear stress and sediment resuspension. Turbidity is related to the statistical frequency distribution of bottom shear stress in a way that implies that a few high-stress events may be responsible for resuspension. Such short-period processes will usually be masked by conventional methods of data handling.

895 - Lettau, B., W. A. Brower, Jr., and R. B. Quayle. 1976. Marine climatology. MESA New York Bight Atlas Monograph 7. MESA New York Bight Project and New York Sea Grant Institute, Albany, NY. 239p.

Abstract: Atlas Monograph 7 presents what we know about the climate of New York Bight in two ways: first, selected data illustrate the seasonal variation of meteorological parameters; second, isopleth maps and statistical graphs give a detailed climatic profile. The Bight coasts have a temperate, rainy climate with warm summers, no dry season, and rapid changes in weather from season to season, at times from day to day. Unique to the Atlantic coast between New England and North Carolina, is the major cyclonic storm track that passes through the outer Bight, following the New Jersey coastline. Because such storms intensify and move very rapidly, they present distinct weather forecasting problems for Bight coastal areas, especially since not all conditions favorable for storm formation actually produce a storm. According to Brower and Quayle, the atlas data provide the best possible climatological picture of the Bight's near-coastal zone, a region of sharp gradients and complex climates.

896 - Levin, M. A., R. E. Wolke, and V. J. Cabelli. 1972. Vibrio anguillarum as a cause of disease in winter flounder (Pseudopleuronectes americanus), Can. Jour. Miorobiol. 18:1585-1592.

897 - Lewis, B. W., and V. G. Collins. 1977. Remotely sensed and laboratory spectral signatures of an ocean-dumped acid waste. NASA-TN-D-8467. 37p.

Abstract: An ocean-dumped acid waste plume was studied by using a rapid scanning spectrometer to remotely measure ocean radiance from a helicopter. The results of these studies are presented and compared with results from sea-truth samples and laboratory experiments. An ocean spectral reflectance signature and a laboratory spectral transmission signature were established for the iron-acid waste pollutant. The spectrally and chemically significant component of the acid waste pollutant was determined to be ferric iron. The remotely sensed ocean spectral reflectance of the plume and the laboratory spectra transmissivity as a function of iron concentration are used to estimate the concentration of the acid waste in the plume by means of a simple mathematical model. The method compares the shape (color) of the spectral reflectance curves, calculated for various iron concentrations from laboratory transmissivity data, with the remotely measured spectral radiance curves of the plume. The calculated curve which most nearly matches the shape (color) of the plume curve gives the estimated concentration of iron in the plume at the point of measurement. In the limited number of stations where good remote spectral data were obtained, the agreement with sea-truth iron concentration was quite good. These studies indicate that remotely sensed monitoring of acid waste is feasible.

898 - Lewis, B. W. 1978. Relation of laboratory and remotely sensed spectral signatures of ocean-dumped acid waste. In: Joint Conference on Sensing of Environmental Pollutants, 4th. New Orleans, LA. 6-11 November 1977. Proceedings. A79-15023 04-45. Washington, DC. pp. 654-658.

Abstract: Results of laboratory transmission and remotely sensed ocean upwelled spectral signatures of acid waste ocean water solutions are presented. The studies were performed to establish ocean-dumped acid waste spectral signatures and to relate them to chemical and physical interactions occurring in the dump plume. The remotely sensed field measurements and the laboratory measurements were made using the same rapid-scanning spectrometer viewing a dump plume and with actual acid waste and ocean water samples, respectively. Laboratory studies showed that the signatures were produced by soluble ferric iron being precipitated in situ.

Relationships were developed between the field and laboratory data both for spectral signatures and color changes with concentration. The relationships allow for the estimation of concentration of the indicator iron from remotely sensed spectral data and the laboratory transmission concentration data without sea-truth samples.

899 - Lewis, E. J., T. K. Sawyer. 1979. Distribution of Acanthamoeba (Amoebida: Acanthamoebidae) in sediments from an offshore sewage disposal site. Jour. Protozool. 26(Pt. 1).20 A.

900 - Light, A. 1977. Marine incineration - a clean answer to chlorinated wastes. Chem. Age. London. 115(3027):11-15.

901 - Linnenbom, V. J., and W. L. Brundage, Jr. 1973. The 1969 survey of deep water dump (DWD) Area A. Appendix A. In: P. E. Wilkniss. Environmental condition report for Deep Water Dump Area A. U.S. Naval Research Laboratory. Washington, DC. NRL Report 7553. pp. 7-11.

902 - Litchfield, C. D., M. A. Devanas, J. Zindulis, M. Meskill, J. Freedman, and C. McClean. (In Press). Influence of cadmium on the microbial populations and processes in sediments from the New York Bight Apex. In: G. S. Mayer, A. Calabrese, F. A. Gross, D. C. Malins, D. W. Menzel, J. S. O'Connor, and F. J. Vernberg (editors), Ecological Stress in the New York Bight: Science and Management. Estuarine Research Federation. Columbia, SC.

Abstract: The effects of cadmium on the ecology of bacteria from selected New York Bight Apex sediments were studied during a two-year period. Cadmium chloride supplementation to media and mineralization experiments indicated an inhibition in the numbers of aerobic heterotrophic bacteria but either no effect or slight stimulation of the rates of decomposition of urea and glycolic acid. This pattern was observed repeatedly during the sampling period and indicated that a rather constant percentage of cadmium-resistant bacteria in sediments accounts for the majority of the microbial mineralization capacity. Microbial strains demonstrating resistance to cadmium on initial isolation were cultured, purified, and identified, and their antibiotic patterns were established. Seventy-two percent of the isolates were resistant to multiple classes of antibiotics, indicating a widespread distribution of both cadmium and antiobiotic resistance in non-enteric sedimentary bacteria. Additional studies with isolate A-4-20 confirmed that both the uptake of amino acids and their incorporation into protein by cadmium-resistant bacteria were unaffected by the presence of cadmium at concentrations tested during field studies.

903 - Loder, T. C. 1975. Effects of baled solid waste disposal in the marine environment. A descriptive model. ACS Symp. Ser. 18. pp. 467-486.

904 - Long Island State Park and Recreation Commission. 1976. Sludge Impact Study As Part of Task 8.7/1 of OPR Work Program. Long Island State Park and Recreation Commission. 33p.

Abstract: The recent sludge situation on Long Island provides a useful model through which the negative effects of an oil spill resulting from OCS operations can be assessed and mitigated. Task 8. 7, subtask 1, of the Office of Parks & Recreation work program within the OCS development impact study provides for a review of the negative effects of oil spills given their magnitudes and strand times. It is felt that an examination of the sludge situation in this context will provide useful insight into possible negative effects of an oil spill on Long Island's recreation industry.

Depending upon your source of information, this industry contributes anywhere from $500 - $800 annually to the region's economy. According to the Long Island Association of Commerce and Industry, "Tourism represents one of the region's largest revenue and job opportunity sources - - $790 million annually." Of this figure, the LIACI estimates some $150 million as being "summer month" and "ocean - oriented."

From these statistics, the loss to Long Island's beaches and related business for any one of the thirteen summer weekends could be in excess of $10 million. One can get a feeling as to the accuracy of such a statement when they realize the scope of Long Island's recreation industry.

Long Island is the second (after NYC) most popular recreation tourist region in the State. The ocean is the number one appeal of Long Island as far as Long Island's desirability as a recreation destination is concerned.

905 - Longwell, A. C. 1976a. Chromosome metagenesis in developing mackerel eggs sampled from the New York Bight. NOAA TM ERL MESA-7. Boulder, CO. 61p.

906 - Longwell, A. C. 1976b. Chromosome mutagenesis in developing mackerel eggs sampled from the New York Bight. In: M. G. Gross (editor), Amer. Soc. Limnol. and Oceanogr. Spec. Symp. 2:337-339.

Abstract: A third of the 30,689 embryo cells were scored in 452 eggs from the New York Bight from 14 different stations, a third of the 30,689 division figures scored were abnormal. The majority of abnormalities came from the early embryo stage just after gastrulation and formulation of the three primitive cell layers.

907 - Longwell, A. C., and J. B. Hughes. 1979. Cytologic, cytogenetic and embryologic state of atlantic mackerel eggs from surface waters of the New York Bight in relation to pollution. NOAA/NMFS. NE Fisheries Center. Milford Laboratory. Milford, CT. 26p.

Abstract: Cytologic, cytogenetic, and embryologic measures were made on about 10,000 eggs of the Altantic mackerel Scomber scombrus at the cleavage through tail-free embryo stages, as collected in the zooplankton from surface waters of the New York Bight. Eggs were sampled in May of '74 and '77 at a total of about 80 different Bight sites.

Development rates as calculated by mitotic index, viability as calculated from early indicators of cell death, and chromosome-mitotic abnormalities of the mackerel embryos varied widely over the sample sites in both '74 and '77. In '77 gross embryo malformations were frequent, and these too varied from station to station.

The '74 stations were grouped for statistical study according to proximity to one another, water temperature, and patterns of water circulation in the Bight. Group estimates of egg moribundity based on cell state, mitotic-chromosome irregularities, and division arrest at gastrulation and just after were then compared. These showed statistically significant lower egg viability in presumed more impacted Bight areas relative to less impacted areas more distant from the coast and waste dumping grounds.

Along with the '77 collection of mackerel eggs for cytological study, water from the microlayer (or surface) and from the subsurface was sampled for heavy metal analysis. Large quantities of zooplankton were also sampled for analyses of 65 aromatic and chlorinated hydrocarbons. The '77 samples provide some evidence for statistically significant associations between cytologic, cytogenetic, and embryologic measures of mackerel egg health at the several sample sites and concentrations of heavy metals in surface water and zooplankton, and toxic hydrocarbon concentrations in zooplankton at these sites. The measures of egg state showing associations fall into the following categories: embryo moribundity (cellular evidence); chromosome-mitotic abnormalities, mitotic index: cell differentiation problems; gross embryo malformation: total egg number collected.

Samples from a '78 cruise will provide additional information on associations between water and zooplankton pollution and mackerel egg health.

908 - Longwell, A. Crosby, and J. B. Hughes. 1980. Cytologic, cytogenetic, and developmental state of Atlantic mackerel eggs from sea surface waters of the New York Bight, and prospects for biological effects monitoring with ichthyoplankton. Rapp. P.-V. Reun. Cons. Int. Explor. Mer. 179:275-291.

Abstract: Buoyant, early-developing stages of fish eggs are expected to be a most sensitive component of plankton responding to various environmental cytotoxins, mutagens, and teratogens. The risk of contaminant exposure must be increased by their position near the ocean surface, as well as by the location of important spawning grounds in polluted coastal areas.

Rather standard procedures have been modified for simple, direct examination of the mitoses and cells of fish embryos as sampled in plankton. Also it has been determined that the yolk-sac membrane of such fish eggs is suitable for detailed chromosome analysis. Accordingly, cytologic, cytogenetic, and embryologic measures were made on about 10,000 early-stage eggs of the Atlantic mackerel *Scomber scombrus*. Eggs were sampled in May 1974 and May 1977 from surface waters at a total of about 80 sites in the New York Bight.

Development rates as calculated by mitotic index, viability as calculated from early indicators of cell death, and chromosome-mitotic abnormalities of the mackerel embryos varied widely over the sample sites in both 1974 and 1977. Early stage embryos that have ceased to undergo chromosome and cell divisions, and embryos with grossly irregular and arrested mitoses, have little chance for any further divisions, and embryos with grossly irregular and arrested mitoses have little chance for any further normal development. In 1977, gross embryo malformations were frequent and these too varied from station to station. Generally, there were significant correlations between measures on chronologically related development stages, and in 1974 between all stages, excluding cleavage.

The 1974 stations were grouped for statistical study according to proximity to one another, water temperature, and patterns of water circulation in the Bight. Group estimates of egg moribundity based on cell state, mitotic-chromosome irregularities, and division arrest at gastrulation and just after were then compared. These showed statistically significant lower egg viability in presumed more impacted Bight areas relative to less impacted areas more distant from the coast and dumping grounds.

Along with the 1977 collection of mackerel eggs for cytological study, microlayer (or surface) and subsurface waters were sampled for heavy metal analysis. Large quantities of plankton were also sampled for analyses of 65 toxic aromatic and chlorinated hydrocarbons. This study provides some evidence for statistically significant associations between cytologic, cytogenetic, and embryologic measures of mackerel egg health at the several sample sites and heavy metal levels of surface water and plankton, and toxic hydrocarbon levels of plankton at these sites. The measure of egg state showing associations fall into the following categories: embryo moribundity (on the basis of cellular and mitotic indicators), chromosome-mitotic abnormalities, development rate, cell differentiation problems, gross embryo malformation, and total egg number sampled.

Temperature and salinity also associated with some of the 1977 egg variables, but not necessarily in the direction that developmental optima for mackerel would have indicated. Atlantic mackerel, furthermore, are coastal spawners with normal development proceeding over a considerable range of temperature and salinity. Data are presently interpreted as indicating that spawned eggs of the southern contingent of Atlantic mackerel are being adversely impacted by hydrocarbon and heavy metal pollution in a portion of their spawning grounds in the stressed New York Bight. Variable temperature and salinity most likely affected overall expression of the measured impact. Samples from a 1978 cruise will provide additional information on associations between surface water, plankton pollution, and mackerel egg health.

This application of procedures of cytology and cytogenetics to ichthyoplankton is the first demonstration of the feasibility of measuring in open ocean waters a technically sublethal effect of sea surface pollution directly on a reproductive phase of a fishery resource species. The mackerel studies are believed to demonstrate the general usefulness of cytological and cytogenetic analysis of pelagic fish eggs as collected in plankton to monitor biological effects of sea surface water pollution. This combined embryo-cell-chromosome approach has the advantage of being a sensitive biological gauge of sublethal effects of pollution. At the same time, it can routinely supply informaiton on impacts, natural and unnatural, on the early life history stages of a commercial resource. Data on fish stocks developed in other studies of these resources can aid in interpretation of monitoring results thereby contributing to the overall value of any biological effects ocean monitoring.

909 - Longwell, A. Crosby. 1981. Cytological examination of fish eggs collected at and near 106-Mile site. In: Assessment Report on the Effects of Waste Sumping in 106-Mile Ocean Waste Disposal Site. NOAA/ODP Dumpsite Evaluation Report 81-1. Rockville, MD. pp. 257-276.

910 - Lorenzen, S. 1971-1972. Die nematodenfauna in verklappungsgebiet für industrieabwasser. Nordwestlich von Helgoland. Part I. Araeolaimida und Monhysterida. Zool. Anz. Vol. 187, pp. 223-248. Part II. Desmoidorida und Chromadorida. Zool Anz. Vol. 187, pp. 283-302. Part III. Cyatholaimidae, revisions of Pomponeria Cobb. Veraff. inst. meeresforsch brenierh. Vol. 13 pp. 285-306.

911 - Lorenzen, S. 1974. Die nematodenfauna der sublitoralen region der Deutschen Bucht, insbesondere im Titan-Abwasser-gebiet bei Helgoland. Veroff. Inst. Meeresforsch. Bremerh. 14. pp. 305-327.

912 - Lowrie, A., Jr., and B. C. Heezen. 1967. Knoll and sediment drift near Hudson Canyon. Science. 157:1552-1553.

913 - Luch, G. 1973. An examination in the Wessor estuary in connection with industrial wastewater introduction (titanium dioxide factory). Ext. Die Kuste. 24:20-27.

914 - Ludwig, H. F., and B. Onodera. 1964. Scientific parameters of marine waste discharge. In: E. A. Pearson (editor), Advances in Water Pollution Research. Proceedings of the first international conference held in London, September 1962. Pergamon Press. New York, NY. Vol. 3. pp. 37-56.

915 - Ludwig, H. F., and P. N. Storrs. 1970. Effects of waste disposal into marine waters. A survey of studies carried out in the last ten years. Water Res. Vol. 4:709-720.

916 - Ludwig, H. F., J. R. Thoman, J. R. and P. N. Storrs. 1972. Trends in marine waste disposal. In: B. L. Edge, (editor), Coastal Zone Pollution Management. Syposium proceedings. Clemson Univ. Clemson, SC. pp. 105-130.

917 - Ludwig. H. F. 1975. Criteria for marine waste disposal in South-east Asia. In: E. A. Pearson, and E. de Fraja Frangipane. Marine Pollution and Marine Waste Disposal. (Supplement to Progress in Water Technology.) Proceedings of the 2nd International Congress. San Remo. 17-21 December, 1973. Pergamon Press. New York, NY. pp. 99-108.

918 - Ludwig, H. F. 1975. Aesthetics- a key factor in the technology of marine waste disposal. In: E. A. Pearson, and E. de Fraja Frangipane (editors), Marine Pollution and Marine Waste Disposal. (Supplement to Progress in Water Technology.) Proceedings of the 2nd International Congress. San Remo. 17-21 December, 1973. Pergamon Press. New York, NY. pp. 163-173.

919 - Ludwig, H. F. 1979. Design of marine disposal systems for industrial wastes. In: S. H. Jenkins, (editor), Marine Municipal and Industrial Wastewater Disposal. Proceedings of a Conference. Sorrento, Italy. 23-27 June 1975. (Progress in Water Technology Volume 4). Pergamon Press. Oxford. pp. 223-230.

920 - Lyons, W. B., H. E. Gaudette, N. D. Chasteen, and T. R. Fogg. 1978. Early diagenesis of nearshore and continental shelf anoxic sediment; the amount, nature, and possible role of dissolved organic carbon. Geol. Soc. of Amer. Abstr. 10(2):74.

Abstract: The role of dissolved organic carbon in diagenetic processes of recently deposited estuarine and hemipelagic anoxic sediments is not clearly understood. Pore water data of this study from Long Island Sound, Great Bay, N.H., and Jeffrey's Basin, New England Continental Shelf suggest that dissolved organic carbon (DOC) may be extremely important in a number of post-depositional processes. Our data demonstrate that the DOC concentrations of these pore fluids are as much as 20 times higher than those of the overlying waters. This suggests to us that DOC in pore waters is an excellent indicator of intense diagenetic processes. Dissolved organic carbon compounds in the anoxic pore waters may lead to inhibition of authigenic $CaCo_3$ metal sulfide formation. Dialysis experiments and electron paramegnetic resonance (EPR) spectroscopy analyses suggest DOC - trace metal associations are extremely important in the speciation and mobility of these metals in the sedimentary environment. Our data also indicate that the majority of DOC in estuarine pore waters may be nonrefractory in character and a condensation product of biochemical monomers.

921 - Lumsdaine, J. A. 1976. Ocean dumping regulation: An overview. Ecology Law Quarterly. 5(4):753-792.

922 - MacDonald, R. W. 1979. Development of meaningful criteria for ocean dumping. In: S. C. Byers, and R. O. Brinkhurst (editors), Report on Ocean Dumping R and D, Pacific Region. Department of Fisheries and Oceans 1978-1979. Pacific Mar. Sci. Rep. 79-4. Institute of Ocean Sciences. Sidney, British Columbia. p. 7-15.

923 - MacGregor, K. A., M. S. Klein, J. S. Bazzolo, and M. E. Delaney. 1980. Municipal solid waste disposal in estuarine and coastal marshlands. U.S. EPA-600/2-80-212. Cincinnati, OH. 183p. NTIS No. PB81 129223.

924 - Mackay, D. W., and G. Topping. 1970. Preliminary report on the effects of sludge disposal at sea. Effluent and Water Treatment Jour. 10:641-649.

925 - Mackay, D. W., W. Halcrow, and I. Thornton. 1972. Sludge dumping in the Firth of Clyde. Mar. Pollut. Bull. 3(1):7-11.

926 - MacLean, S. A., C. A. Farley, M. W. Newman, and A. Rosenfield. 1981. Gross and microscopic observations on some biota from Deep Water Dumpsite 106. In: B. H. Ketchum, D. R. Kester, and P. K. Park (editors), Ocean Dumping of Industrial Wastes. Plenum Press. New York, NY. pp. 421-437.

Abstract: Euphausiids were infested on the appendages with phoronts of an aposome ciliate and also had foci of melanization in the gills. The isopod, Idotea, showed patchy lack of pigmentation, particularly in the telson, and edematous gills. Heteropod mollusks had larval cestode infestions, neoplasialike lesions, and other cell abnormalities. Examination of leptocephali revealed possible traumatic lesions and developmental anomalies. Microbial testing for the presence of mutagens promises to be a very useful method for monitoring pollution in the ocean.

927 - MacLean, Sharon, A., et al. 1981. Histopathology of organisms sampled at and near 106-Mile site. In: Assessment Report on the Effects of Waste Dumping in 106-Mile Ocean Waste Disposal Site. NOAA/ODP Dumpsite Evaluation Report 81-1. Rockville, MD. pp. 277-293.

928 Macomber, R. T., R. L. Mairs, and D. T. Stanczuk. 1974. A study of the surface fluid flow characteristics of the New York Bight and the development of research data display techniques. Earth Satellite Corporation. Washington, DC. 63p.

Abstract: Man has increasingly used the coastal oceans as a depository for wastes produced from accelerating coastal development and industrialization. As large metropolitan centers have developed, the quantities of waste materials to be disposed of have increased to a point where they are making a recognizable impact on the marine environment. What impact such environmental changes may have, how irreversible the processes they initiate may be, and how significant these processes may be over long periods of time have yet to be determined.

Besides having a large amount of marine traffic with its attendant pollution problems, the New York Bight area has been used for deliberate coastal marine waste disposal for well over a century. The environmental impact of these wastes and their movement, dispersion, and accumulation have not been studied in detail, nor were systematic investigations seriously considered before 1968.

The National Oceanic and Atmospheric Administration (NOAA) conceptualized a bold plan in 1972 to develop a better understanding of the New York Bight ecosystem. This program, called the Marine Ecosystems Analysis (MESA) Program, has for two years been studying physical, chemical, and biological interactions within the New York Bight.

Numerous federal agencies, state governments, and local municipalities are involved in the management of the New York Bight. These agencies require various types of information which summarize environmentally related characteristics and phenomena, which can contribute realistically to routine decision making, and which can be data inputs to practical coastal management.

In response to the need for (1) a demonstration of remote sensing data acquisition applied to studies of the coastal marine ecosystem in the New York Bight area, and (2) a demonstration of the remote sensing based information products which can be provided to area users, Earth Satellite Corporation (EarthSat) entered into a contract agreement with the applicability of remote sensor records characteristics, and to develop remote sensor-derived products for potential users in the New York Bight area.

EarthSat's activities along the Atlantic Coast of New Jersey, using ERTS-1 and complementary aircraft imagery, combined with the representation of study results in a format compatible with potential user needs have provided early benefits to SPOC, and to MESA project as a whole.

929 - Macrae, J. 1977. Getting rid of sludge - lose a few myths and gain an asset. Municip. Eng. 154(24):807-809.

930 - Madeley, J. R. 1973. Disposal of liquid industrial wastes to the marine environment. Jour. Inst. Fuel. 46. p. 63 and 65-69.

931 - Maggi, P., et P. Lassus. 1974. Effects bilogiques des effluents de quelques frabriques d'oxyde de titane. ICES. C.M. 74/E:50.

932 - Mahoney, J. 1968. Feasibility for ocean based solid waste disposal. Report. Harvard University. School of Public Health. Department of Environmental Health Services. 7p.

933 - Mahoney, J. B., F. H. Midlige, and D. G. Deuel. 1973. A fin rot disease of marine and euryhaline fishes in the New York Bight. Trans. Amer. Fish. Soc. 102:596-605.

Abstract: A fin rot disease reached epizootic proportions in 1967 among at least 22 species of marine and euryhaline fishes in the New York Bight and has continued to occur annually to the present time. Fin necrosis was the consistent external sign of the disease; other external pathological changes included skin hemorrhages, skin ulcers and, occasionally, blindness. Bluefish, summer flounder, winter flounder, and weakfish were the principal species affected. Results of these studies indicate a bacterial cause. Bacteria of three genera, *Aeromonas*, *Vibrio*, and *Pseudomonas* are likely implicated. Similar epizootics among freshwater fishes have been associated with unsanitary conditions in aquaria and with water pollution in nature. The epizootic primary center, lower New York Harbor, is grossly polluted with sewage and industrial wastes. We suspect pollution had a role in the disease.

934 - Mahoney, John B. 1978. The seasonal maxima of *Ceratium tripos* with particular reference to a major New York Bight bloom. NOAA/NMFS TRS No. 16. Highlands, NJ. 28p.

Abstract: Measurements of spring bloom maxima of *C. tripos* in 1976. May's mean water column density was 240,000 cells per liter. Normal level is 46 cells per liter. Evidence suggests the bloom migrated into the New York Bight and then collapsed after the nutrients were exhausted.

935 - Mahoney, John B. 1979. Plankton dynamics and nutrient cycling. Part II. In: R. L. Swanson and C. J. Sinderman (editors), Oxygen Depletion and Associated Benthic Mortalities in New York Bight, 1976. NOAA Professional Paper No. 11. U.S. Dept. of Commerce. Rockville, MD. pp. 219-229.

Abstract: *C. tripos* remained a significant portion of the bottom floc in the oxygen-deficient area between Manasquan and Barnegat at least until mid-July. Around the end of July, no *C. tripos* were seen in bottom samples from off Sandy Hook and it had almost disappeared from the bottom floc off Manasquan, but it was still evident in the Barnegat area. The floc from the Manasquan area at this time was about equally black and white under the microscope, but the Barnegat floc retained much of the earlier yellow-brown appearance. By mid-August, the bottom samples from off Barnegat had no *C. tripos* and little floc; what was present appeared generally decayed. Bottom samples from off Atlantic City still contained abundant yellow-brown floc.

Vaughan (1977) surveyed the *C. tripos* population in southern New Jersey coastal waters between Great Bay and Cape May from July 21 to its complete disappearance around August 20. The decline of *C. tripos* abundance he observed had a pattern similar to that seen in the more northerly regions, but occurred later. In addition, during late July and the first half of August, Vaughan (personal communication) found *C. tripos* present nearly always, even inshore, as individual, intact cells retaining cytoplasmic contents. Because a preservative was used, he was not able to determine whether the cells were alive when collected. However, the cells were at least not disrupted or aggregated in a detrital mass as they were to the north when the first samples were examined on July 4. The combined observations indicate that the decomposition proceeded earlier or at a more rapid rate in the Sandy Hook-Manasquan area and progressed southward.

Some *C. tripos* concentrations, alive and apparently vigorous, were found around middepth between the middle of July and early August off Sandy Hook, Manasquan, and Barnegat. They were all at stations 20 km or greater from the shore. During the same period, intact but nonmotile cells and fragments of cells, but no live cells, were found at various depths less than 20 km from shore. This suggests that *C. tripos* survived better in offshore waters. If so, a possible explanation is that by May and June, the *C. tripos* population within 20 km of the shoreline between the Bight Apex and Barnegat Inlet was light limited.

Based on microscopic observations, some bacterial presence was associated with aggregates of phytoplankton material in the water column, but it was greatest at or near the bottom where most of the floc was also found. Also, general bacterial presence seemed to be associated with the presence of *C. tripos* (the notable exception was in the August 16 Atlantic City bottom sample in which abundant bacteria but no *C. tripos* were seen). Unmistakable bacterial decomposition of *C. tripos* cells was observed several times; fungus attached on *C. tripos* was also seen. Protozoans and small nematodes were fairly numerous in some samples and may have been feeding on the floc material, although this was not determined. Vigorous activity and clustering around the floc by ciliate protozoans seen in a number of samples did suggest feeding behavior. All these forms probably contributed to the decomposition process. The cell wall of *C. tripos* is cellulosic. In the Gulf of Maine and Georges Bank, Waksman *et al.* (1973) found extensive populations of bacteria able to use cellulose and hemicelluloses, although cellulose-decomposing marine bacteria were less abundant than species not having this capability. Barghoorn and Linder (cited in Zobell 1945) found several species of marine fungi able to use cellulose. Marine ciliates have been known to feed on bacteria, diatoms, or other protozoa (Lackey 1936).

The presence of numerous *O. luteus* (many individuals apparently in a senescent state) as a floc constituent in locations from Sandy Hook to Sea Girt, N.J., between July 22 and August 4 is interesting. This species bloomed intensely throughout the southern half of Lower New York Bay between June 6 and 13, 1976. Tidal action gradually washed the bloom water to the ocean. If we assume that the *Olisthodiscus* concentrations in the bottom floc originated in the bay, then inshore along the New Jersey coast the bottom received large quantities of phytoplankton from the bloom in Lower New York Bay. The *Olisthodiscus* bloom may not have added measurably to the 1976 oxygen depletion, because it was small compared to the *C. tripos* bloom. Perhaps a more important implication is that material from chronic seasonal blooms of phytoflagellates would contribute to annual bottom water oxygen sag in at least the coastal area near the bay. Segar and Berberian (1976) determined that oxidation of phytoplankton material below the thermocline was a major cause of the low oxygen values they observed in the bottom waters of the Bight Apex.

936 - Mahoney, J. B., and F. W. Steimle, Jr. 1979. A mass mortality of marine animals associated with a bloom of *Ceratium tripos* in the New York Bight. In: D. L. Taylor and H. H. Seliger (editors), Proc. Sec. Int. Conf. on Toxic Dinoflagellate Blooms. Elsevier-North Holland, Amsterdam. pp. 225-230.

Abstract: A depletion of oxygen in bottom waters of the New York Bight developed during the summer of 1976. At its maximum areal extent in September, the oxygen depletion extended over 13,000 km^2. The low oxygen levels and, probably, the subsequent production of hydrogen sulfide, caused extensive mortalities of marine animals, especially, the surf clam, *Spisula solidissima*; 5-7 years may be required for the recovery of shellfish stocks. The event was apparently related to the decline and ultimate collapse of an immense bloom of the dinoflagellate, *Ceratium tripos*, which had developed in the Bight during the late winter and spring. A time/locale coincidence between advanced bloom decomposition and zones of maximum oxygen depletion suggests a relationship between the two phenomena. Comparison of features of this bloom with those of prior maxima of *C. tripos* in the New York Bight and other locales indicates that its intensity was extraordinary but other aspects such as season and duration of occurrence are not unique.

937 - Malone, Thomas C. 1976a. Phytoplankton productivity in the Apex of the New York Bight: Environmental regulation of productivity/chlorophyll *a* In: M. G. Gross (editor), Amer. Soc. Limnol. and Oceanogr. Spec. Symp. 2:260-272.

Abstract: Phytoplankton productivity within a 600-km^2 area of New York Bight Apex ranged from a December minimum of 0.1 g C $m^{-2}d^{-1}$ to a June maximum of 6 g C m^{-2} d^{-1}. Netplankton productivity peaked in February (1.7 g C $m^{-2}d^{-1}$) and June (2.2 g C $m^{-2}d^{-1}$); nanoplankton productivity was highest in June (2.1 g C $m^{-2}d^{-1}$) and July (3.7 g C $m^{-2}d^{-1}$). Productivity above 2 g C $m^{-1}d^{-1}$ was restricted to within 20 km of the mouth of the Hudson-Raritan esturay. Netplankton and nanoplankton accounted for 41% and 59% of the annual phytoplankton production (370 g C m^{-2}).

The supply of dissolved inorganic nitrogen from the Hudson estuary exceeded phytoplankton demand except during June, July, and August when regeneration within the Apex was an important source of nitrogen. No evidence of nitrogen-limited phytoplankton growth was found. Evidence shows that chlorophyll a specific phytoplankton productivity was regulated by light and temperature. Ocean dumping had no observable effect on phytoplankton assimilation numbers or on the distribution of environmental factors regulating phytoplankton growth in the Apex.

938 - Malone, Thomas C. 1976b. The 1976 *Ceratium tripos* bloom in the New York Bight: Causes and consequences. NOAA TR NMFS Circ. NEFC/NMFS. Highlands, NJ. (410):1-14

Abstract: An extensive bloom of the dinoflagellate *Ceratium tripos* occurred throughout the New York Bight between January and July 1976. Population size peaked during April-June and declined rapidly during July. A floc consisting primarily of decaying *C. tripos* cells was observed to cover the bottom during July between Sandy Hook and Atlantic City between 5 and 50 km offshore. The distribution of the floc roughly coincided both temporally and spatially with the development of a subthermocline oxygen minimum layer and extensive fish kills.

Prior to the onset of thermal stratification (January-March), the *C. tripos* population was uniformly distributed throughout the water column and was growing photosynthetically. As the water column began to stratify in April, the population aggregated in a layer 1-3 m thick near the base of the thermocline between the 0.1 and 10% light depths. If photosynthetic growth was occurring during May-June, it was at a very low rate (about 0.02 g $C/m^{-2}d^{-1}$ at the 1% light level). The possibility of no growth or heterotrophic growth cannot be dismissed, especially in the Apex of the New York Bight and along the New Jersey coast.

The *C. tripos* bloom resulted in a gradual accumulation of a large quantity of particulate organic matter which did not enter pelagic food chains. Respiration of this biomass and its decay below the thermocline were probable major factors in the development of oxygen-poor bottom waters in June and July. Localization of the oxygen minimum layer off the New Jersey coast probably reflects the bottom topography of the New York Bight and the distribution of the *C. tripos* biomass within the Bight.

The occurrence of *C. tripos* blooms per se is not unusual. The bloom was unique only in terms of the size of the population produced, its areal extent, and its duration. Because high densities of *C. tripos* developed throughout the New York Bight during January-March with maximum densities between midshelf and the shelf break, it is unlikely that the bloom occurred in response to local nutrient enrichment (related to the disposal of domestic and industrial wastes) during the period of the bloom. However, causes of the bloom and its collapse cannot be determined, based on existing information.

939 - Malone, T. C. 1977a. The 1976 New Jersey fish kill: Summary of workshops. In: Oxygen Depletion and Associated Environmental Disturbances in the Middle Atlantic Bight in 1976. NOAA TR No. 3. NEFC/NMFS. Highlands, NJ. pp. 9-29.

Abstract: Mass mortalities of finfish and shellfish populations occurred off the New Jersey coast between Sandy Hook and Atlantic City from late June through September 1976. This has had immediate impacts on commercial and sport fisheries and could have long term effects as a consequence of larval mortality and reductions in the size of spawning populations. The fish kills were apparently caused by the development of an oxygen minimum layer and local hydrogen sulfide production below the thermocline in waters 20 to 40 m deep between 5 and 40 km offshore.

Because of these fish kills and other unusual episodes in the New York Bight during the first half of 1976 (e.g. the <u>Ceratium tripos</u> bloom), workshops were organized to collate existing environmntal data, to document the environmental impact of the oxygen minimum layer, and to determine why exceptionally low oxygen concentrations developed off the New Jersey coast during summer 1976. The following is an integrated summary of these workshops.

940 - Malone, T. C. 1977b. Plankton systematics and distribution. MESA New York Bight Atlas Monograph 13. MESA New York Bight Project and New York Sea Grant Institute. Albany, NY. 45p.

Abstract: ATLAS MONOGRAPH 13 describes the distribution and abundance of phytoplankton and zooplankton species in New York Bight. Phytoplankton species composition in the Bight is strongly influenced by estuarine processes; zooplankton species composition is more strongly influenced by oceanic processes. Using published and unpublished data collected over the past 75 years, this paper summarizes the known plankton ecology in the Bight but concludes that this knowledge is of limited value. To more thoroughly understand plankton distribution and the effects of man's activities on plankton population in the Bight, it will be necessary to conduct more systematic observations using standardized methodologies.

941 - Malone, T. C. 1977c. Light-saturated photosynthesis by phytoplankton size fractions in the New York Bight. Marine Biology. 42:281-292.

Abstract: The relationships between netplankton and nanoplankton assimilation numbers, temperature, and major nutrient concentrations were studied and evaluated in the context of seasonal patterns in the biomass of these phytoplankton size fractions. Netplankton and nanoplankton blooms typically occur during late winter (2° to $8^{\circ}C$) and summer (18° to $24^{\circ}C$), respectively. Variations in nanoplankton and netplankton assimilation numbers were not statistically related to the development or collapse of specific blooms based on weekly sampling, but assimilation numbers were higher during the bloom periods than during transition periods of rapid temperature change (8° to $18^{\circ}C$). Differences in the assimilation numbers between size fractions could account for the dominance of the nanoplankton fraction during the summer bloom period but not for the dominance of netplankton during the winter bloom period. Nanoplankton and netplankton assimilation numbers were exponential functions of temperature between 8° and 24° and 8° and $20^{\circ}C$, respectively. Below $8^{\circ}C$ the assimilation numbers of both fractions were higher than expected on the basis of temperature. Above 20° netplankton assimilation numbers declined with temperature. Netplankton and nanoplankton assimilation numbers were occasionally correlated with dissolved inorganic nitrogen concentrations from less than 1.0 to more than 15 μg- at 1^{-1}. Under these conditions, nanoplankton growth rates (calculated from assimilation number and carbon: chlorophyll) were higher and increased more rapidly with dissolved inorganic nitrogen than netplankton growth rates.

942 - Malone, Thomas C. 1977d. Environmental regulation of phytoplankton productivity in the Lower Hudson Estuary. Estuar. and Coastal Mar. Sci. 5:157-171.

Abstract: Phytoplankton productivity and mean photic zone chlorophyll a concentrations ranged from 0.01 g C $m^{-2}d^{-1}$ (December) to 2.22 g C $m^{-2}d^{-1}$ (July) and from 0.8 μg Chll 1^{-1} (December) to 6.4 μg Chll 1^{-1} (July), respectively. Dissolved inorganic nutrient concentrations were high throughout the year, and (with the possible exception of silicate during the summer) temporal variations were not correlated with concurrent variations in phytoplankton productivity. Chlorophyll a specific phytoplankton productivity was regulated by light and temperature. Nanoplankton growth accounted for increases in phytoplankton productivity above 0.25 g C $m^{-2}d^{-1}$ within the lower Esturay. Except during the summer when phytoplankton growth rates were high, increases in phytoplankton biomass were related to the advection of phytoplankton into the Hudson Estuary from Raritan Bay or the Apex of the New York Bight. Although major nutrient concentrations were consistently high, netplankton growth rates were apparently less than flushing rates and, therefore, too low to generate blooms in the lower Estuary.

943 - Malone, T. C. 1978. The 1976 Ceratium tripos bloom in the New York Bight: Causes and consequences. NOAA TR NMFS Circular 410. NEFC/NMFS. Highlands, NJ. 14p.

944 - Malone, Thomas C., and M. B. Chervin. 1979. The production and fate of phytoplankton size fractions in the plume of the Hudson River, New York Bight. Limnol. and Oceanogr. 24(4):683-696.

Abstract: Seasonal effects of water column stability, phytoplankton growth, and copepod grazing on the biomass of phytoplankton size fractions are described in the context of variations in particulate organic carbon (POC). Net plankton biomass was more variable in time (between weeks and months), while nanoplankton biomass was more variable in space (between surface and bottom layers and between stations). Biomass was highest and most uniformly distributed during February-March when net plankton diatoms dominated and doubling times approximated the flushing time of the plume. Biomass declined rapidly with depth and distance from the mouth of the estuary during June-July when nanoplankton dominated and doubling times were short relative to the flushing time of the plume. Variations in POC were well correlated with phytoplankton biomass when net plankton dominated and biomass was limited by sinking and mixing. When nanoplankton dominated, biomass was limited mainly by grazing and correlations between POC and biomass were poor. It appears that most of the biomass produced by net plankton diatoms sinks from the surface layer before the summer thermocline develops, while most of the biomass produced by nanoplankton during the summer enters pelagic food chains before sinking from the surface layer.

945 - Malone, Thomas C., Wayne E. Esais, and Paul Falkouski. 1979. Plankton dynamics and nutrient cylcing. Part I. In: R. L. Swanson and C. J. Sinderman (editors), Oxygen Depletion and Associated Benthic Mortalities in New York Bight, 1976. NOAA Professional Paper No. 11. U.S. Dept. of Commerce. Rockville, MD. pp. 193-216.

Conclusions: Although data were not collected synoptically in time or space, coastal observations correlated well with those in the Apex and outer Bight. Temporal variations in *C. tripos* cell density at Fire Island Inlet reflect the early stages of the bloom before stratification, and mean cell densities along the New Jersey shore appeared to reflect at least the latter stages of the bloom during the period of thermal stratification.

The *C. tripos* bloom apparently began throughout the New York Bight in January; maximum cell densities developed in the midshelf to shelf break region in late March before the onset of thermal stratification. The temporal and spatial distributions of cells indicate that the population was increasing most rapidly in the outer Bight during March or that the outer Bight received a larger initial inoculum of cells than the inner Bight. The large area over which the bloom occurred indicates that it did not develop in response to local nutrient enrichment of the coastal zone during the actual period of the bloom. This is supported by the observation that *C. tripos* cell densities were lowest in the Apex where local nutrient enrichment is greatest. The causes of the bloom, whether related to increased growth or decreased mortality rates, must have involved processes operative on spatial scales on the order of the continental shelf and time scales on the order of months to years.

The temporal and spatial development of the bloom during April and May suggest an onshore transport of cells once the population began to aggregate below the thermocline. By mid-June a large population of cells was present below the thermocline in a relatively flat region of the New Jersey shelf between the 20- and 40-m isobaths. Much of this population was below the euphotic zone in considerable organic loading over the past two decades, and the development of oxygen minimum layers and local anoxia have occurred previously during summer.

Based on the effect of *C. tripos* on the content of POC in the water column and on the development of large, subthermocline populations, it is likely that *C. tripos* made a very significant contribution to the oxygen demand required to account for the the oxygen minimum layer. In the latter context, a flocculent suspension of organic matter at least 1 cm thick coated the bottom during July between Sandy Hook and Atlantic City from 5 to 50 km offshore. The floc consisted primarily of phytoplankton cells dominated by *C. tripos*. Microscopic examination indicated a steady increase in the decomposition of *C. tripos* cells during July.

A computer simulation model was used to explore the combined effects of benthic respiration and *C. tripos* respiration on the rate of oxygen depletion below the thermocline. *C. tripos* respiration rates were calculated from the expression $R = aW^b$ (Banse 1976) where a and b are temperature dependent constants and W and R are the weight of the cell in picograms (pg) of carbon and the respiration rate in picograms of carbon/cell/hour, and the respectively. The carbon content of *C. tripos* was calculated from both CHN analysis and regression of netplankton chlorophyll *a* on netplankton chlorophyll *a* on netplankton carbon. Values ranged from 20,000 to 30,000 pg/cell; a mean value of 25,000 was chosen for calculating respiration rates. Using $Q_{10} = 2.3$, the carbon specific respiration rate of a single cell was calculated to be 0.003/h at 10°C (= 1.4×10^{-4} ul O_2/cell/h). In 1977, *C. tripos* respiration rate of $1.39 \pm 0.17 \times 10^{-4}$ ul O_2/cell/h at 10° C, agreeing well with the rate calculated according to the expression given by Banse (1976).

There is no evidence that the *C. tripos* bloom influenced the growth of netplankton diatoms or nanoplankton populations. The distribution and abundance of these groups were similar to previous years' observations.

The model output indicated that within two months the oxygen concentration in the bottom 5-m layer reaches a steady-state concentration that is 45 percent of the initial oxygen concentration. The simulated rate of oxygen depletion below the thermocline is extremely sensitive to changes in eddy diffusivity, and small changes in diffusivity are sufficient to cause simulated anoxia.

These calculations show the potential metabolic influence of *C. tripos*. The water column integrated *C. tripos* respiration rate exceeds the benthic oxygen consumption rate by a factor of 20. Therefore, *C. tripos* biomass is a large potential source of BOD. Oxidation of the *C. tripos* biomass (3,255 mg-at C/m^2) within 20 m of the bottom would require 8,463 mg-at O_2/m^2 or 71 percent of the initial oxygen content. Thus, the combined effects of respiration and subsequent death and decay of the biomass were more than sufficient to produce anoxia.

Nanoplankton productivity per se was probably not a major factor in the 1976 oxygen depletion even though it normally accounts for most of the input of POM to the region. With the exception of winter-spring diatom blooms, which apparently go ungrazed, there is no evidence that a significant portion of phytoplankton producton normally accumulates below the thermocline during summer. In effect, the *C. tripos* bloom provided a mechanism by which large quantities of POC were accumulated over several months. The change in the relative abundance of phytoplankton species and the effects of this change on the distribution and quantity of POC in the subthermocline water column resulted in exceptionally high BOD in 1976. Unforntunately, we do not understand this type of species succession very well, and the basic question of why the *C. tripos* bloom occurred in the first place remains unanswered.

946 - Malone, Thomas C. (In Press). Factors influencing the fate of sewage-derived nutrients in the Lower Hudson estuary and New York Bight. In: G. F. Mayer, A. Calabrese, F. A. Cross, D. C. Malins, D. W. Menzel, J. S. O'Connor, and F. J. Vernberg (editors), Ecological Stress in the New York Bight: Science and Management. Estuarine Research Federation. Columbia, SC.

Abstract: The lower Hudson River estuary receives large quantities of sewage wastes and associated plant nutrients. Most of the nutrient load is transported into adjacent coastal water where it is taken up by phytoplankton within a relatively small area ($<1,500\ km^2$). A better understanding of the effects of freshwater flow of the Hudson River on this nutrient flux is required before urban use of fresh water is increased above current levels. Arguments based on the dynamics of phytoplankton productivity and nutrient uptake lead to the conclusion that secondary sewage treatment is having a beneficial effect on water quality in the lower estuary and Apex of the New York Bight, but that tertiary treatment could have a negative impact on commercial and sport fisheries of the region.

947 - Manheim, F. T. (In Press). Who is doing what in marine dumping. In: I. W. Duedall, B. H. Ketchum, P. K. Park and D. R. Kester (editors), Wastes in the Ocean, Vol. I: Industrial and Sewage Wastes in the Ocean. John Wiley and Sons, Inc. New York, NY.

Abstract: In 1978, major categories of United States waste products being discharged to the oceans by vessels were: dredged material, 65 million metric tons; solid sewage waste, 5.0 million metric tons; industrial and construction wastes, 2.4 million metric tons. The amount of treated (liquid) sewage waste discharged to the ocean and estuaries from the U.S.A. coastal communities has been estimated at somewhat less than 20 million tons per day.

The principal Federal agencies engaged in inhouse or contracted research on waste disposal in the sea (in order of allocation of funds): National Oceanographic and Atmospheric Administration, Environmental Protection Agency, and Department of Energy. The outlays by the Departments of Defense and Interior, the National Science Foundation, and other agencies are smaller. Of the total estimated Federal budget of $186 million for pollution-related work in 1980, some $30 million was allotted for research on marine waste disposal categories. A very rough estimate of $100 million is spent annually by industry for the disposal of 2.2. million metric tons of industrial waste. Costs for the disposal of dredged material mostly by the U.S. Army Corps of Engineers may be on the order of $75 million.

The operation of large sectors of industry, commerce, and coastal urban communities is still dependent on waste material handling or disposal in the marine environment, in spite of the Marine Protection Research and Sanctuaries Act of 1972 (Public Law 92-532). Innovation in disposing of wastes in the coastal area is needed.

948 - Margalef, R. 1975. Assessment of the effects on plankton. In: E. A., Pearson and E. de Fraja Frangipane (editors), Marine Pollution and Marine Waste Disposal. (Supplement to Progress in Water Technology.) Proceedings of the 2nd International Congress. San Remo. 17-21 December 1973. Pergamon Press. New York, NY. pp. 301-306.

949 - Marindin, H. L. 1888. Tidal levels and flow of currents in New York Bay and Harbor. U.S. Coast and Geodetic Survey Rep. App. 9. p. 405-408.

950 - Marine Ecosystem Analysis Program. 1974. Bibliography of the New York Bight. Part 1-List of Citations. Part 11-Indexes. Rockville, Md. U.S. Dept. of Commerce, NOAA. Part 1-184p. Part 11-493p. Bibliography concerning maritime law, administration, air pollution, meteorology, and solid waste management. Covers from 1951-April 1973.

951 - Marine Ecosystems Analysis Program. 1976. Evaluation of proposed sewage sludge dumpsite areas in the New York Bight. NOAA TM ERL MESA-11. Boulder, CO. 212p. NTIS No. PB 253727/AS.

952 - Marine Ecosystems Analysis Program. 1977. MESA-New York Bight project development plan and technical development plan for Fiscal Year 1977. U.S. Dept. of Commerce. NOAA. MESA. Stony Brook, NY.

Abstract: The New York Bight Project was initiated to determine the condition of the ecosystem in the Bight and to identify where significant environmental problems exist or may develop in the future.

953 - Marine Ecosystems Analysis Program. 1978. NOAA/MESA New York Bight project annual report for Fiscal Year 1977. Boulder, CO. 131p.

Abstract: The annual report for fiscal year 1977 presents an overview of MESA New York Bight Project activities between October 1, 1976, and September 30, 1977. Specifically, it summarizes research efforts sponsored by the Project and reviews significant technical, operational, and administrative achievements during the period. The report is produced as a means for formally reporting progress made by the Project in meeting its goals and objectives. This document is intended as an administrative tool for the MESA Program Office, the Environmental Research Laboratories (ERL), and the National Oceanic and Atmospheric Administration (NOAA). In a similar manner, it provides information to the U.S. Congress and Office of Management and Budget on the use of funds allocated for research into ocean dumping and marine ecosystems analysis. Finally, the report documents Project status to groups and individuals with interests in the Bight, including Project investigators, the New York Bight Project Synthesis Steering Committee, the MESA New York Bight Advisory Committee, users of MESA-generated information, and members of the concerned public.

954 - Marine Research, Inc. 1975a. Analysis of operation "Deep Six" benthic invertebrates. U.S. EPA Cruises (D. W. Lear) Mid-Atlantic Bight Dumpsites. Marine Research Inc. Falmouth, MA.

955 - Marine Research, Inc. 1975b. Analysis of operation "Midwatch" benthic invertebrates. U.S. EPA Cruises (D. W. Lear) Mid-Atlantic Bight Dumpsites. Marine Research Inc. Falmouth, MA.

956 - Marine Research, Inc. 1976a. Analysis of operation "Dragnet" benthic invertebrates. U.S. EPA Cruises (D. W. Lear) Mid-Atlantic Bight Dumpsites. Marine Research Inc. Falmouth, MA.

957 - Marine Research, Inc. 1976b. Analysis of operation "Touchstone" benthic invertebrates. U.S. EPA Cruises (D. W. Lear) Mid-Atlantic Bight Dumpsites. Marine Research Inc. Falmouth, MA.

958 - Marine Research, Inc. 1976c. Analysis of operation "Mogul" benthic invertebrates. U.S. EPA Cruises (D. W. Lear) Mid-Atlantic Bight Dumpsites. Marine Research Inc. Falmouth, MA.

959 - Marmer, H. A. 1935. Tides and currents in N.Y. Harbor. U.S. Coast and Geodetic Survey. Spec. Publ. 111. 198p.

960 - Martin, C. 1972. Toxicity of incinerator residues for marine organisms. Ecological aspects of the disposal of incinerator residues in the sea. In: M. W. First (editor), Municipal Waste Disposal by Shipborne Incineration and Sea Disposal of Residues. Report. Harvard University School of Public Health. Boston, MA. pp. 106-125.

961 - Martin, J. M., M. Meybeck, and F. Salvadori. 1976. Pollution chimique des estuaires: Etat actuel des connaissances. Revue bibliographique arretee en juin 1974. Publ. Centre nat. Exploit. Oceans. Ser. Rapp. Sci. Techn. No. 22. 286p.

962 - Martinsen, R., J. Ginter, C. Tuthill, J. Spagnoli, and A. C. Jensen. 1972. Environmental impact of dumping sewage sludge in the New York Bight. Marine Sciences Research Center. State Univ. of New York, Stony Brook. NY. Unpublished report. 52p.

Abstract: Sludge dumping modifies water quality and habitat. It creates public health problems. There are concerns for physical and social environments. The approach to this problem involves: (1) gradual decrease in dumping by developing uses for sludge. (2) removal of all toxic elements in product of treatment facilities. (3) a long range-plan for the reuse of all sludge generated. (4) cleansing of adjacent estuaries by virtue of benefits accured by passage of Environmental-Bond Issue.

963 - Matthews, W., (Chairman.) 1975. Marine litter. In: Pollutants. Nat. Res. Coun. U.S. National Academy of Science. Washington, DC. pp. 405-438.

964 - Mauchline, J. 1963. The biological and geogaphical distribution in the Irish Sea of radioactive effluent from Windscale Works, 1959-60. United Kingdom Atomic Energy Authority, Harwell. 70p.

965 - Maugh, T. H. 1979. Incineration, deep wells gain new importance. Science. 204:1188-1190.

966 - Maul, G. A., R. L. Charnell, and R. H. Qualset. 1974. Computer enhancement of ERTS-1 images for ocean radiances. Remote Sensing of Environment. 3:237-253.

967 - Mayer, D. A., D. V. Hansen, and S. M. Minton. 1979. Water movement on the New Jersey shelf, 1975 and 1976. In: R. L. Swanson and C. J. Sinderman (editors), Oxygen Depletion and Associated Benthic Mortalities in New York Bight, 1976. NOAA Professional Paper No. 11. U.S. Dept. of Commerce. Rockville, MD. pp. 149-163.

Summary: Our principal objective was to explore the differences between currents observed in 1975 and in 1976 as a possible cause of, or condition to, the important differences in D.O. concentrations observed during these 2 years. Insufficient flushing of the coastal waters could be an important factor in the depletion of the D.O. reservoir. A distinct difference was observed in the flow beneath the pycnocline in the area of anoxia off New Jersey. The weakened flow and reversal of direction in 1976 altered the usual pattern of material transports in the Bight. Much is still unknown about mechanisms of shelf circulation and its variability, but one major influence certainly was the period of southerly and southwesterly winds from about J.D. 125-215 (early May through early August). Although these winds were not particularly unusual on any given day, their direction was persistently opposite the normal southwestward flow of shelf waters, and their cumulative effect over more than 2 months must have been considerable. Similar reversals of surface currents off New Jersey in summer were reported during the 1960s (Bumpus 1969). Bumpus primarily attributed these reversals to the low river discharge at the time. Anoxia or mass mortalities were not associated with these events.

A likely reason that the southwesly winds evoked so little response in water below the pycnocline off Long Island is that the water is about half again as deep off Long Island as off New Jersey. The pressure gradient response to wind stress in shallow water is expected to be nearly inversely proportional to depth, hence the shoaler waters off New Jersey are expected to be more responsive to local winds. Another possible cause of the weaker perturbation of the current off Long Island is the spatial pattern observed in the wind stress. Wind stress was generally parallel to the shelf contours off New Jersey, but crossed the coast at a considerable angle off Long Island. Csanady (1976) explained that the alongshore component of wind stress is more important than the cross-shelf component in its influence on currents.

During the period of observations, upwelling occurred primarily in association with the current perturbation off New Jersey, but not off Long Island. Before this time, however, upwelling occurred consistently off Long Island and may have contributed indirectly to the low D.O.'s observed by advection of nutrients into the region. Notable here is that during late spring and early summer 1976, circulation below the thermocline was favorable off New Jersey but not off long Island for concentration of Ceratium tripos. The average onshelf flow (nearly 1.5 cm/s) observed off New Jersey during the 2 months before initial discovery of the benthic mortalities corresponds to a virtual displacement of about one-third the total shelf width per month. This is enough to have swept a large amount of these organisms from the outer shelf onto the inner shelf off New Jersey.

We infer from the relatively minor response of water below the pycnocline, first, that even though hurricane Belle's winds were intense, its rapid rate of passage was such that its impact on vertical mixing below the thermocline in the Bight was not especially significant. To the extent that wind-induced mixing may result in asymmetric, upward, turbulent entrainment of water across the thermocline, rather than symmetrical vertical mixing, then very little downward flux of oxygen into the bottom water would occur under even more extreme conditions than those of hurricane Belle. Second, we infer that reduced vertical mixing as a result of anomalously low summer wind speeds is not a likely cause of the low D. O. across the pycnocline. This implies that, once the summer stratification has become established, the only physical mechanism for oxygen renewal is by advection or by mixing along isopycnal surfaces that typically slope upward in the offshelf direction, but at very small angles. There is, of course, a period of time in spring and again in autumn when establishement or destruction of the pycnocline is critically dependent upon the occurrence and timing of atmospheric events.

968 - McAllister, R. F. 1969. Demonstration of the limitation and effects of waste disposal on an ocean shelf. F.W.P.C.A. Demonstration Project Grant W.P.D. 165-01 (R-1)-67. Florida Ocean Sciences Inst. Inc. Report No. AR69-2. 115p.

969 - McAuliffe, C. D. 1976. Environmental aspects of offshore disposal of drilling fluids and cuttings. Society of Petroleum Engineers of Aime: 46th Annual California Regional Meeting. In Society of Petroleum Engineers of Aime. SPE 5864. 16p.

970 - McCain, Bruce B., and Donald C. Malins. (In Press). Effects of petroleum hydrocarbons on selected demersal fishes and crustaceans. In: G. F. Mayer, A. Calabrese, F. A. Cross, D. C. Malins, D. W. Menzel, J. S. O'Connor, and F. J. Vernberg (editors), Ecological Stress in the New York Bight: Science and Management. Estuarine Research Federation. Columbia, SC.

Abstract: Investigations dealing with the effects of petroleum hydrocarbons on marine fishes and crustaceans are described. In one study conducted in our laboratory, three species of flatfishes (of the family Pleuronectidae) were exposed to crude oil-contaminated sediments for periods ranging from one to four months. English sole (Parophrys vetulus) was the only species to undergo significant adverse effects, including weight loss, liver structure aberrations, and hematological changes. In another study conducted in our laboratory, the feeding behavior of spot shrimp (Pandalus platyceros) exposed to seawater-soluble fraction of the crude oil was monitored. Concentrations of as low as 100 ppb caused significant disruption of shrimp feeding behavior. Investigations by other researcher pertinent to possible adverse effects of hydrocarbons on species of fishes and crustaceans similar to those found in the New York Bight are also reported.

971 - McDermott, D. J., et al. 1978. Metal contamination of flatfish around a large submarine outfall. Jour. Water Poll. Control Fed. 48(8):1913.

972 - McDonough, Kathleen B. 1976. A benthic index of environmental quality for the New York Bight Apex and Raritan Bay. Masters Thesis. State Univ. of New York. Stony Brook, NY. 89p.

Abstract: There are much data available quantifying the abundance and diversity of benthic communities in the New York Bight Apex and Raritan Bay, areas heavily impacted by ocean disposal (dumping, ocean outfalls, riverine inputs). A proposed index of environmental quality (EQ Index) is developed to evaluate these data. Each benthic invertebrate in a sample is given a tolerance value based on an evaluation of the organism's ability to colonize a defined zone with an unpredictable harsh environment. The product of the tolerance value and the abundance of each organism is summed for each sample and divided by the total number of organisms in the sample, resulting in an index of environmental quality at each station sampled. A grid with the resulting index numbers delineates an area in both the New York Bight Apex and Raritan Bay within which the benthic community structure has been altered by ocean disposal.

973 - McFarlane, C. F. 1974. Ocean waste disposal practices in metropolitan areas of California. Government Reports Announcements. 74(12):131.

974 - McFarlane, C. F. (Undated). Ocean waste disposal practices in metropolitan areas of California. U.S. EPA Ocean Disposal Program Report No. 4460 C1635.

975 - McGraw, Donald J. 1969. Continuation of the investigation of waste disposal in the New York Bight. U.S. Dept. Health, Education and Welfare. Public Health Service. 11p.

976 - McHugh, J. L. 1975. Management of New York's hard clam fishery. In: Proceedings of a Workshop. 28 March 1975. New York State Sea Grant and New York State Department of Conservation. Albany, NY. pp. 44-47.

Summary: Describes the problems associated with management of New York's hard clam fishery and the approaches that should be taken to overcome these problems.

977 - McHugh, J. L. & A. D. Williams. 1976. Historical statistics of the fisheries of the New York Bight area. New York Sea Grant Institute. Albany, NY. Publication. NYSSGP-RS-76-013. 75p.

Abstract: A compilation of available information on fish catches of this area since 1880 is divided into 3 categories: domestic commercial landings off New York and New Jersey, recreational, and foreign catches. The migratory habits of many species valuable to domestic fisheries make it useful to include certain catch statistics from the entire Southern New England and Middle Atlantic Bight regions. Comparability and accuracy of weight estimates of foreign and domestic catches are discussed. The ICNAF quotas for the immediate coastal area are charted for catches of species for which adequate scientific information is not available to estimate allowable catches. The major provisions of the Fishery Conservation and Management Act of 1976 set out U.S. fishery policy and authority, describe how the Act relates to foreign fishing and existing international agreements, establish national standards for fishery conservation and management, and provide mechanisms for implementation.

978 - McHugh, J. L. 1977. Fisheries and fishery resources of New York Bight. NOAA TR NMFS Circular 401. NEFC/NMFS. Highlands, NJ. 50p.

Abstract: The history of total fish and shellfish landings in the two states (New York and New Jersey) that form the landward boundaries of New York Bight is a history of change. Resource after resource has produced maximum landings, then declined. Total landings dropped from about 315,000 metric tons in 1956 to about 23,000 in 1967 and have risen only moderately since that time. The rise and fall of the industrial fisheries, mostly menhaden, was responsible for most of this decline, and this has masked trends in the food fisheries.

Altogether about 132 species or groups of species of fishes and invertebrates have been reported as landed in New Jersey or New York since 1880. Fifty of these are discussed and illustrated with figures and tables of landings.

Edible finfish species as a group reached peak landings in 1939 and declined fairly steadily to about one-third that level in the 1970s. Molluscan and crustacean shellfish production reached two peaks, in 1950 and 1966, the second considerably higher than the first. This recovery of shellfish landings in 1966 would not have occurred where it not for the rapid development of the surf clam fishery in the 1950s.

The timing of the declines makes it clear that foreign fishing was not the cause, for foreign fishing probably could not have affected the fisheries of New York Bight before the mid-1960s. Actually, total catches of resources taken only by domestic fishermen have declined more sharply than total domestic catches of species shared with foreign fleets. Foreign fishing is but a symptom of the troubles of the domestics fisheries, some of which are imagined. The ills of the domestic fisheries are economic and sociopolitical, and they will not yield easily to scientific solutions.

979 - McHugh, J. L. 1978. United States clam industry: Where is it going? In: Proceedings of Northeast Clam Industries: Management for the Future. 27-28 April 1978. Hyannis, MA. pp. 7-24.

Abstract: Clam production and industry along Atlantic Coast, Great South Bay is largest producer. Inshore clams continue but surf clams resource is collapsing. Clam management and regulation. Clam culture techniques and harvesting.

980 - McHugh, J. L., and J. J. C. Ginter. 1978. Fisheries. MESA New York Bight Atlas Monograph no. 16. New York. Sea Grant Institute. Albany, NY. 129p.

Abstract: Statistics of fisheries, both domestic and foreign, in the New York Bight area are presented. Maps and graphs provide an historical picture of past trends. Details of trends in commercial fishery landings are given for 14 major food finfish, 6 species or groups of species used mostly for industrial purposes, and 6 crustacean and molluscan shellfish resources. Brief accounts of 17 minor species are included. Foreign pressure in the international waters of the Bight began to increase steadily in the 1960s. The problem improved in 1977 when the U.S. extended its domestic fishery jurisdiction to 200 mi, thus assuming total responsibility for fishery management over the U.S. continental shelf. Foreign fishing within the 200-mi economic zone is now permitted only by special permit. The Mid-Atlantic Regional Fishery Management Council was formed to develop management plans for major fisheries 3 mi from the coast between Montauk Point, Long Island, and Chesapeake Bay; to oversee catch limits and assess surpluses; and to approve permits allowing foreign fleets access to designated surpluses. New York, New Jersey, Pennsylvania, Delaware, Maryland, and Virginia form the basis of membership in the council.

981 - McHugh, J. L. 1979. Status of the fisheries of the Middle Atlantic Bight region. Marine Sciences Research Center, State Univ. of New York. Stony Brook, NY. Special Report 31. Ref. 79-15. 56p.

Abstract: Maximum annual domestic commercial fish and shellfish landings of about 15 billion pounds, maximum annual foreign catch of 700 million pounds, and recreational catch of 300 million pounds. To improve fishery development-strategies, fishery techniques, market development. Need to improve coordination between government (federal, state and local) and between government, industry, and recreational interests.

982 - McIntyre, A. D., and R. Johnston. 1975. Effects of nutrient enrichment from sewage in the sea. In: Discharge of Sewage from the Sea Outfalls. Pergamon Press. Oxford. pp. 131-.

983 - McIntyre, A. D. 1977. A review of the effects of the disposal of sewage sludge to sea. In: Sewage Sludge Disposal Data and Technical Notes. Tech. Note No. 6. Department of the Environment. London. pp. 16-28. 43p.

984 - McIntyre, A. D. 1980. A note on the input of contaminants from the atmosphere to the sea. ICES. C.M. 80/E:47.

985 - McKinney, T. F., A. E. Cok, G. L. Freeland, and D. J. P. Swift. 1974. Side-scan sonar mapping of sediment facies and dump material in the New York Bight Apex. Geol. Soc. of Amer. Abstr. 6(1):54.

Abstract: Closely spaced (2000') side-scan sonar records and 1/2- and 1 nm-spaced grab sampling have been used to produce a detailed bottom facies map of part of the New York Bight Apex. The distribution of bottom facies can be related to the Y-shaped, N-S trending headward portion of the Hudson Shelf Valley. Bottom facies include: A) silty mud and fine muddy sand along the floor of the valley, generally below 25m; B) artifact gravel, well displayed by sonar mapping, in the higher segments between the branches of the Y and forming the Castle Hill cellar dirt deposit; C) coarse sand and/or gravel mostly on the irregular shallow bottom west of the Y to the New Jersey shore; D) fine to medium sand - a) as zones alternating with zones of coarse sand and trending ENE along coarser material; and b) as a broad bank facies to the east of the Y.

In addition, the sonographs reveal the following smaller scale features: 1) degraded sandwave-like forms, oriented NW to N on the bank east of the shelf valley; 2) elongate patches and bands of fine to medium sand across the coarser sediments; locally two sets of patterns suggest both longitudinal and transverse elements are present; and 3) dump lineations and trails across both sediment zones and patches.

986 - McLaughlin, Dennis B., and J. A. Elder. 1976. A conceptual representation of the New York Bight ecosystem. In: M. G. Gross (editor), Amer. Soc. Limnol. and Oceanogr. Spec. Symp. 2:249-259.

Abstract: Large scientific enterprises need effective techniques for organizing and relating research findings from different fields. Graphical representations (or models) of the ecosystem can be particularly useful organizational tools. These, conceptual models help lay the groundwork for detailed mathematical or empirical descriptions of ecological processes; they also demonstrate the scope of the problem being studied. The New York Bight ecosystem model discussed here is based on a descriptive technique developed by H. T. Odum and widely applied by others. The model is constructed from a few basic components which fall into the following categories: energy or mass storage compartments; energy or mass sources; energy or mass flow regulators. These components are pieced together into a comprehensive representation of physical, chemical, and biological processes in the Bight. Selected aspects of the representation are examined and, where possible, mass and energy fluxes are estimated.

987 - McLaughlin, J. J. A., G. S. Kleppel, M. P. Brown, R. J. Ingram, and W. D. Samuels. (In Press). The importance of nutrients to phytoplankton production in New York Harbor. In: G. F. Mayer, A. Calabrose, F. A. Cross, D. C. Malins, D. W. Menzel, J. S. O'Connor, and F. J. Vernberg (editors), Ecological Stress in the New York Bight: Science and Management. Estuarine Research Federation. Columbia, SC.

Abstract: A four-year study was conducted to identify the importance of nutrients in regulating phytoplankton growth and productivity in New York Harbor.

While nitrogen (occurring primarily as ammonia in yearly concentrations of 25-88 μg-at/liter) and phosphorus appear to be present well in excess of concentrations capable of limiting phytoplankton growth, silicon delivery to the estuary and the New York Bight Apex decreases sharply in May and June. The phytoplankton community at this time changes from diatom-dominated to chlorophyte- and dinoflagellate-dominated. Primary productivity reflects this change, initially decreasing between April and May, then increasing toward summer as the chlorophytes become dominant. Data from laboratory experiments suggest that diatom growth during the late spring decline could be restored by the addition of silicon, trace elements, and vitamins.

During peak photosynthesis periods (February-March, May-July) productivity seems to be regulated by light intensity, but during the fall, when productivity is low, unidentified constituents of the water appear to inhibit photosynthesis.

Water quality management strategies for the harbor must address all aspects of the system (planktonic, benthic, pelagic regimes) and consider the influence of the Bight Apex and Long Island Sound as sources of biomass.

988 - Mearns, A. J., and M. J. Sherwood. 1974. Environmental aspects of fin erosion and tumors in Southern California dover sole. Trans. Amer. Fish. Soc. 103(4):799-810.

989 - Mearns, A. J., and D. R. Young. 1977. Chromium in the Southern California marine environment. In: C. G. Giam (editor), Pollutant Effects on Marine Organisms. Lexington Books. D.C. Heath and Co. Lexington, MA. 213p.

990 - Mearns, A. J., and M. J. Sherwood. 1977. Changes in the prevalence of fin erosion off Los Angeles and Orange Counties. Annual Report, Southern California Coastal Water Research Project. El Segundo, CA. pp. 143-146.

991 - Mearns, Alan. J. 1981. Ecological effects of ocean sewage outfalls: Observations and lessons. Oceanus. 24(1):44-54.

Abstract: Overview and summary of findings from the Southern California Coastal Water Research Project (SCCWRP), universities, state and federal agencies from the 1960's. Discussion of waste loading, outfall structure and function, effects and noneffects, far field effects rates of impact and recovery, and research needs.

992 - Mearns, Alan J., and Jack Q Word. (In Press). Forecasting effects of sewage solids on marine benthic communities. In: G. F. Mayer, A. Calabrese, F. A. Cross, D. C. Malins, D. W. Menzel, J. S. O'Connor and F. J. Vernberg (editors), Ecological Stress in the New York Bight: Science and Management. Estuarine Research Federation. Columbia, SC.

Abstract: A portion of the solids from marine municipal discharges settles to the sea bottom where they cause major but potentially reversible changes in the biomass and trophic structure of macrobenthic communities. Analysis of data from macrobenthic surveys around five southern California coastal discharge sites indicates that there are significant exponential relationships between the mass emission rate of solids and the size of altered macrobenthic communities. Size was expressed both in terms of biomass (metric tons, wet weight) and area (km^2) occupied by communities dominated by surface and subsurface deposit-feeding organisms.

The relationships may be useful in forecasting changes in the size and structure of benthic communities in response to changes in solids emissions and in providing guidance for the design and execution of responsive monitoring surveys.

993 - Mearns, A. J., et al. (In Press). Summary report of the panel on the effects of nutrient and carbon loadings on communities and ecosystems. In: G. F. Mayer, A. Calabrese, F. A. Cross, D. C. Malins, D. W. Menzel, J. S. O'Connor, and F. J. Vernberg (editors), Ecological Stress in the New York Bight: Science and Management. Estuarine Research Federation. Columbia, SC.

Abstract: Waters of the Hudson River Estuary and the New York Apex experience several unpleasant and costly problems which are caused, in part, by excessive inputs of nitrogen and carbon from shore-based sewage discharges. The principal problem is the summertime depression, and localized depletion, of dissolved oxygen in the estuary and in the bottom waters of the Apex. In the estuary, decay of organic matter (carbon) from sewage and runoff, and from phytoplankton blooms stimulated by nitrogen from sewage, reduces dissolved oxygen to levels insufficient for diverse and abundant populations of normal estuarine and coastal organisms. Moreover, much of the nitrogen is not used in the estuary and flushes out into surface waters of the Apex where it is assimilated by coastal phytoplankton in an area covering up to 1,300 sq. km. in the summer. The organic matter produced by this excess growth, coupled with certain oceanographic conditions, may contribute to episodes of oxygen depletion in limited portions of Apex bottom waters. Oxygen-demanding materials from sludge dumping do not appear to contribute greatly to this problem. Moreover, the large-scale anoxic episode of 1976 was not caused by these inputs.

A key to control lies in knowing which factors limit and stimulate phytoplankton blooms and how to control them. A large-scale reduction in nitrogen input, especially in the summer, might help to alleviate the low oxygen episodes that occured in 1976. Reduction in inputs of organic matter, which is underway with advanced treatment, may help, but only in a minor way. Other nutrients and toxic materials may be encouraging growth of some species and inhibiting others, depending on location and time of year, and it is still unclear what changes would result from their control.

994 - Mellor, G. L., and A. F. Blumberg. 1978. Numerical simulation of the New York Bight coastal waters. New Jersey Sea Grant Program. New Jersey Marine Sciences Consortium. 13p.

Abstract: The construction and implementation of a fully three-dimensional numerical model capable of predicting the dynamics and thermodynamics of the New York and Middle Atlantic Bights are presented. Preliminary results indicate that the model is operational and plans for its verification and use are discussed.

995 - Mendia, L. 1979. Criteria for industrial wastewater disposal in the marine environment. Progr. Water Technol. 4:271-284.

996 - Menzies, R. J. 1973. Results of the biological survey in Deep Water Dump Area A in 1972. Appendix F. In: P. E. Wilkniss, (editor). Environmental Condition Report for Deep Water Dump Area A in 1972. U.S. Naval Research Laboratory. Washington, DC. NRL Report 7553. pp. 63-90.

997 - Merentutkmuslaites. 1973. Lausunto kemira Oy: n Uuorikemian tetaan jatevesien mereen johtamiseen liittyvissa kysymyksissa.

998 - Meyers, T. D. 1974. An observation of rapid thermocline formation in the Middle-Atlantic Bight. Estuar. and Coastal Mar. Sci. 2:75-82.

999 - Meyers, T. R. 1978. Prevalence of fish parasitism in Raritan Bay, New Jersey. Proc. Helm. Soc. Wash. 45(1):120-128.

1000 - Michael, Allan D. (In Press). The potential contribution of petroleum hydrocarbons to changes in benthic communities of the New York Bight. In: G. F. Mayer, A. Calabrese, F. A. Cross, D. C. Malins, D. W. Menzel, J. S. O'Connor, and F. J. Vernberg (editors), Ecological Stress in the New York Bight: Science and Management. Estuarine Research Federation. Columbia, SC.

Abstract: Contaminant levels in the lower Hudson River estuary and New York Bight have affected the distribution and abundance of benthic invertebrates. One of the major contaminants, petroleum hydrocarbons, includes persistent and toxic components such as polyaromatic hydrocarbons. Input levels, mostly, from nonpoint sources, are high and difficult to control. Since a large number of contaminants are present in the New York Bight, it is difficult, if not possible, to identify individual contaminants that have caused changes in benthic communities. However, oil spills have occurred where levels of other contaminants were extremely low and where the effects on benthos can be traced directly to hydrocarbons. These effects include reductions in the numbers of species and individuals and increases in the number of opportunistic species. Changes in benthic community patterns observed in parts of the New York Bight and the Hudson River estuary are similar to those caused by oil spills and theoretically could be attributed solely to petroleum hydrocarbons.

1001 - Michael, J. D. 1975. The need for international co-operation in coastal resource quality management. In: E. A. Pearson, and E. de Fraja Frangipane (editors), Marine Pollution and Marine Waste Disposal. (Supplement to Progress in Water Technology.) Proceedings of the 2nd International Congress. San Remo. 17-21 December 1973. Pergamon Press. New York, NY. pp. 27-30.

1002 - Michaelis, H. 1973. Untersuchungen über das Makrobenthos der Wesermundung. Forschungsstelle Norderney, Jahresbericht 1972. 24:103-170.

1003 - Miller, H. C. 1973. Ocean dumping--prelude and fugue. Jour. of Maritime Law and Commerce. 5(1):51-75.

1004 - Miller, J. N., K. E. Chave, M. H. Allen, A. L. Char, E. H. Chave, R. Moberly, and J. F. Walters. 1980. Physical, chemical and biological characteristics of deep water dredge spoil disposal. Third International Symposium on Dredging Technology. Paper G2. pp. 389-402. (Also pub in Inter. Dredg. and Port Constr. Series II. 7(1980)12. pp. 14-20.

1005 - Miller, S. S. 1977. Ocean incineration anew. Environ. Sci. Technol. 11(3):236-237.

1006 - Milstein, C. B., E. V. Garlo, and A. E. Jahn. 1977. A major kill of marine organisms in the Middle Atlantic Bight during Summer 1976. Ichthyological Associates, Inc. 56p.

Abstract: During summer 1976, widespread and persistent anoxic conditions in bottom waters of the Middle Atlantic Bight caused mortalities of marine organisms. Although the cause of anoxia is uncertain, it has been ascribed to a combination of factors including a major bloom of Ceratium, unusual physicl conditions, and the contributions of nutrients and organic material from a variety of sources in the Bight. Information presented to date indicates that anamolous meteorological events in early 1976 led to the early establishment of a vertically stable water column which in turn may have enhanced the development and maintenance of the anoxia.

1007 - Mironov, O. G. 1965. The role of marine birds in the transmission of radioactive substances from sea to land. Okeanologiia. 5(4):715-717. (In Russian).

1008 - Mitchell, F. K. 1974a. Properties of ocean sludge. In: S.C.C.W.R.P. Annual Report, 1974. Southern California Coastal Water Research Project. El Segundo, CA. pp. 159-162.

1009 - Mitchell, F. K. 1974b. Evaluating the impact of sludge discharge to Santa Monica Bay, California. In: A. Freiberger (editor), Pretreatment and Ultimate Disposal of Wastewater Solids. Proceedings of a Research Symposium. Cook College, Rutgers University. New York. May 21 & 22. U.S. Environmental Protection Agency. Region III. Philadelphia, PA. EPA/902/9 74 002. pp. 417-433.

1010 - Mitchell, F. K., and H. A. Schafer. 1975. Effects of ocean sludge disposal. In: Southern California Coastal Water Research Project. Coastal Water Research Project: Annual Report for year ended 30 June, 1975. Southern California Coastal Water Research Project. El Segundo, CA. pp. 153-162.

1011 - Mitchell, N. T., and J. G. Shepherd. 1981. The UK disposal of solid radioactive waste into the Atlantic Ocean and its environmental impact. Proc. Conf. on Environ Impact Nucl. Power. pp. 119-154.

A review of ocean disposal of radioactive wastes.

1012 - Mitre Corporation. 1979a. Disposal of dredged material within the New York district. Present practices and candidate alternatives. Vol. I. McLean, VA. 215p.

Summary: Volume I contains a discussion of the present dredged material disposal program, its impacts, general conditions in the New York Bight, and an evaluation of the 21 most feasible disposal alternatives identified during the project.

1013 - Mitre Corporation. 1979b. Disposal of dredged material within the New York district. Present practices and candidate alternatives. Vol. II. McLean, VA. 92p.

Summary: Preliminary evaluation of the feasibility of the use of upland disposal sites for dumping of dredged material.

1014 - Mitre Corporation. 1979c. New York dredged material disposal alternatives workshop bibliography (1899-1977). Mitre Corp. McLean, VA. 43p.

1015 - Mizenko, D., and J. L. Chamberlin. 1977. Anticyclonic Gulf Stream eddies and water circulation at Deepwater Dumpsite 106 during 1976. (Unpublished manuscript).

1016 - Mizenko, D., and J. L. Chamberlin. 1979. Anticyclonic Gulf Stream eddies off the northeastern United States during 1976. In: J. R. Goulet, Jr., and E. D. Haynes (editors), Ocean variability in the U.S. fishery conservation zone, 1976. NOAA TR NMFS Circ. 427. Washington, DC., pp. 259-280.

1017 - Mizenko, David, and J. Lockwood Chamberlin. 1981. Gulf stream anticyclonic eddies and shelf water at 106-Mile site during 1977. In: Assessment Report on the Effects of Waste Dumping in 106-Mile Ocean Waste Disposal Site. NOAA/ODP Dumpsite Evaluation Report 81-1. Rockville, MD. pp. 207-232.

Abstract: This report summarizes for 1977 the movements of Gulf Stream anticyclonic eddies in the Slope water region off the New England and Middle Atlantic coasts and through 106-Mile Site. Bisagni (1976) prepared a similar report for 1974-75 and Mizendo and Chamberlin (1979) for 1976. Bisagni (1976) also includes informaton on the formation, structure, and dynamics of anticyclonic eddies.

1018 - Moller, H. 1978. Geographic and seasonal variation in fish pests in North Sea and Baltic. ICES. C.M.78/E:9. 14p.

1019 - Moller, Heino. 1981. Field guide to the diagnosis of fish disease and important fish parasites in North Sea and Baltic Sea. Berichte aus dem Institut für Meereskunde an der Christian-Albrichts-Universität. Kiel. Nr. 86. 65p. ISSN 0341-8561.

1020 - Moller, Heino. 1981. Untersuchungen über den gesundeitszustand der unterweser-fische. 56p. GKSS, Forschungszentrum Geesthacht GMBH.

1021 - Mommaerts, J. P., W. Baeyens, and G. Decadt. 1980. Synthesis of research on nutrients in the Southern Bight of the North Sea. ICES. C.M. 80/E:60.

1022 - Moore, D. C., and R. Johnston. 1979. Report on the St. Abb's and Bell Rock Sewage Sludge Dumping Grounds (1978). Dept. of Agriculture and Fisheries for Scotland. Aberdeen Marine Laboratory. Aberdeen, Scotland. 18p+ Appendices.

1023 - Moore, D. C. 1980. A baseline study of the physico-chemical and benthic charcteristics of two sewage sludge dumping grounds outside the Firth of Forth, Scotland. ICES. C.M. 79/E:67.

1024 - Moore, D. C., and R. Johnston. 1980. Report to Lothian Regional Council on the St. Abb's Head and Bell Rock sewage sludge dumping grounds (1979). Dept. of Agriculture and Fisheries for Scotland. Aberdeen Marine Laboratory. Aberdeen, Scotland. 20p. + Appendices.

1025 - Morgan, C. W., J. M. Bishop, and F. F. Mulher. 1976. Oceanography of the New York Bight, August 1974. U.S. Coast Guard Oceanographic Unit. Oceanogaphic Report No. CG 373-71. 89p.

Abstract: The physical oceanography of the shelf and slope of the New York Bight (Block Island to Cape May) in August of 1974 is described. Temperature, salinity, and density data, presented in surface contours and section profiles, showed the shelf/slope front, a cold core on the shelf, and a salinity core on the slope. Geostrophic currents in the slope water were inferred from the density structure, and showed two anticyclonic eddies with maximum geostrophic clocities of approximately 40 cm s^{-1}. Temperature and salinity profiles indicated shelf/slope mixing related to the eddies.

1026 - Morgan, Charles W., and Joseph M. Bishop. 1977. An example of Gulf Stream eddy-induced water exchange in the Mid-Atlantic Bight. Jour. of Physical Oceanography. 7:472-479.

Abstract: Temperature and salinity data collected by USCGC EVERGREEN in August 1974 provide an interesting example of eddy-induced horizontal water exchange across the continental slope of the Mid-Atlantic Bight. A tongue of shelf water (MSW), defined by a salinity of 34‰, extending seaward of the 100-fathom curve behind the eddy contained a measured volume of shelf water. The volume transport of MSW through this tongue was estimated. Assuming a typical eddy residence time of 3 months in the Mid-Atlantic Bight and an eddy frequency of 3 years^{-1}, the annual offshore transport of MSW across the 100-fathom curve due to eddies is also estimated. This line of reasoning suggests that eddies may be responsible for an important portion of the mixing of shelf water with the slope water region in the Mid-Atlantic Bight.

1027 - Morooka, Hitoshi. 1961. Determination of sea-water pollution. III. A standard of pollution caused by city waste. Bull. Jap. Soc. Fish. 27(3):220-224. Also: Sport Fish Abstr. No. 4522.

1028 - Morse, John W., and John deKanel. 1976. Determination of phosphate fluxes across the sediment-water interface of the New York Bight. Final Research Report to NOAA, MESA NY Bight Project Office. Dept. of Ocn. Florida State University. Tallahasee, FL. 10p.

Abstract: Only a small portion of the total dissolved phosphate can be acounted for as distinct species in the interstitial waters of the site which was investigated. Although there is no direct evidence it appears probable that the dominant mode of exchange of dissolved species across the sediment water interface is bioturbation. Small-scale lateral heterogeneities exist which indicate that much more intensive sampling is necessary to understand the distribution of phosphate in New York Bight sediments.

1029 - Mueller, J. A., J. S. Jeris, A. R. Anderson, and C. Hughes. 1976a. Contaminant inputs to the New York Bight. NOAA TM ERL MESA-6. Boulder, CO. 347p.

Abstract: This report presents an estimate of the location and magnitude of contaminant inputs into the New York Bight, indicates their relative importance, and identifies data gaps.

Four sources of contaminant inputs were evaluated in the study: barge dumps and atmospheric fallout as direct Bight inputs, and wastewater and runoff as coastal inputs to waters ultimately draining to the Bight. The sources were further subdivided into their various constituent: dredge spoils, sewage sludge, acid wastes, chemical wastes and rubble for the barge dumps; gaged stream flow, urban runoff, and groundwater outflow for the runoff, and municipal and industrial wastewater inputs. The wastewater inputs were evaluated only downstream of the gaged stream stations because all inputs above these points are reflected in the gaged runoff values. In addition to flow or volume for each source, 23 separate contaminants were investigated to estimate the inputs of solids, organic matter, nutrients, heavy metals, and microbes.

A large quantity of recent (1970 to 1974) data was obtained from numerous agencies, including Federal, State, county, municipal, and academic. These data were rigorously analyzed on a flow-weighted basis to obtain the average mass loads into the New York Bight for each contaminant. A summary of the contribution by source and location is presented as well as an estimate of the annual or seasonal variability for selected sources. To evaluate the significance of the mass loads generated in the study, they were compared to an estimate of the background mass loads entering the Bight as a result of a net current across the ocean boundaries.

Because of their usefulness in water quality modeling, the raw dredge spoil, wastewater discharge, and gaged runoff data for each source are included as appendixes in this report.

1030 - Mueller, James A., Andrew R. Anderson, and John S. Jeris. 1976b. Contaminants entering the New York Bight: Sources, mass loads, significance. In: M. G. Gross (editor), Amer. Soc. Limnol. and Oceanogr. Spec. Symp. 2:162-170.

Abstract: Major contaminant inputs to the New York Bight originate from the New York metropolitan area and the Hudson River drainage basin, principally from wastewater, runoff, and barged discharges.

Major sources of the microbial load are unchlorinated municipal wastewater discharges and urban runoff. Sewage sludge barge dumps constitute an insignificant microbial load on Bight waters.

The seasonal variability of municipal wastewater and gauged runoff and annual variability of barge discharges range from a maximum of 1.5-2.0 to a minimum of 0.5-0.7 times the average mass loads. Between 1960 and 1975 the sludge volume as well as fraction digested shows a definite increasing trend. Few data are available on atmospheric and urban runoff inputs.

The mass loads reaching the Bight are highly related. Poor quality of the dredged materials is caused by contaminants settling from poorly treated wastewater and urban runoff. Increasing levels of wastewater treatment produce more municipal and industrial sludge for disposal. For conservative substances, such as heavy metals, various control measures may redistribute the load among the sources but cause no net decrease in the total.

1031 - Mueller, J. A., and A. R. Anderson. 1977. Contaminant inputs to New York Bight phase II. Urban runoff. Manhattan College. Environmental Engineering and Science Program Report. 105p.

Abstract: An indirect method of estimating urban stormwater runoff and overflow loads and concentrations from combined sewer systems is presented in this report. A mathematical framework consisting of mass and flow balances for the sewer system and its attendant treatment plant is used with treatment plant sampling data and rainfall data as input. The mass balance model is applied to the drainage area of the 26th Ward Water Pollution Control Plant in Brooklyn, N.Y., which drains to Jamaica Bay. Data from this plant on suspended solids, volatile suspended solids, and total and soluble BOD_s were available for a thirty-year period (1945-1975) and seven of these years (1951, 54, 57, 60, 66, 69) were analyzed in depth. Mean annual values of load and concentration for these constituents are compared with others' directly sampled runoff and overflow data. Sensitivity of model results to variations in several input parameters were checked.

For areas having mainly combined sewer systems, the approach is suggested as an economical and accurate alternative to sampling programs or presently available deterministic runoff quality mathematic models for areawide assessment over a large time scale.

1032 - Mueller, J. A., and Andrew R. Anderson. 1978. Industrial wastes. MESA New York Bight Atlas. Monograph 30. New York Sea Grant Institute. Albany, NY. 39p.

Abstract: The offshore water in the bend of the Atlantic coastline from Long Island on 1 side to New Jersey on the other is the New York Bight. This 15,000 mi^2 of the Atlantic coastal ocean reaches seaward to the edge of the continental shelf, 80-120 mi offshore. Lying just offshore New York City, it is one of the world's most intensively used coastal areas for recreation, shipping, fishing and shellfishing, and for dumping sewage sludge, construction rubble, and industrial wastes. Its potential is being considered for resources like sand and gravel and oil and gas. The character and importance of industrial wates discharges in the New York Bight region are evaluated. The magnitude of Bight area industrial wastes is considered as compared to that of the nation. Also considered are geographic distribution and input locations; mass loads from industrial waste discharges by source, by geographic location, and in comparison to waste inputs from other sources; industrial treatment methods; and official policies. Much wastewater in the Bight area is now discharged through primary and secondary treatment plants. New federal regulations presenting treatment levels and phasing out some ocean dumping should improve water quality. Industrial discharges contribute significant portions of the contaminant input loads to New York Bight. Future decisions on waste management must consider interrelationships among all contaminant sources.

1033 - Muir, W. C., and G. M. Horwitz. 1978. Ocean disposal in the Mid-Atlantic Bight. IEEE Mar. Technol. Soc. OCEANS' 78. The Ocean Challenge. Washington, DC. 6-8 Sept. 743p.

1034 - Muir, William C. (In Press). History of ocean disposal in the mid-Atlantic Bight. In: I. W. Duedall, B. H. Ketchum, P. K. Park, and D. R. Kester (editors), Wastes in the Ocean, Vol. I: Industrial and Sewage Wastes in the Ocean. John Wiley and Sons, Inc. New York, NY.

Abstract: During the 1960s and 1970s a variety of industrial and municipal wastes were dumped in the ocean on the mid-continental shelf bordering Maryland and Delaware. U.S. Environmental Protection Agency (U.S. EPA) Region III managed four disposal sites in this area between 1972 and 1980.

With the passage of the Marine Protection, Research, and Sanctuaries Act of 1972 and the subsequent regulations, the U.S. EPA's Region III began a systematic review of each dumper for the need to dump, the feasible alternatives to ocean disposal, and the impacts of the dumping on the marine environment. As a result of the Region's efforts and the eventual assistance of the dumpers themselves, all four disposal sites have been successfully phased out. Further, in all instances environmentally acceptable land-based alternatives were developed. In several cases these alternatives were less costly than ocean disposal, once implemented.

1035 - Mukherji, Prithviraj, and Dana R. Kester. (In Press). Acid-iron disposal experiments in summer and winter at deep water dumpsite-106. In: I. W. Duedall, B. H. Ketchum, P. K. Park, and D. R. Kester (editors), Wastes in the Ocean. Vol. I: Industrial and Sewage Wastes in the Ocean. John Wiley and Sons, Inc. New York, NY.

Abstract: The concentrations and distributions of total and particulate iron, copper, cadmium, and lead associated with acid-iron wastes were measured at Deep Water Dumpsite 106 Mile site (DWD-106) during July 1977 and February 1978 to study the fate of these wastes in the ocean after dumping. Concentrations were measured by obtaining discrete samples which were analyzed by atomic absorption spectrophotometry.

In July the waste material was confined to the upper 20 m of the water column and within the mixed layer for a period of up to 27 h. The waste mixed to greater depths during February when the seasonal thermocline was absent and the permanent pycnocline was below 100 m. Total iron was used as a tracer to follow the waste plume. The results were related to two models for plume dispersion by diffusion processes during the initial 5 h after the dump. A model based on a constant diffusion velocity (as opposed to a constant eddy diffusivity) fit the data best, yielding a diffusion velocity of 1.1. cm^{-1}.

1036 - Muller, I. W. 1953. Die einleitung von Abwassern ins Meer. [Discharge of waste waters into the sea.] Gesundheits Inq. 74:286-288.

1037 - Murchelano, Robert, and John Ziskowski. 1974. Fin erosion in winter flounder. Mar. Pollut. Bull. 6(2):26-29.

Abstract: Diseased fish showing erosion of the fins have been recorded from polluted waters in several parts of the world. This article gives the results of a year's survey of fin erosion among fish in oceanic and estuarine waters of New York Bight. The pristine Great Bay was chosen as a control for the heavily contaminated Sandy Hook-Raritan Bay system. Almost all fin erosion was confined to demersal flat fish, particularly the winter flounder, a species which spends a long time in polluted estuaries near Sandy Hook.

1038 - Murchelano, R. A. 1975. The histopathology of fin rot disease in winter founder from the New York Bight. Jour. Wildlife Diseases. 11:263-268.

1039 - Murchelano, R. A., and J. Ziskowski. 1976. Fin rot disease studies in the New York Bight. In: M.G. Gross (editor), Amer. Soc. Limnol. and Oceanogr. Spec. Symp. 2:329-336.

Abstract: Surveys from February 1974 through June 1975 showed the prevalence of fin rot disease in winter flounder Pseudopleuronectes americanus from the New York Bight Apex was 3.9%, compared to 0.7% outside the Apex. Prevalence of the disease in winter flounder from Apex areas of low carbon deposits was 2.9%; whereas, in Apex areas of high carbon deposits, disease prevalence was 5.1%. The prevalence of fin rot disease in summer flounder Paralichthys dentatus from the Apex was 0.6% and in summer flounder from Sandy Hook-Raritan Bay was 3.4%. No summer flounder with fin rot were noted in Great Bay, a control area outside the Apex. Although the numbers of diseased fish are small, incidence of fin rot disease in summer flounder from Sandy Hook-Raritan Bay increased monthly from June-November 1974. Attempts to induce fin rot disease in winter flounder in cages submerged in the sewage sludge area of the Bight Apex produced active fin lesions on the candal fins more often than on the dorsal and anal fins. These flounder were in substantially worse condition than fish in cages at the control site.

1040 - Murchelano, R. A., and J. Ziskowski. 1977. Histopthology of an acute fin lesion in the summer flounder (Paralichthys dentatus) and some specultions on the etiology of fin rot disease in the New York Bight. Jour. of Wildlife Diseases. 13:103-106.

Abstract: The histopathology of acute fin rot disease in summer flounder, Paralichthys dentatus, from the New York Bight is described. Grossly, caudal and dorsal fin lesions appeared ragged or frayed with no evidence of resolution. Microscopically, there was epidermal and dermal necrosis, congression, edema, focal and diffuse hemorrhage, and Zenkers necrosis of underlying muscle. Gram-negative bacteria were presented in the fin tissues as well as in heart muscle and liver parenchyma. The inflammatory response consisted mostly of macrophages. The significance of the acute disease in summer flounder is related to the etiology of fin rot disease in winter flounder from the Bight.

1041 - Murchelano, R., and J. Ziskowski. 1980a. Some observatins on an ulcer disease of red hake, Urophycis chuss, from the New York Bight. ICES. C.M. 80/E:23.

1042 - Murchelano, R., and T. Azarowitz. 1980b. Fish disease surveys in the western North Atlantic. ICES. C.M. 80/E:24.

1043 - Murchelano, R. A., and J. Ziskowski. 1980c. Fin rot disease - a sentinel of environmental stress. ICES. C.M. 80/E:25.

1044 - Murchelano, Robert A. (In Press). Some pollution-associated diseases and abnormalities of marine fish shellfish: a perspective for the New York Bight. In: G. F. Mayer, A. Calabrese, F. A. Cross, D. C. Malins, D. W. Menzel, J. S. O'Connor, and F. J. Vernberg (editors), Ecological Stress in the New York Bight: Science and Management. Estuarine Research Federation. Columbia, SC.

Abstract: In recent years, there has been a tendency to associate environmental degradation with diseases of marine fishes and shellfishes. A brief review is provided of pollution-associated diseases caused by viruses and bacteria and of a number of pollution-associated diseases whose etiology is uncertain. Microbial diseases discussed include lymphocystis, herpeslike virus, Baculorvirus, vibriosis, and shell disease: diseases of uncerain etiology include fin rot, skeletal anomalities, and "black gill." The prevalence of these diseases varies geographically, and in any one area usually has not been statistically compared to an acceptable reference area. The prognosis for the animal also varies; some diseases are fatal and others have little effect on their hosts. The only pollution-associated diseases that are numerically prevalent in the New York Bight are shell disease, fin rot, and "black gill" disease.

1045 - Murchelano, Robert A., and John Ziskowski. Fin rot disease in the New York Bight (1973-1977). (In Press). In: G. F. Mayer, A. Calabrese, F. A. Cross, D. C. Malins, D. W. Menzel, J. S. O'Connor, and F. J. Vernberg (editors), Ecological Stress in the New York Bight: Science and Management. Estuarine Research Federation. Columbia, SC.

Abstract: The results of research on the prevalence and etiology of fin rot disease of New York Bight winter flounder, Pseudopleuronectes americanus, are presented. Except for 1975 and 1976, significant differences ($P<0.05$) in disease prevalence were found between the Apex and Sandy Hook/Raritan Bays, the Apex and control area. The Apex had the highest prevalence of the disease (7.4%) for the five-year period from 1973 to 1977. It was not possible to determine whether the etiology of the fin rot is infectious or noninfectious. Bacteria do not appear to be involved in the pathogenesis of the disease, and the concentrations of trace metals and polychlorinated biphenyls (PCB's) in tissues were not elevated consistently. Based on bacteriologic, histopathologic, and immunologic studies, the disease probably is not fatal; however, behavioral modification may result in some mortalities. Negligible mortality and the absence of an extensive winter flounder fishery in the New York Bight both make it unlikely that there is a major effect on the population dynamics of the species. The disease is unlikely to reduce the marketability of fish since muscle tissue is not affected and since the fish are not sold in the round. The most important aspect of the presence of fin rot disease in Bight winter flounder rests in its value as an indicator of environmental stress.

1046 - Murphy, Lynda S., and Peter R. Hoar. 1981. Effect of pollutants on marine phytoplankton at 106-Mile site. In: Assessment Report on the Effects of Waste Dumping in 106-Mile Ocean Waste Disposal Site. NOAA/ODP Dumpsite Evaluation Report 81-1. Rockville, MD. pp. 309-319.

1047 - Murphy, Lynda S., Peter R. Hoar, and Rebecca A. Belastock. 1981. The effects of industrial wastes on marine phytoplankton. In: B. H. Ketchum, D. R. Kester, and P. K. Park (editors), Ocean Dumping of Industrial Wastes. Plenum Press. New York, NY. pp. 399-410.

Abstract: Nineteen clones in three marine phytoplankton species were used in bioassays of DuPont Grasselli waste. We determined that low doses of waste stimulate growth; higher concentrations were inhibitory. The transition from stimulation to inhibition in culture occurs at approximately the highest concentration that is sustained in the environment. Clones from polluted estuaries are less sensitive to the waste than are clones of the same species from other environments.

1048 - Murphy, Lynda S., Robert R. L. Guillard, and Jerome Gavis. (In Press a). Evolution of resistant phytoplankton strains through exposure to marine pollutants. In: G. F. Mayer, A. Calabrese, F. A. Cross, D. C. Malins, D. W. Menzel, J. S. O'Connor, and F. J. Vernberg (editors), Ecological Stress in the New York Bight: Science and Management. Estuarine Research Federation. Columbia, SC.

Abstract: We evaluated the responses of 17 clones of marine diatoms, isolated from a range of environments, to two industrial wastes now disposed of at deep-ocean dumpsites. The clones showed a wide range of sensitivity to the stress, and there were no correlations with latitude of isolation, distance from shore, temperature regime, salinity, or general environmental stability. Three neritic clones from heavily polluted waters were much less sensitive than were other neritic oceanic clones. Further, we surveyed cupric iron sensitivity in 15 clones of phytoplankton isolated from different neritic environments. All extremely resistant clones came from regions known or presumed to be heavily polluted. These data suggest that some phytoplankters can develop strains resistant to a marine pollutant.

1049 - Murphy, Lynda S., Elin M. Haugen, and Jeffrey F. Brown. (In Press b). Phytoplankton: Comparison of laboratory bioassay and field measurements. In: I. W. Duedall, B. H. Ketchum, P. K. Park, and D. R. Kester (editors), Wastes in the Ocean, Vol. I. Industrial and Sewage Wastes in the Ocean. John Wiley and Sons, Inc. New York, NY.

Abstract: A laboratory bioassay was developed to predict the effects on the phytoplankton community of the several wastes disposed of at deep-ocean dumpsites. The bioassay measured change in chlorophyll fluorescence and cell number of specified clones in a controlled environment. For all phytoplankton clones tested, the pharmaceutical and the American Cyanamid wastes were more toxic than the Dupont-Edge Moor waste, which was more toxic than the Dupont-Grasselli waste. Centric diatoms and dinoflagellates were more sensitive to the wastes than were pennate diatoms and some representatives of the monad classes, but some degree of resistance was shown in isolates of other classes established from polluted estuaries.

These bioassays predict that changes in community structure should occur at the concentration existing in the wake of the barge during dumps. Initial studies at the Puerto Rico dumpsite showed an immediate, short-term pulse effect on the community structure, with dinoflagellates decreasing and monads increasing.

1050 - Murray, Lindsay A., and M. G. Norton. 1979. The composition of dredged spoils dumped at sea from England and Wales. U.K. Min. of Agr., Fisheries and Food (MAFF), Lowestoft. Fish. Res. Tech. Rept. No. 52. 10p.

Conclusions: This survey of the composition of dredged spoils dumped at sea around England and Wales shows that wide variations of the concentrations of metals and 'persistent' oil occur between different sites. Some spoils from industrialised estuaries which are heavily used by shipping contain concentrations of metals up to 500 times greater than those in sediments from the least contaminated estuaries or ports.

The presence of the metal and 'persistent' oil contaminants may lead to environmental effects over and above those associated with the physical impact of spoil dumping, through effects on the water column and on biota in contact with the sediment. Although laboratory and field studies have suggested that effects on water quality and metal uptake by biota are likely to be only of minor significance, the contamination of spoils may also reduce the likelihood of recolonisation of the dumped sediment by the benthos. The association in some areas of deposition with certain diseases of fish and shellfish is less certain. The nature of the receiving area will, however, determine the extent to which these effects occur. Field studies in specific dumping areas are thus required to establish in greater detail whether the chemical contamination of dredged spoils leads to an adverse environmental impact.

1051 - Murray, L. A., M. G. Norton, R. S. Nunny, and M. S. Rolfe. 1980a. The field assessment of effects of dumping wastes at sea: 6. The disposal of sewage sludge and industrial waste off the River Humber. U.K. Min. of Agri, Fisheries and Food (MAFF), Lowestoft Fisheries Tech. Rept. No. 55. 35p.

Conclusions: The dumping of approximately 200,000 t a^{-1} of sewage sludge and industrial waste represents only one of a number of inputs of material to the area around the Spurn Head dumping site. The major input of organic substances and trace metals to the area is the discharge of the River Humber and it is estimated that, depending on the substance, dumping contributes between 1 and 24% of inputs from all sources (the Humber, coastal discharge, and atmospheric deposition). Copper and mercury are the only two metals for which dumping is estimated to contribute in excess of 10-20% of total inputs.

The dumping area is subject to strong tidal flows and is exposed to very active wave conditions during storms which may at times generate near-bed flows up to three times those of tidal flows. Residual water movements determined by sea-bed drifter releases appear to be dominantly to the south or south-west. Thus the area is one where rapid dispersion of dumped waste is expected with little probability of local long-term accumulations of dumped material.

This picture is confirmed by the sediments of the area. In the northern part of the survey area, which includes the dumping ground, the substrate is a stable coarse gravel deposit. Further south, gravels and mobile sands occupy areas of the sea floor with some mud deposits from erosion of local clay outcrops on the sea bed. Small amounts of fine material are present in most gravel areas as interstitial deposits, but no evidence could be found of the long-term accumulation of fine sediments.

The organic and trace metal analyses of sediments indicate that sewage-derived particulates occur on the sea bed within the interstices of the coarse substrates found on the dumping ground and inshore of it. From the distribution of some metals, it was difficult to isolate the influence of dumping from that of the Humber Estuary. However, the enrichment of sediments by organic carbon and mercury appeared to be due to dumping activity. Outside the zone of local enrichment, transport pathways could not be identified and there was no evidence of areas where organic and metal-rich material had accumulated. Sludge-derived material thus becomes mixed with suspended particulates from the Humber Estuary and from erosion of the sea bed, all of which become very widely dispersed.

Due to the sampling difficulties of the gravelly substrate, it has not been possible to describe the macrobenthos of the dumping area in other than qualitative terms. Although no gross effects could be attributed to dumping, it is not known whether the changes in sediment composition attributable to dumping have had any minor effects on the benthos. However, one faunal association dominated by the horse mussel, *Modiolus modiolus*, is found widely distributed over the survey area, including the dumping ground.

1052 - Murray, Lindsay A., M. G. Norton, R. S. Nunny, and M. S. Rolfe. 1980b. The field assessment of effects of dumping wastes at sea: 7. Sewage sludge and industrial waste disposal in the Bristol Channel. U.K. Min. of Agr., Fisheries an Food (MAFF), Lowestoft. Fisheries Res. Tech. Rept. No. 59. 40p.

Conclusions: The dumping of approximately 400,000 t a^{-1} of sewage sludges (wet weight) and small quantities of industrial waste into the central Bristol Channel represents only one of a number of inputs of material to that area. Large amounts of organic substances and some metals are also discharged to the Severn Estuary and Bristol Channel from sewage and industrial outfalls; others enter from rivers, atmospheric deposition, and the dumping of dredged spoil. Estimates of the relative magnitude of these inputs during the period of these surveys (1975-77) have been made and suggest that sewage sludge dumping is mainly of local significance, amounting to up to 51% of the direct inputs of organics and trace metals to the central Channel.

The dumping area and the central Channel are characterised by strong tidal flows and also, during westerly storms, by strong wave-induced near-bed movements. Initial residual water movements from the dumping ground appear to be to the north-east. Under some conditions this movement, as deduced from seabed drifter releases, would appear to be maintained in bottom waters into Swansea Bay and toward Nash Point. Under other conditions, however, landward movement may be slow, with movement upstream suggested, particularly to Bridgwater Bay. Dumped waste would thus be expected to be dispersed along the tidal axis with some residual movement to the north-east. In the longer term, however, particulate matter would become widely dispersed within the central and inner Channel. Limited confirmation of the short-term movements of dumped sludge particles was obtained from the measurement of faecal bacteria in the sediment.

A study of the natural sediment distribution and dynamics showed the dumping area to be in a region of muddy gravels. To the north and east, mobile sands predominated while into Swansea Bay finer sands and silts were found. Sites of deposition of suspended sediment, whether of sewage or other origins, were identified in the interstices of the gravels near the dumping area, in Swansea Bay, and further upstream in Bridgwater Bay. The areas of accumulation locally were considered to be of limited capacity and subject to reworking during storms.

Analysis of the organic carbon and trace metal concentrations in the sediment showed that elevated concentrations of these substances occurred in several zones. One, arising from dumping of sewage sludge, was located along an east-west transect through the dumping area. The orientation of this main area of enrichment was slightly different from that expected from the north-easterly residual bottom currents measured at the dumping ground. Although this may indicate that the observations of currents at the dumping ground may not accurately reflect those in the general area, the presence of mobile sands to the north-east would make movement in this direction difficult to detect. Other zones of enrichment were identified as arising from the dumping of dredged spoil in outer Swansea Bay and from industrial discharges to the bay.

Because of the difficulty of sampling the gravelly substrate quantitatively for benthos, it has not been possible to describe the macrobenthos in other than qualitative terms. No gross effects of sewage sludge dumping have been identified; the fauna was relatively sparse and consistent with the substrate type. In the immediate vicinity of the dumping area, however, the increased organic content of the sediments may have resulted in slight changes in species composition. Dredged spoil dumping on the other hand does appear to have depleted the bottom fauna locally in the disposal area in Swansea Bay.

1053 - Murray, Talbot, E. 1981. Summary of waste inputs to 106-Mile site during 1976 and 1977. In: Assessment Report on the Effects of Waste Dumping in 106-Mile Ocean Waste Disposal Site. NOAA/ODP Dumpsite Evaluation Report 81-1. Rockville, MD. pp. 33-37.

1054 - Musick, J. A., C. A. Wenner, and G. R. Sedberry. 1975. Archibenthic and abyssobenthic fishes of Deepwater Dumpsite 106 and adjacent area. In: May 1974 Baseline Investigations of Deepwater Dumpsite 106. NOAA/ODP Dumpsite Evaluation Report 75-1. Rockville, MD. pp. 229-269.

1055 - Musser, T. W. 1974. The ocean disposal permit program and spill prevention. Proceedings of the 1974 National Conference on Control of Hazardous Material Spill sponsored by Amer. Inst. Chem. Eng. & U.S. EPA (Ind. Waste Treatment Research Lab).

1056 - Mytelka, A. I., et al. (In Press). Summary report of the panel on management of industrial wastes. In: G. F. Mayer, A. Calabrese, F. A. Cross, D. C. Malins, D. W. Menzel, J. S. O'Connor and F. J. Vernberg (editors), Ecological Stress in the New York Bight: Science and Management. Estuarine Research Federation. Columbia, SC.

Abstract: A management system for the New York Bight must be capable of achieving and protecting water quality by preventing pollutants which adversely affect marine life and the ecosystem from entering Bight waters. Although the continental shelf beyond the New York and New Jersey coasts has not yet been damaged severely by waste disposal, the Bight Apex is degraded. Improved environmental conditions will permit industrial and recreational uses of the Bight in areas closer to population centers than is now possible. Interdisciplinary research is needed to understand and control the effects of various pollutants on the Bight. Resulting information must be generally available. National effluent limitations and industrial pretreatment regulations must be developed and enforced. Public and private treatment facilities also should be required to monitor their discharges for designated toxic materials. New and innovative approaches to attain industrial pretreatment and direct discharge standards also must be encouraged.

1057 - Nadeau, R., and Robert A. Davis. 1976. Polychlorinated biphenyl (PCBs) in the Hudson River. Bull. Environ. Contam. Toxicol. 16(4):436-444.

Abstract: Polychlorinated biphenyls (PCB's), as recently as 1966, have been recognized as an environmental contaminant. PCB's are primarily used in industry and are even more persistent than DDT. These compounds are essentially nonalterable by microbial or physical-chemical activities and are incorporable into living protoplasm. The bonding between the chlorine atom and the biphenyl structure is resistant to shearing by natural degradation processes; therefore PCB's are not easily metabolized by enzyme systems presently found in nature.

The atomic structure of PCB's gives them specific chemical characteristics very desirable for certain industrial uses, i.e. dielectric fluids in capacitors and transformers.

PCB's not only are incorporable into living biomass in natural ecosystems, but are transferable within food webs, the end result being a much higher concentration of these compounds occurring in specific tissues of sumit carnivores (biomagnification).

It has been estimated that 4 to 5 x 10^3 tons/year of PCB's are lost into the nation's fresh and coastal waters.

1058 - Naeve, E. 1974. Sources of pollution in the Mediterranean and its effects on living resources and fishing. Rev. Int. Oceanogr. Med. pp. 35-36.

1059 - Nakayama, O., and M. Ohno. 1981. Bactericidal effect of seawater containing digested night soil and marine plankton against intestinal bacteria. Bull. Jap. Soc. Sci. Fish. 47:165-169.

1060 - Nash, N. 1975. Sludge disposal and the coastal metropolis. In: T. M. Church (editor), Special Symposium on Marine Chemistry in the Coastal Environment. Amer. Chem. Soc. Symp. No. 18:410-415.

Abstract: Ocean disposal is a serious matter which deserves calm and scientific study, a modicum of private and public pressure, and a minimum of emotion and scare tactics. A professional study of its effects on marine and, perhaps, human life and a dispassionte evaluation of the ocean's capacity to accept its share of man's wastes are needed. The land and air, the only other possible repositories, cannot receive all of it; New York City produces about 90 million ft^3 of sludge/yr. that will increase to 190 million by 1981, the year that Region 2 of the Federal EPA hopes will mark the end of ocean disposal. In New York's view, that volume represents a badge of merit, for sludge volume is earned by longterm sewage collection and treatment. One alternate is wet air oxidation of sludge, to be incorporated in a new plant already designed. Another is high-temperature incineration with solid waste, but time is needed to field-test these processes.

1061 - Nassau County (NY) Dept. of Health. 1973. Report of a comprehensive water quality study of South Shore bays - Nassau County, NY. Nassau County Dept. of Health. Mineola, NY. 204p.

Summary: A comprehensive water quality study of the South Shore Bays of Nassau County, conducted by the Nassau County Department of Health, is reported herein. Levels and trends in water quality were evaluated in relation to assigned State standards. Sources of pollution and their individual and collective water quality impacts were also evaluated. Alternate solutions were reviewed in terms of their efficacy in preserving and restoring the quality of the bay waters. Necessary actions for planning a pollution abatement program are recommended to restore water quality to assigned standards. Data from studies conducted by several agencies as well as extensive data collected by the Health Department were used.

1062 - Nassau County (N.Y.) Department of Health. 1976a. Offhore sampling results. Nassau County Dept. of Health. Mineola, NY. 14p.

Distributions of: total and fecal coliform bacteria in the water column and sediment, volatile solids, chemical oxygen demand, total chromium, copper, lead, nickel and zinc for May 1975 and September 1975.

1063 - Nassau County (NY) Department of Health. 1976b. Nassau County water quality assessment report - 1976 report year. Nassau County Dept. of Health. Mineola, NY. 100p.

Content: The report consists of a summary and five sections. Sections 1 through 5 contain the monitoring results of each of the following programs: Surface Water Bacteriological and Chemical Monitoring Program - Sampling Results from October 1975 thru September 1976. (197p), Bathing Water Quality Bacteriological Monitoring Program from April 1976 to September 1976. (136 p), Atlantic Ocean Macrobenthic Study from Results May 1975. (56 p), Impact of Ocean Disposal of Sewage Sludge on Atlantic Ocean Water and Sediment Quality from September 1975 29 p., and Groundwater Chemical Monitoring Program from October 1975 thru September 1976.

1064 - Nassau County (NY) Dept. of Health. 1978a. Nassau County water quality assessment report for 1977 report year. Mineola, NY. 80p.

Report Content: The 1977 Report consists of a summary and four sections. Contents are as follows: Surface Water Bacteriological and Chemical Monitoring Program from October 1976 thru September 1977, Bathing Water Quality Bacteriological Monitoring Program from April 1977 thru September 1977, Atlantic Ocean Special Studies: Impact of Ocean Disposal of Sewage Sludge on Atlantic Ocean Water and Sediment Quality - Conclusions from Sampling Results of May 1975 to September 1977, Impact of Treated Effluent Discharged from the Cedar Creek Park Water Pollution Control Plant on Atlantic Ocean Water and Sediment Quality, and Groundwater Chemical Monitoring Program from October 1976 thru September 1977.

1065 - Nassau County (NY) Dept. of Health. 1978b. Ocean sludge monitoring cruise - October, 1978 Preliminary Report. Nassau County Dept. of Health. Mineola, NY. 11p.

Abstract: Water column samples (49), collected in six-liter Van Dorn bottles, were obtained from one meter above the bottom and one meter below the surface. These samples were subjected to routine bacteriological analysis (total and fecal coliform) and chemical analysis (salinity and dissolved oxygen). In addition, water columm samples will be examined microscopically for plankton and known sewage artifacts.

Sediments samples (49), collected with a Shipek Grab, were divided for bacteriological, chemical, and bio-geological analysis.

1066 - Nassau County (NY) Environmental Management Council. 1974. Nassau County environmental plan report Aug. 1974. Mineola, NY. [Report includes consideration of Marine Environment conditions and activities].

1067 - National Institute for Water Research, C.S.I.R. and Natal Town and Regional Planning Commission, South Africa. 1969. The disposal of effluents into the sea off the Natal coast. Natal Town and Regional Planning Commission. Vol. 14. 139p.

1068 - National Marine Fisheries Service. 1972a. The effects of waste disposal in the New York Bight. Summary final report. U.S. Dept. Commerce. NOAA Informal Rep. No. 2. Highlands, NJ. 70p.

1069 - National Marine Fisheries Service. 1972b. The effects of waste disposal in the New York Bight. Section 3: Zooplankton studies. Highlands, NJ. 54p.

1070 - National Research Council. 1978. Multimedium management of municipal sludge. National Academy of Sciences. Washington, DC. 187p.

1071 - Nauke, M. 1974a. Red mud tilting in the North Sea. Sonderdrück aus Meerestechnik Band 5 Nr. 5. pp. 149-153.

Abstract: The construction and intended extension of aluminium-producing installations at the German North Sea Coast caused problems of the removal of residues which are known as red mud. The United Aluminium Works (Vereinigte Aluminiumwerke AG) of Berlin and Bonn have applied for permission to tilt 800,000 tons of red mud yearly into the North Sea, starting in 1973. The Federal Minister for Traffic concerned has instructed in 1971 several Federal Intitutes to investigate the effects which are to be expected. These investigations include trial tiltings of a total amount of 15,000 tons into the North Sea, besides numerous experiments in the laboratories.

1072 - Nauke, M. 1974. Die Schwermetallgehalte der Sedimente im Klarschlamm - Verklappungsgebiet vor der Elbmundung. [Content of heavy metals in the sediments in the area of sewage sludge disposal off the Elbe Estuary.] Dt. Hydrogr. Z. 27:203-213.

1073 - Neff, J. M. 1979. Polycyclic aromatic hydrocarbons in the aquatic environment: sources, fates and biological effects. Applied Science Publ. London. 262p.

1074 - Nelson, S. B., and R. C. Tipper. 1971. Ocean dumping - a rational approach to an international problem. Engng. in the Ocean Environment Conf., San Diego, Calif., 21-24 Sept 1971. Inst. Electrical and Electronics Engineers (IEEE). New York, NY. pp. 317-320.

1075 - Nelson, S. B. 1972. Basic considerations for managing ocean dumping. Ann. Tech. Mtg. Inst. Environmental Sci. 18th PRoc. Inst. Envir. Sci., held in New York. 1-4 May 1972. pp. 36-53.

1076 - Nelsen, Terry A. 1979. Suspended particulate matter in the New York Bight Apex: Observations from April 1974 through January 1975. NOAA TM ERL MESA-42. Boulder, CO. 78p.

1077 - Nespital, W. 1973. Zur schadlosen beseitigung von dunnsaure im meer. [Harmless marine disposal of residuary acid]. Wasser und Boden. 25(9):277-281.

1078 - Newell, R. C. 1972. The effect of chemical waste on marine organisms. Effluent and Water Treatment Jour. 12(6):307-311.

1079 - New England Aquarium. 1979. Investigating the impact of drilling mud and its major components on bivalve species of Georges Bank. Boston, MA. DOE/EV/04580-T1. 90p.

1080 - New Jersey V City of New York. 1931. (City dumping garbage in ocean as nuisance). 283 US 473 51 Sup Ct. 519-522.

1081 - New Jersey V City of New York. 1933. (Remedy against city for pollution of state waters and beaches). 290 US 54 Sup Ct. 136-137.

1082 - New Jersey V City of New York. 1935. (Offshore sewage sludge disposal). 296 US 259, 56 Sup Ct. 188.

1083 - New Jersey Dept. of Environmental Protection. 1977. Ocean dumping of sludge - program through phase out in 1981. N.J. DEP Staff Report. 40p.

1084 - New York City. 1979. Water quality management plan - Summary 208 areawide waste treatment management planning program. 19p.

Abstract: Plan is the improvement of water quality in NY Harbor through the year 2000. Specific goals are: 1) Attain Federal quality goal of "fishable, swimmable waters, where attainable". 2) Open new beaches and restore quality of present beaches. 3) Restore some shellfishing with improved conditions. 4) Improve water quality in local tributaries. 5) Strengthen fiscal support for water quality management.

1085 - New York (State) Ocean Science Laboratory. 1974. An interdisciplinary study of the estuarine and coastal oceanography of Block Island Sound and adjacent New York coastal waters: Ground truth. New York Ocean Science Laboratory. Montauk, NY. Tech. Report. 0027. 107p.

Abstract: A program was established to study the characteristics of the waters of Block Island Sound and the New York Bight as they relate to ERTS "A" imagery. Temperature, salinity, density, optical characteristics, and tidal currents were analyzed. The annual temperature regime is largely governed by solar radiation and correlates with the mean month temperatures in the atmosphere, lagged 1 mo. The annual salinity regime is regulated by the stream discharge entering Long Island Sound and New York Harbor. The average coefficient for Block Island Sound in the visible spectral band was 0.335, compared to a mean value for the New York Bight of 0.663. Observations on spatial and temporal distribution of particulate phosphate (PO_4^{-3}), reactive PO_4^{-3}, and soluble organic PO_4^{-3}, nitrate-N, silica, and chlorophyll _a_ indicate the waters of Block Island Sound and those along the southern shore of Long Island were deficient in N. With the exception of particulate and soluble organic P, no significant correlations were found between any measured parameter and chlorophyll _a_. In Block Island Sound, correlations of 0.815 were found between particulate P and chlorophyll _a_. No correlation was found for these parameters in the New York Bight. In the latter area, a correlation of 0.88 existed between soluble organic P and chlorophyll _a_ at the entrance to New York Harbor. A high correlation between phytoplankton and suspended particles was found in Block Island Sound (0.858) indicating phytoplankton may contribute largely to the suspended material in this region. The lower correlation between these parameters in the New York Bight (0.586) indicated that other factors were adding to the suspended load in this area. The phytoplankton population was highest at the New York Bight and Block Island Sound stations, a result of organic enrichment caused by the disposal of sewage sludge in the New York Bight.

1086 - Nichels, J. E. Geer, and A. Penna. 1979. New York Bight diver observation program, results of 1978 survey. Tech. Rept. No. 1. American Littoral Society. Highlands, NJ. 32p.

Abstract: The American Littoral Society, in cooperation with the Marine Ecosystem Analysis project of the National Oceanographic and Atmospheric Administration, conducted a Diver Observation Program using sport divers to monitor and record information on water conditions and marine life over rough bottoms and wrecks off the New Jersey Coast from April to November 1978.

Thirty-eight divers visited over 120 different sites from the shore out to 18 miles, covering about 680 square miles. They recorded physical conditions as well as comments and observations on unusual behavior of marine life and other general information.

1087 - Nichols, Maynard M. 1979. The problem of misplaced sediment. In: H. D. Palmer and M. G. Gross (editors), Ocean Dumping and Marine Pollution. Dowden, Hutchinson and Ross, Inc. Stroudsburg, PA. pp. 147-161.

Abstract: Disposal of estuarine dredged material is a major depositional process with important sedimentologic and hydraulic consequences. Large-scale dredging in U. S. East Coast estuaries has cut channels far below natural equilibrium depths and induced rapid sediment accumulation. In turn, the increased frequency of dredging has created a massive disposal problem and a need for ocean dumping.

An estuary channel attains maximum stability by adjusting its bed geometry and hydraulic regime. When a channel is deepened, tidal currents are reduced and sediment deposition accelerates. Deepening causes more salty water to penetrate landward and shifts the zone of maximum shoaling upstream. Landward density currents accelerate the potential for return of dredged material from seaward reaches, while higher stratification enhances the potential for entrapment. Therefore, dredging is partly self-perpetuating.

Long-continued open water disposal along channel shoulders of the Upper James Estuary, Virginia, has built up mounds close to the surface where they are subject to wave erosion and current scour. Mounds of misplaced sediment with slopes greater than 6° allow return of sediment to the channel and they restrict tidal flow. By reducing the cross-sectional area, they restrict tidal flow. By reducing the cross-sectional area, they supress the tide range and tidal discharge in landward reaches. Consequently, dispersion of pollutants is reduced and accumulation of fine-grained sediment is promoted. As a consequence of dredging and disposal, many estuaries are losing their capacity to absorb more misplaced sediment.

1088 - Nielson, S. O. 1974. Nuclear waste disposal in the oceans. Science. 185(4157):1183.

1089 - Niintmaki, J. 1972. Vahingonkorvausesitys Likkihappo Oyin Vuorikernian tehtarden. Jatevesien vaikutusalueen ammattikalastajille.

1090 - Nitkowski, M. F., S. Dudley, and J.T. Graikoski. 1977. Identification and characterization of lipolytic and proteolytic bacteria isolated from marine sediments. Mar. Pollut. Bull. 8(12):276-279.

Abstract: Lipolytic and proteolytic bacteria were isolated from sediments at two sampling stations in the New York Bight Apex and one sampling station each in Sandy Hook Bay and Great Bay, New Jersey. The stations in the Bight Apex and Sandy Hook Bay have received industrial wastes and sewage for several decades, while Great Bay has received little of such materials. Proteolytic counts were 2-4 times higher and lipolytic counts generally 4 times higher in the polluted areas. Of the isolates taken from casein and lipid plates, 76% were Gram-negative rods; 80% of the latter were identified as *Vibrio* and *Pseudomonas*. The vibrios comprised more than 60% of the isolates from Station 4 (Great Bay) and Station 1 (Bight Apex), and were tested for their ability to break down casein, lipid, starch, and chitin. From Station 1, 75% of the *Vibrio* were active in degrading one or more substrates in addition to the substrate of the initial isolation medium; from Station 4, 52% of the *Vibrio* were active.

1091 - Norris, D. P., L. E. Birke, Jr., R. T. Cockburn, and D. S. Parker. 1971. A comprehensive environmental approach to marine waste disposal planning. Paper presented at the 44th Annual Conference of the Water Pollution Control Federation. San Francisco, CA. October, 1971. 26p. (mimeogr.)

1092 - Norris, D. P., L. E. Birke, Jr. , R. T. Cockburn, and D. S. Parker. 1973. Marine waste disposal - a comprehensive environmental approach to planning. Jour. Water Poll. Control Fed. 45:52-70.

1093 - North, W. J. 1958. The effects of waste discharges on kelp. Annual progress report. University of California, Institute of Marine Resources. IMR Ref. 58-11. 43p.

1094 - Norton, M. G. 1976. The operation of the dumping at sea act 1974. Chemistry and Industry. October 2. pp. 829-834.

Abstract: The Ministry of Agriculture, Fisheries and Food (MAFF) has exercised a responsibility for the control of marine pollution from the disposal of wastes from ships by dumping since the early 1960s, when a voluntary control scheme was started by inviting companies, or other organizations involved in the dumping of wastes, to apply for MAFF's approval of their activities. These applications were considered and, where no significant adverse effects on the marine environment were expected, a consent for the disposal of the waste to a specific dumping area was granted. By the early 1970s nearly all industrial wastes and sewage sludges dumped to sea were included within this voluntary scheme. In 1972, the Oslo Convention was concluded whereby the countries bordering the North Sea and North-East Atlantic agreed on common measures to control marine pollution caused by dumping. Following this, the UK introduced the Dumping at Sea (DAS) Bill which was enacted on 27 June 1974.

1095 - Norton, M. G. 1978a. Monitoring of waste disposal areas around England and Wales. ICES. C.M. 78/E:7.

1096 - Norton, M. G. 1978b. The control and monitoring of sewage sludge dumping at sea. Jour. Water Poll. Control. Fed. 3:402-407.

1097 - Norton, M. G., and M. S. Rolfe. 1978. The field assessment of effects of dumping wastes at sea: 1. A introduction. U.K. Min. of Agr., Fisheries and Food (MAFF), Lowestoft. Fish. Res. Tech. Rept. No. 45. 9p.

Foreword: As a consequence of the Dumping at Sea Act (1974) and the ratification by the UK of the Oslo Convention for the Prevention of Marine Pollution by Dumping from Ships and Aircraft and the London Convention on the Dumping of Wastes at Sea, it is necessary for the Ministry of Agriculture, Fisheries and Food to survey areas of sea used for the disposal of nonradioactive wastes by dumping from ships. The surveys assist in the control of marine pollution and are carried out by the Marine Environment Protection Division of the Ministry's Directorate of Fisheries Research. The group of scientists concerned is based at the Directorate's Burnham Laboratory. This present report acts as an introduction to a series setting out the results of those surveys.

1098 - Norton, M. G. 1980a. Marine disposal by dumping. In: J. Ryan (editor), Proceedings of a seminar: 'Todays and Tomorrow's wastes,' held at Cork, Ireland, May 1979. National Board for Science and Technology. Dublin. 172p.

1099 - Norton, M. G. 1980b. Monitoring of areas receiving dumping wastes around England and Wales. ICES.C.M. 80/E:28. 11p.

1100 - Norton, M. G., and R. Lloyd. 1981. The role of laboratory testing in the United Kingdom's controls on dumping wastes at sea. Chemosphere. 10(6):641-657.

Summary: The objective of the UK's pre-discharge controls under national legislation and the Conventions is taken to be the prevention of pollution from dumping. Laboratory tests have an essential role to play in assessing the polluting potential of a proposed discharge. Most emphasis is placed on toxicity tests to supplement basic physical and chemical data on the waste, and the results of these tests allow the possibility of acute toxic effects after discharge to be assessed. Longer term effects may arise from persistence or bioaccumulation potential in the waste. However, where tests are used to investigate such effects, care must be taken in the extrapolation of laboratory data to field conditions. The appropriate literature on the chemical components of the waste may often provide sufficient information on which to base decisions. Other essential components of an effective control programme are enforcement of regulations and monitoring of the effects of dumping. In the latter context examples are given of environmental effects which are not capable of being adequately predicted on the basis of laboratory testing alone.

1101 - Norton, M. B., R. A. Eagle, R. S. Nunny, M. S. Rolfe, P. A. Hardiman, and B. L. Hampson. (In Press). The field assessment of effects of dumping wastes at sea: 8 Sewage sludge dumping in the Outer Thames Estuary. U.K. Ministry of Agriculture, Fisheries and Food, Directorate of Fisheries Research, Lowestoft.

1102 - NL Industries, Inc. 1978. Summary of Studies made Relating to NL Ocean Disposal Site. NL Industries. Sayreville, NJ. 20p.

Abstract: A superior methodology to estimate the impact of an ocean disposal operation on the marine environment consists of extensive biosurveys, supplemented by laboratory studies, of the disposal operation after the site has been in use for a number of years. Such surveys and studies have been made of NL's disposal site, and are reported herein. The conclusion reached is that there is minimal, or no, effect of the acid waste disposal on the marine environment. Therefore, N L's ocean disposal operation meets the requirements of the U.S. EPA Ocean Dumping Criteria, with respect to having minimal, and therefore acceptable, impact on the marine environment.

1103 - Nunny, R. S. 1975. Disposal of wastes at sea. IV. The role of sedimentology in the assessment of effects off the northeast coast of England. ICES. C.M. 75/E:24.

1104 - Nunny, R. S. 1978. A survey of the dispersal of colliery waste from Lynemouth beach, Northumberland. U.K. Min. of Agr., Fisheries and Food (MAFF) Lowestoft. Fish. Res. Tech. Rept. No. 43. 17p.

Conclusions:

1. The colliery waste tipped along the high water mark at Lynemouth is actively reworked by wave action. Littoral drift transports material north and south within the bay, the slightly more dominant direction of movement being to the south. Alternation of destructive and constructive wave conditions causes onshore/offshore movement of beach sediments (definable as quartz-density particles with a diamater greater than approximately 180 µm) between the beach face and the immediate sub-littoral zone (maximum depth 10m).

2. Material does not escape offshore of breaker line unless it is fine enough to be carried in suspension through the decreased turbulence seawards of the breakers, i.e. particles finer than approximately 180 um diameter quartz, 500 µm diameter coal. Although only small quantities of the waste as tipped are fine enough to act in this way, processes of attrition and abrasion along the extensive beach face of colliery waste appear to be producing large amounts of fine sediment capable of moving offshore. It is difficult to make precise estimates of the rate of loss of material from the beach zone, but it is certainly a substantial proportion of the current tipping rate and may, since 1970, have come very close to this rate. The present situation therefore may be approaching a state of equilibrium.

3. The waste material that escapes the breaker zone and settles out offshore is composed of fine quartz sand (indistinguishable from natural sands) and fine coal particles (<500 µm diameter). The extent of the offshore reservoir of coal particles has been established. No mechanisms are apparent whereby coal in this reservoir could be returned to the beaches, as any sea bed material that might be transported shorewards beneath storm waves is fine enough to go into suspension within the nearshore zone and be carried seawards again.

4. There is no evidence that significant amounts of beach materials are being transported out of Lynemouth Bay by littoral drift, as the containing headlands are fringed with sediment-free rock ledges. Thus the sea coal of the Northumberland beaches is unlikely to have been derived from the Lynemouth tip, as there is no obvious transport pathway along the intertidal zone and the sea coal particles found on the beaches are too coarse (500-2000 μm) to have beEn derived from the offshore reservoir; they are more likely to be of natural origin. The slightly more extensive coal deposits at Newbiggin may be an exception; it is possible that during severe storms coal particles are washed across the rock ledges between Lynemouth and Newbiggin in water depths of 0-10m. The capacity of this pathway, if it exists, cannot be very great, and a special study involving underwater observations and sediment traps would be necessary to demonstrate its existence.

5. The results of the study have provided a baseline against which the effects on amenity and fisheries of any proposals for the use of alternative methods of disposal can be evaluated.

1105 - Nusbaum, I., H. E. Miller, and A. W. Reinhardt. 1955. Travel of pollution in shoreline coastal waters. Proc. Amer. Soc. Civil Eng. 81(781).

1106 - Oakley, S. M., P. O. Nelson, and K. J. Williamson, 1981. Model of trace-metal partitioning in marine sediments. Envir. Sci. Technol. 15:474-480

1107 - Oceanus. 1981. The oceans as waste space. Woods Hole Oceanographic Institution, Woods Hole, MA. 24(1):68.

1108 - O'Connor, D. J. 1962. Organic pollution of New York harbor - theoretical considerations. Jour. Water Poll. Control Fed. Vol. 34.

1109 - O'Connor, D. J. 1970. Water quality analysis for the New York Harbor complex. In: A. A. Johnson (editor), Water Pollution in the Greater New York Area. Gordon and Breach Science Publishers. New York, NY. pp. 121-144.

1110 - O'Connor, D. J., R. V. Thomann, and H. J. Salas. 1977. Water quality. MESA New York Bight Atlas Monograph 27. New York Sea Grant Institute. Albany, NY. 104p.

Abstract: Data presented on the water quality of the New York Bight include both reduced data combined for graphical presentations, and raw data. Physical and chemical characteristics measured include temperature, light, salinity, DO, pH, heavy metals, coliform bacteria, phytoplankton chlorophyll, and N and P forms. Other properties examined include the following: geomorphology, hydrology, circulation, biological properties, and the effects of various inputs such as dredge spoil, sewage sludges acid and chemical wastes, cellar dirt, municipal and industrial waste discharges, river discharges, and drainage and atmospheric fallout. Bottom DO percent saturation levels in the disposal areas of the Apex have decreased from 67% in 1949 to 30% in 1974. Surface total iron concentrations in the Apex have increased from 20 µg/1 in 1949 to approximately 90 ug/1 in 1969, and are higher than background open ocean levels. Coliform bacterial influence appears to be confined to an area of 3.2-4.8 km radius from the discharge point of a sewage sludge barge dump. Nitrogen is generally more important than P in limiting phytoplankton growth in the Bight. Turbidity may also be a limiting factor. Man's impact on the water quality of the region is significant with >50% of the total imput of Fe, Ca, Cd, Cr, suspended solids, and total K attributed to barge discharges alone.

1111 - O'Connor, H. B., Jr., and I. W. Duedall. 1975. The seasonal variation in sources, concentrations, and impacts of ammonium in the New York Bight Apex. ACS Symposium Series. Marine Chemistry in the Coastal Environment. Ser. No. 18. pp. 636-663.

Abstract: Two sources of NH_4^+ input to the New York Bight have been identified: (1) Hudson River-Raritan Bay discharges into the Apex of the Bight through the Rockaway Point, New York-Sandy Hook, New Jersey, transect and (2) barge-dumped sewage sludge from the greater New York metropolitan region.

The Bight-ward flux of ammonium through the transect has been calculated for one 24-hour period in June and was found to be 5 to 10 times greater than the ammonium input from barge-dumped sludge from a typical two-day period in July. During the April bloom, the rate of phytoplankton uptake and the effect of freshwater dilution were found to decrease the NH_4^+ concentration a similar amount in the transect.

The Hudson River plume was observed to be a persistent year-round feature near the New Jersey shore and was responsible for the advection of large amounts of chlorophyll a into the Bight.

1112 - O'Connor, H. B. Jr., C. F. Wurster, C. D. Powers, D. C. Biggs, R. G. Rowland. 1978. Polychlorinated biphenyls may alter marine trophic pathways by reducing phytoplankton size and production. Science. 201(43-7):737-739.

Abstract: Polychlorinated biphenyls at concentrations of 1 to 10 micrograms per liter reduced phytoplankton biomass and size in natural estuarine phytoplankton communities grown within dialysis bags in situ in an estuarine marsh. In polychlorinated biphenyls-contaminated waters, these changes could increase the number of trophic levels and divert the flow of biomass from harvestable fish to jellyfish and other gelantinous predators.

1113 - O'Connor, H. B., Jr., C. D. Powers, C. F. Wurster, K. D. Wyman, G. M. Nau-Ritter, and R. G. Rowland. (In Press). Fates and biological effects on plankton of particle-sorbed PCBs in coastal waters. In: G. F. Mayer, A. Calabrese, F. A. Cross, D. C. Malins, D. W. Menzel, J. S. O'Connor, and F. J. Vernberg (editors), Ecological Stress in the New York Bight: Science and Management. Estuarine Research Federation. Columbia, SC.

Abstract: The adsorption and desorption of polychlorinated biphenyls (PCB's) (Aroclor 1254) to and from organic and inorganic particulates characteristic of coastal wasters were examined, as were the effects of PCB's on biological processes in natural phytoplankton assemblages and common estuarine copepods. Partitioning of PCB's in aqueous particle suspensions appeared related to the amount of organic matter present. Particle-desorbed PCB's reduced growth rate, carbon fixation, and chlorophyll _a_ content of natural phytoplankton assemblages incubated for several days in situ and were more toxic than an equivalent amount of PCB's added directly to water. When a single dose of PCB's (1 or 10 µg/liter) were added directly to water containing a natural phytoplankton community, a dose-dependent suppression of cell biomass, photosyntesis, and chlorophyll _a_ concentration was observed. Biomass within larger size classes, predominantly diatoms, was greatly reduced. PCB uptake by two copepod species (_Acartia clausi_ and _A. tonsa_) from water was asymptotic, reaching maximum concentrations in 30 hours. Mortality of _A. tonsa_ in water containing 1 µg PCB's/liter was greater than controls and was independent of the presence of PCB-contaminated food. However, in water containing 5, 10, or 20 µg PCB's/liter, mortality was significantly greater in the presence of phytoplankton that had accumulated PCB's

1114 - O'Connor, J. M., and M. Eisenbud. (In Press). Environmental health. Section I. MESA Atlas Monograph 32. New York Sea Grant Institute. Albany, NY.

1115 - O'Connor, J. M., J. B. Klotz, and Theo. J. Kneip. (In Press). Sources, sinks and distribution of organic contaminants in the New York Bight ecosystem. In: G. F. Mayer, A. Calabrese, F. A. Cross, D. C. Malins, D. W. Menzel, J. S. O'Connor, and F. J. Vernberg (editors), Ecological Stress in the New York Bight: Science and Management. Estuarine Research Federation. Columbia, SC.

Abstract: Data are given on the concentrations of chlorinated pesticides, polychlorinated biphenyls (PCB's), and polynuclear aromatic hydrocarbons (PAH's) in the sediments, water, and biota of the Hudson River, New York harbor, and New York Bight. Major sources of organic contaminants to the estuarine/coastal system are: for pesticides, Hudson River transport and municipal wastewater; for PCB's, Hudson River transport, atmospheric fallout, and surface runoff; and for PAH's, atmospheric fallout, surface runoff, and oil spills. Major sinks for all contaminants are the harbor, Lower Bay, and the Bight.

Of the three classes of compounds, the most common contaminants in marine and estuarine biota are the PCB's. Pesticides and PAH's, although present in many organisms, occur at relatively low levels. Except for PCB's in striped bass from the Hudson River, no instances of contaminants exceeding U. S. Food and Drug Administration (FDA) Action Levels were recorded. The implications of these data for ecosystem management and reclamation are discussed.

1116 - O'Connor, Joseph M., and Joseph W. Rachlin. (In Press). Perspectives on metals in New York Bight organisms: Factors controlling accumulaton and body burdens. In: G. F. Mayer, A. Calabrese, F. A. Cross, D. C. Malins, D. W. Menzel, J. S. O'Connor, and F. J. Vernberg (editors), Ecological Stress in the New York Bight: Science and Management. Estuarine Research Federation. Columbia, SC.

Abstract: Despite the vast array of data on metals bioconcentration factors (BCFs) and levels of toxic metals in marine organisms, the data generally are not applicable to impact assessment. This is because adequate data are not available for defining metal species in seawater and marine sediments, and because laboratory-derived data for BCFs and toxic effects generally relate to metals species not measured in field studies.

Metals partitioning from food and sediment and the potential metals regulation in marine organisms were evaluated with the use of selected data from the New York Bight and other coastal systems. It was concluded from these data that marine organisms regulate essential and nonessential metals, although regulation by various species differed. Metals regulation is accomplished by several mechanisms, ranging from selective uptake and selective elimination to sequestration by metal-binding proteins. A general scheme for internal metals regulation is provided, along with recommendations for future field and laboratory research.

1117 - O'Connor, J. S. 1973. Dredging and spoiling on Long Island. Marine Sciences Research Center. State Univ. of New York. Stony Brook, NY. Technical Report Series No. 19. 34p.

Abstract: Dredging and spoiling activity in the shore zone of Nassau and Suffolk Counties, New York, is summarized for the years 1961 through 1971. Statistics on dredge/spoil activity (volume dredged and area spoiled) are compiled and reviewed in terms of private and governmental actions, geographic distribution, and temporal trends. Most of the dredge/spoil activity has been conducted at the margins of Great South and Peconic Bays, principally for maintenance dredging of navigation channels and development of the shore zone for residential and commercial purposes. A marked reduction in dredge/spoil activity occurred in 1968. The government of Suffolk County has initiated most of the dredge/spoil activity.

The environmental impacts of large-scale spoil actions are described as major and persistent. Spoil sites are typically developed rather quickly wherever population pressures and land values are high.

The existing Permit System for dredging and spoiling is discussed. Recommendations for improving the planning and monitoring of dredge/spoil activity are made in the context of environmental impact and existing regulations by various levels of government.

1118 - O'Connor, J. S. 1975. Contaminant effects on biota of the New York Bight. In: Proc. 29th Annual Session Gulf and Caribbean Fisheries Institute. pp. 50-63.

Abstract: The Marine Ecosystems Analysis (MESA) Project of NOAA was initiated in 1973 to assess the impacts of man's activities and the natural influences on the New York Bight. Several examples of contaminant impacts on marine resources have been identified rather quickly: (1) high prevalence of diseases in several species of finfish and shellfish, (2) major alterations in the distribution and abundance of bottom living organisms, (3) widespread distribution in exceptionally high numbers of coliform and fecal coliform bacteria, indicative of pathogenic bacteria (which findings have led to closure of clam fishing operations in extensive areas around and landward of ocean dumping sites), (4) presence of transfer-resistant (R^+) bacteria which are resistant to broad spectra of heavy metals and antiobiotics, and (5) noxious concentrations of suspended particulate material, flotsam and surface slicks, particularly on beaches used very intensively for bathing and sport fishing.

1119 - O'Connor, Joel S. 1976a. Evaluating the significance of degradation in a large ecosystem - The New York Bight experience. Proceedings of the Symposium on Biological Evaluation of Environmental Impact. Annual Meetings of the AIBS and ESA. 2-3 June 1976. New Orleans, LA. 30p.

Abstract: Problems of waste disposal management and suggestions for evaluating the scientific significance of environmental impacts. Evaluation requires wide perspectives, from scientists, public decision makers, and general public. Each generation should act as trustee for succeeding generations. Scientists are responsible for strictly objective perspectives.

1120 - O'Connor, Joel S. 1979. A perspective on natural and human factors. In: R. L. Swanson and C. J. Sinderman (editors), Oxygen Depletion and Associated Benthic Mortalities in New York Bight, 1976. NOAA Professional Paper No. 11. U.S. Dept. of Commerce. Rockville, MD. pp. 323-331.

Abstract: While human activities are very important sources of nitrogen and carbon loadings to the Bight, the quantities of some natural loadings have not yet been estimated. The natural contributions of carbon and nitrogen from along-shelf transport, shelf onwelling, and sediment regeneration could be major, but estimates are not available.

While the bulk of these anthropogenic loadings enter the Bight through the Sandy Hook/Rockaway Point transect, nitrogenous loadings from the New Jersey coast are more than 20 percent of those from the transect. These sources, largely from sewage treatment plant outfalls all along the New Jersey coast, may well have contributed to the nearshore oxygen depletion events in the past-particularly in 1968, 1971, 1974, and 1977. It is not yet possible to link quantitatively the particular sources of nutrients and carbon to their roles in depleting oxygen from Bight waters.

Other contributors to this report have developed an entirely plausible hypothesis for the severe oxygen depletion in 1976. This hypothesis implies that the anthropogenic loadings of carbon and plant nutrients contribute to oxgyen depletion of the inner Bight, but have not been the major cause of anoxia, even in 1976 when physical conditions favored oxygen depletion. However preponderant the anthropogenic loadings, the resulting summer phytoplankton blooms and detritus are actively grazed upon and dispersed widely while still in surface water. The particulate carbon falling into the bottom layer does not become concentrated enough to cause anoxia. The hypothesis invokes _C. tripos_, which bloomed in 1976 independently of known anthropogenic stimuli, as a mechanism for accumulating beneath the pycnocline unusually large quantities of carbon. Because _C. tripos_ is seldom eaten, essentially all this biomass respires actively until death when the labile fractions of the cells are oxidized, making _Ceratium_ a very effective agent for oxygen depletion. This hypothesis is consistent with the extended duration of stratification, probable physical concentration of _C. tripos_, and increased residence time of New Jersey coastal bottom water, all of which contributed to oxygen depletion in 1976.

This hypothesis seems entirely consistent with realistic ranges of the loadings and processes involved. However, the realistic ranges are great for oxidation rates of POC and DOC in bottom waters. If sinking rates of nanoplankton and detritus are toward the high end of presently realistic values and if substantial quantities of this material are not incorporated in planktonic food webs before being oxidized, this fraction of POC would be more significant than the hypothesis allows. Also, despite the overwhelming concentrations of DOC, there seems to be very little quantitative evidence about its sources or oxidation rates. Turnover rates would not need to be very large to make DOC more significant than has been presumed. Substantially more precise and systematic evaluations are needed to determine the relative significance of nutrient and carbon loadings to oxygen depletion. Mathematical models of carbon/oxygen/nutrient dynamics in the Bight are being developed to refine the assessments in this volume.

The extent of human contributions to this coastal eutrophication and seasonal oxygen depltion remains arguable. Annual variation in the degree of oxygen depletion is pronounced. Reductions in any significant, relatively constant BOD loading would, on the average, reduce the likelihood of anoxic events. Consequently, curbs on human waste loadings would reduce BOD in bottom waters of the inner Bight and, hence, the probability and severity of benthic mortalities. Other alternatives to the same ends are not obvious to me.

1121 - O'Connor, J. S., and H. M. Stanford. 1979. Chemical pollutants of the New York Bight: Priorities for research. NOAA. Boulder, CO. 217p.

Abstract: The New York Bight Marine Ecosystem Analysis Project of the NOAA is an 8-yr interdisciplinary field program involving studies of the New York Bight and of the effects of human influence on the ecosystem. The background and purpose of the investigation, methods, and a summary of findings are provided. Included are subpanel reports on metals, petroleum and halogenated hydrocarbons (including production and transport), halogenated compounds (e.g., pesticides), artificial radionuclides, and industrial synthetic organic chemicals. The potential impact of the contaminants upon humans is emphasized. Detailed figures and tables are provided which map the New York Bight, the decision framework for ranking contaminants, and gas chromatograms from packed and capillary columns.

1122 - O'Connor, J. S. (editor). (In Manuscript to nearly In Press). Perspectives of the New York Bight. NOAA, MESA Special Report Series. MESA New York Bight Project. State Univ. of New York. Stony Brook, NY.

Abstract: Coverage of this report series (15 papers) includes the historical development of the New York Bight region; comparative human influences on other populated coastal areas; the consequences of continued industrialization; definition of pollutant discharges; shore-zone impacts; impacts on fishes, shellfishes, and their food chains; the implications of current and possible future contaminant levels; the probability of marine environmental crises; summaries of recreation and aesthetics, human health and well-being, resources and ecosystem well-being, and economic develpment; and an assessment of the future of the Bight.

1123 - O'Connor, Thomas P., and P. Kilho Park. 1981. Waste characteristics, waste dynamic and biological effects. In: Assessment Report on the Effects of Waste Dumping in 106-Mile Ocean Waste Disposal Site. NOAA/ODP Dumpsite Evaluation Report 81-1. Rockville, MD. pp. 11-24.

1124 - O'Connor, Thomas P., and P. Kilho Park. 1981. Assessment findings, recommendations and future studies. In: Assessment Report on the Effects of Waste Dumping in 106-Mile Ocean Waste Disposal Site. NOAA/ODP Dumpsite Evaluation Report 81-1. Rockville, MD. pp. 25-29.

1125 - O'Connor, Thomas P., and P. Kilho Park. (In Press). Consequences of industrial waste disposal at the 106-Mile ocean waste disposal site. In: G. F. Mayer, A. Calabrese, F. A. Cross, D. C. Malins, D. W. Menzel, J. S. O'Connor, and F. J. Vernberg (editors), Ecological Stress in the New York Bight: Science and Management. Estuarine Research Federation. Columbia, SC.

Abstract: Industrial wastes are dumped beyond the edge of the continental shelf at the 106-Mile ocean waste disposal site. These wastes are mainly liquids that, in some cases, form a precipitate floc after mixing with seawater. Rather than descending to the sea floor, wastes are distributed primarily as plumes within the upper mixed layer. The method of dumping yields a plume with a maximum initial concentration of about 200 ppm waste. Within about four hours, the maximum concentration is reduced to approximately 10 ppm with a waste plume width of about 1 km. Further dilution and plume growth proceed slowly, leading to the conclusion that storm events may be necessary to dilute extensively the waste. The average flow through the dumpsite seems to be sufficient, when compared to the frequency of dumping, for each dump to be an independent event so that later dumps do not occur in previously created plumes. It is estimated that in the average, the plumes move toward the southwest along the contour of the continental shelf.

Laborarory studies have shown that oceanic strains of phytoplankton are less tolerant of waste than coastal strains of the same species, and that there is considerable variation in response to waste among species. Although no tested phytoplankton species have been found to be affected at 10 ppm waste, it is possible that within a waste plume there will be a readjustment of phytoplankton communities favoring more resilient species. Tests of this are proceeding, as are tests on the sublethal responses of zooplankton to wastes, histopathological and chemical analyses of field-collected organisms, studies of the stability of waste-derived organic compounds, and studies of the partitioning of waste constituents between dissolved and particulate phases.

1126 - Oertel, George F. 1979. Depositional characteristics of sediments at low energy ocean disposal site, Savannah, Georgia. In: H. D. Palmer and M. G. Gross (editors), Ocean Dumping and Marine Pollution. Dowden, Hutchinson and Ross, Inc. Stroudsburg, PA. pp. 97-107.

Abstract: Since the early 1900's the bar-channel system of the Savannah navigation channel has been dredged for maintenance and improvement. Between 1967 and 1977, approximately 10.8×10^6 yds^3 (8.3×10^6 m^3) of this dredge material was deposited at a predetermined ocean disposal site. The prominent sources of material dredged form the navigation channel are fine-grained river deposits, littoral sands, Pleistocene river gravels, and submarine Miocene outcrops. Material that is released from hopper dredges is hydraulically sorted as it settles through the water column. Coarse material is deposited in relatively thick mounds corresponding to the disposal path of the dredge, thin layers of fine material accumulate in deeper areas adjacent to mounds.

After deposition, the fine coarser material is relatively stable during nonstorm conditions. During storm conditions, coarse material in the upper parts of the mounds are reworked by waves and storm currents.

1127 - Officer, Charles B., and John H. Ryther. 1981. Swordfish and Mercury: A case history. Oceanus. 24(1):34-41.

Abstract: Overview of mercury pollution, poisoning (Minamata disease, Japan) in the 1970's to present, including Swedish studies, with discussion on levels in fish and the resultant FDA standard.

1128 - Ohlhorst, C. W., R. W. Johnson, and E. R. Meyer. 1977. Remote sensing and laboratory techniques for monitoring ocean dumping. Amer. Geoph. Union. Fall Meeting. San Francisco, CA. Dec. 5-9. 11p.

Abstract: Results of field experiments conducted in the Atlantic Coastal Zone indicate that plumes resulting from dumping of acid waste and sewage sludge have distinguishable spectral characteristics when the radiance of the pollutant is normalized (ratioed to) background ocean water. Acid waste spectra peak between 550-650 nm while sewage sludge spectra have peak values at wavelengths of about 700 nm or greater. Results indicate that identification of acid waste and sewage sludge plumes may be independent of geographical location in the Atlantic Coastal Zone. Radiance ratio curves obtained in the laboratory qualitatively agree with those obtained from field experiments. Results from the July 25, 1977, Galveston Deep Ocean Dump Site experiment show the radiance ratio curve of the biodigested industrial waste to be fairly flat and similar to the radiance ratio curves of sewage sludge line dumps and sewage sludge spot dumps that have been in the water for several hours.

1129 - Ohlhorst, C. W., and G. S. Bahn. 1978. Quantitative mapping of particulate iron in an ocean dump using remotely sensed data. American Congress on Surveying and Mapping and American Society of Photogrammetry. Annual Spring Convention. Washington, DC. 26 Feb.-3 Mar. 1978. Paper. 17p.

1130 - Oliff, W. D., D. J. Livingstone, and V. C. Stone. 1969. Factors determining dilution in the marine environment and affecting the return of effluent to shore. Water Pollution Control (Britain). 68(5):560.

1131 - Olivier, J. P. 1975. Sea disposal of packaged radioactive wastes. Transactions American Nuclear Society. 20:693-695.

1132 - Olson, T. A., and F. J. Burgess. (editors). 1967. Pollution and Marine Ecology. Proceedings of a Conference held at Texas A&M University, March 1966. Interscience Publishers. New York, NY. 364p.

1133 - O'Neill, T. M. 1974. Ocean disposal of waste off New Jersey. Commentary. 1(2):32.

Abstract: Ocean disposal of waste prevents more productive use of the disposal areas and eliminates the possibility of reuse of the waste materials. The ecology of the continental shelf, the continental slope, and the deep ocean is described. Special geologic features of the New Jersey coast are included. Amounts and types of wastes disposed of in the New York Bight are listed. The effects on the marine environment of ocean dumping of sewage sludge, dredge spoil, pesticides, petroleum products, Hg, and radioactive materials are discussed. Alternatives to ocean disposal are considered for each type of waste. Actions to regulate ocean dumping are covered.

1134 - Oostdam, B. L. (In Press). Sewage sludge dumping in the mid-Atlantic Bight in the 1970s: Short-, intermediate- and long-term effects. In: I. W. Duedall, B. H. Ketchum, P. K. Park, and D. R. Kester (editors), Wastes in the Ocean, Vol. I: Industrial and Sewage Wastes in the Ocean. John Wiley and Sons, Inc. New York, NY.

Abstract: The fate of treated sewage sludge dumped by the City of Philadelphia off the coast of Delaware and Maryland was considered on short-, intermediate-, and long-term time scales. Short-term (minutes to hours) studies of the water column before, during, and after dumping operations show the importance of the thermocline both as a barrier to settling and as a surface enhancing dispersal. Fractionation of trace elements in seawater-sewage sludge mixtures may decrease copper concentrations, relative to backgrounds but increase concentrations of zinc. Intermediate-term (days to months) events indicate the importance of the net southward bottom drift to the dispersal of sewage sludge. The fine floc material consisting of natural aggregates possibly mixed with settled sewage sludge disperses widely. Long term effects (>1 year) deal with the environmental changes at newly established interim dumpsite and the recovery of an abandoned old dumpsite. Attempts were made to evaluate these effects from study of changes in trace elements concentrations in sediments and in the surf clams Spisula solidissima. During 1974 concentrations of titanium, vanadium, and zinc were higher, but lead and cadmium lower in the interim dumpsite than in the old dumpsite, respectively. No consistent increase in the concentrations of cadmium could be demonstrated from extensive data on the ocean quahog, Actica islandica and on the sea scallop, Placopecten magellanicus for 1975 and 1976. Increased inputs of trace element from the dumping of sewage sludge did not result in increased levels of trace elements in these organisms.

1135 - O'Reilly, J. E., J. P. Thomas, and C. A. Evans. 1976. Annual primary production (nanoplankton, netplankton, dissolved organic matter) in the Lower New York Bay. Paper No. 19. In: W. H. McKeon and G. J. Lauer (editors). Fourth Symposium on Hudson River Ecology. Hudson River Environmental Society, Inc. New York, NY. 39p.

1136 - Orr, M. H., L. Baxter, Jr., and F. R. Hess. 1980. Remote acoustic sensing of the particulate phase of industrial chemical wastes and sewage sludge. Woods Hole Oceanographic Institution Report. Woods Hole, MA. No. 79-38. 153p.

1137 - Orr, Marshall H. 1981a. Acoustic detection of the particulate phase of industrial chemical wastes released at 106-Mile site. In: Assessment Report on the Effects of Waste Disposal site. NOAA/ODP Dumpsite Evaluation Report 81-1. Rockville, MD. pp. 157-202.

1138 - Orr, Marshall H. 1981b. Qualitative dispersion characteristics of the particulate phase of Dupont Edge Moor Plant Acid-Iron Waste and Grassell: Plant wastes for summer and winter. In: Assessment Report on the Effects of Waste Dumping in 106-Mile Ocean Waste Disposal Site. NOAA/ODP Dumpsite Evaluation Report 81-1. Rockville, MD. pp. 203-204.

1139 - Orr, Marshall H. 1981c. Qualitative dispersion characteristics of sewage sludge released in the presence of a shallow seasonal thermocline. In: Assessment Report on the Effects of Waste Dumping in 106-Mile Ocean Waste Disposal Site. NOAA/ODP Dumpsite Evaluation Report 81-1. Rockville, MD. pp. 205-206.

1140 - Orr, Marshall H., and Lincoln Baxter, Jr. (In Press). Dispersion of particles after disposal of industrial and sewage wastes. In: I. W. Duedall, B. H. Ketchum, P. K. Park, and D. R. Kester (editors), Wastes in the Ocean, Vol. I: Industrial and Sewage Wastes in the Ocean. John Wiley and Sons, Inc. New York, NY.

Abstract: A high-frequency acoustic backscattering system has been used to study the dispersion of particulates released or formed during the disposal of industrial chemical wastes and sewage sludge at the deepwater dumpsite mile-site 106 (DWD-106) located 196 km southeast of New York Harbor, New York. Particles associated with waste plumes serve as waterborne tracers which can be acoustically tracked. In addition the data sets can be used to determine the temporal variability of the particle vertical and horizontal distribution. The acoustic systems provide real-time data to guide chemical and biological sampling of contaminated and uncontaminated sections of the water column.

The acoustic data have shown that the vertical and horizontal distribution of the particulates released at DWD-106 is seasonally dependent, reflecting both the temporal variability of the magnitude of density gradients associated with the seasonal mixed layer and the water mass variability of the area. In a qualitative sense, the acoustic backscattering technique is beneficial to an environmental impact monitoring experiment. However, one of the objectives of developing the technique has been to measure the horizontal and vertical dispersion coefficients of particle plumes. The quantitative analysis of the acoustic data has been limited by navigation inadequacies, the inability to measure vertical shear, the lack of knowledge concerning waste particle size distributions and densities (hence settling rates), and the influence of oceanic turbulence on the particle field.

1141 - Ortner, Peter B., and Lynda S. Murphy. 1981. Biological effects of a dump at 106-Mile site. In: Assessment Report on the Effect of Waste Dumping 106-Mile Ocean Waste Disposal Site. NOAA/ODP Dumpsite Evaluation Report 81-1. Rockville, MD. pp. 245-255.

1142 - Osborn, J. 1979. Marine chlorophyll _a_ analysis. U.S. National Aeronautics and Space Administration. Technology Utilization Division. NASA Tech. Brief. 4(1):54-55.

Abstract: Experiments were conducted with aircraft multispectral scanners over the James River in Virginia and the New York Bight ocean area and were compared with onsite chlorophyll _a_ measurements. A statistical stepwise regression analysis ·-as conducted and used to determine calibrated regression equations for comparing sea-truth measurements to remotely sensed data. The same calibration technique was used to quantitatively relate remotely sensed data to water quality parameters and to map synoptic distributions of chlorophyll _a_ in the remotely sensed scenes.

1143 - Oser, R. K. 1973a. Physical oceanography of Deep Water Dump Area A. Appendix G. In: P. E. Wilkniss (editor), Environmental Condition Report for Deep Water Dump Area A. U.S. Naval Research Laboratory. Washington, DC. NRL Report 7553. pp. 91-102.

1144 - Oser, R. K. 1973b. Seafloor bathymetry and sediment properties in Deep Water Dump Area A. Appendix H. In: P. E. Wilkniss (editor), Environmental Condition Report for Deep Water Dump Area A. U.S. Naval Research Laboratory. Washington, DC. NRL Report 7553. pp. 103-116.

1145 - Osterberg, C., and S. Keckes. 1977. The state of pollution of the Mediterranean Sea. Ambio. 6(6):321-326.

1146 - O'Sullivan, A. J. 1971. Pollution by industrial waste and sewage: scientific aspects and some problems of control in the marine environment. Water pollution as a world problem. Report of a conference held at the University College of Wales, Aberystwyth, 11-12 July, 1970. David Davies Memorial Institute of International Studies. London. Europa Publications, London. pp. 143-151.

1147 - O'Sullivan, A. J. 1975. Discharges of wastes to Liverpool Bay, 1971. Nat. Envir. Res. Coun. Publs. Ser. C. No. 14. pp. 13-16.

1148 - O'Sullivan, D. A. 1976. Large ocean going ship to burn wastes at sea. Chem Eng. News. 54(13):12.

1149 - Otter, R. J. 1978. The Interim Paris Commission. Pub. by the Institute of Water Pollution Control, London. As Conference Paper No. 7 of the 1978 Annual conference. 5p.

1150 - Otto, Robert G., and S. Ian Hartwell. 1981. Preliminary studies of responses of fish to complex pharmaceutical and petrochemical wastes. In: B. H. Ketchum, D. R. Kester, and P. K. Park (editors), Ocean Dumping of Industrial Wastes. Plenum Press. New York, NY. pp. 473-484.

Abstract: Investigations were conducted to define the abilities of fish to survive or to detect and avoid exposure to complex organic waste materials from pharmaceutical and petrochemical sources. These wastes are currently barge-dumped in deep waters north of Puerto Rico. The 24-hr and 96-hr LC_{50}'s for juvenile spot <u>Leiostomus xanthurus</u> and killifish <u>Fundulus heteroclitus</u> and avoidance limits for juvenile spot and juvenile menhaden <u>Brevoortia tyrannus</u> were determined. Fish accepted exposures without mortality to waste concentrations substantively greater than are expected to occur in the field situation.

1151 - O'Viatt, C. A. 1968. The effects of incinerator residue on selected marine species. Proceedings of the Annual North Eastern Regional Anti-Pollution Conference. July 22-24. University of Rhode Island. Kingston, RI.

1152 - Oviatt, C. 1972a. Toxicity of incinerator residues for marine organisms. Bioassays. In: M. W. First (editor), Municipal Waste Disposal by Shipborne Incineration and Sea Disposal of Residues. Report. Harvard University School of Public Health. Boston, MA. pp. 4:1-4:16.

1153 - Oviatt, C. 1972b. Toxicity of incinerator residues for marine organisms. Long period in situ assays. In: M. W. First (editor), Municipal Waste Disposal by Shipborne Incineration and Sea Disposal of Residues. Report. Harvard University School of Public Health. Boston, MA. pp. 4:17-4:53.

1154 - Oviatt, C. 1972c. Toxicity of incinerator residues for marine organisms. Additional tests for incinerator residue toxicity. In: M. W. First (editor), Municipal Waste Disposal by Shipborne Incineration and Sea Disposal of Residues. Report. Harvard University School of Public Health. Boston, MA. pp. 4:54-4:70.

1155 - Oviatt, C. 1972d. Test dumps and survey at Cox Ledge Station; 54 M depth. In: M. W. First (editor), Municipal Waste Disposal by Shipborne Incineration and Sea Disposal of Residues. Report. Harvard University School of Public Health. Boston, Massachusetts. pp. 5:102-5:131.

1156 - Oviatt, C., and M. W. First. 1972. Oceanographic studies. In: M. W. First (editor), Municipal Waste Disposal by Shipborne Incineration and Sea Disposal of Residues. Report. Harvard University School of Public Health. Boston, MA. pp. 5:1-5:57.

1157 - Oy Tiesbus Laboratoria Centrallaboratorium Ab. 1968. Vuorikemia Oy: n piaveiriarin vaikutuksista Reposaaren ja Mantyluodon etelapuolisella merialueella v 1965.

1158 - Oy Vesitekniikka Ab. 1972. Tutkimus Vuorikemian tehtaiden jatevesien vaikutuksista purkualueen tilaan.

1159 - Oza, G. M. 1975. The dangers of dumping wastes into sea and river waters of India. Envir. Conser. 2:59-60.

1160 - Ozturgut, E., J. W. Lavelle, O. Steffin, and S. A. Swift. 1979. Environmental investigations during manganese nodule mining tests in the north equatorial Pacific in November 1978. NOAA TM ERL MESA-48. Boulder, CO.

1161 - Pagoria, P. S., and C. Y. Kuo. 1979. Interpretation of the characteristics of ocean-dumped sewage sludge related to remote sensing. Final Report, 18 Jul. 1977 - 17 Feb. 1979. Old Dominion Univ. Research Foundation. Norfolk, VA. NASA-CR-158454. 94p.

Abstract: Wastewater sludge characteristics in general, and characteristics of wastewater sludges generated by the City of Philadelphia in particular, were addressed. The types and sources of wastewater sludges, a description of sludge treatment and disposal processes, examination of sludge generation and management for the City of Philadelphia, and definition of characteristics for typical east coast sludges undergoing ocean disposal were discussed. Specific differences exist between the characteristics of primary and secondary wastewater sludges, especially with the nature and size distribution of the solids particles. The sludges from the City of Philadelphia monitored during remote sensing experiments were mixtures of various sludge types and lacked distinguishing characteristics. In particular, the anaerobic digestion process exerted the most significant influence on sludge characteristics for the City of Philadelphia. The sludges generated by the City of Philadelphia were found to be typical and harbor no unique features.

1162 - Palmer, H. D. 1976. Sedimentation and ocean engineering: Ocean dumping. In: D. J. Stanley and D. J. P. Swift (editors). Marine Sediment Transport and Environmental Management. Wiley-Interscience. John Wiley & Sons. New York, NY. pp. 557-577.

1163 - Palmer, H. D., and D. W. Lear (editors). 1973. Environmental survey of an interim ocean dumpsite, Middle Atlantic Bight. U.S. Environmental Protection Agency Region III. Philadelphia, PA. EPA 903/9-73-010A. 134p.

1164 - Palmer, H. D., J. R. Guala, and J. L. Nolder. 1976. Current meter data reduction, with comments on bedload sediment transport: Middle Atlantic Bight. Westinghouse Electric Corporation. Oceanic Division. Final Report to EPA Region III. Philadelphia, PA. No. WD-6-99-0669-B. 17p

165 - Palmer, Harold D., and M. Grant Gross (editors). 1979. Ocean dumping and marine pollution. Douden, Hutchinson and Ross, Inc. Stroudsburg, Pennsylvania. 268p.

1166 - Pamatmat, M. M. 1971. Oxygen consumption by the seabed. IV. Shipboard and laboratory experiments. Limnol. and Oceanogr. 16:536-550.

1167 - Pararas-Carayannis, G. 1973. Ocean dumping in the New York Bight: An assessment of environmental studies. U.S. Army, Corps of Engineers. (Coastal Engineering Research Center). Fort Belvoir, VA. T.M. No. 39. 159p.

Abstract: Interdisciplinary short-term investigations related to the effects of ocean dumping in the New York Bight were contracted by the Coastal Engineering Research Center (CERC) as directed by Office, Chief of Engineers. Studies made by the Sandy Hook Laboratory of the National Marine Fisheries, the State University of New York at Stony Brook, the Woods Hole Oceanographic Institution, and the Sperry Rand Corporation were reviewed by a special advisory committee established by the Smithsonian Institution at CERC's request. The studies included hydrographic, geological, chemical, biological investigations, and a feasibility study for a remote-controlled electronic sensing system that could assist regulating agencies in detecting the location and dump status of waste disposal vessels operating in the Bight. Circulation patterns were estimated from data obtained by current meters and by seabed and surface drifters. Chemical analyses were made of the concentration of phosphorus (ortho, organic, meta, and total), nitrate, total iron, dissolved oxygen, and chlorophyll-*a* in water samples. Temperature, salinity, turbidity, and pH were measured. Sediment samples were analyzed for organic content and the heavy metals, copper, chromium, lead, silver, nickel and zinc. Selected biological samples were analyzed for heavy metals and mercury. Biological investigations included studies of benthic meiofauna and macrofauna, zooplankton, finfish, and bacteria. The studies include basic data related to the disposal of sewage sludge, dredge spoils and acid-iron wastes, and have helped provide a more detailed environmental description of the Bight dumping grounds and adjacent areas. The findings of these and other related studies are presented and analyzed in this report in terms of impact on ecology, water quality, and total environmental effects.

1168 - Park, P. Kilho, and Thomas P. O'Connor. 1981. Ocean dumping research: Historical and international development. In: B. H. Ketchum, D. R. Kester, and P. K. Park (editors), Ocean Dumping of Industrial Wastes. Plenum Press. New York, NY. pp. 3-23

Abstract: Plans for systematic and comprehensive research and monitoring of the effects of ocean dumping were prepared in 1970, followed by their execution in recent years. Forty-three nations have joined the international convention on the "Prevention of Marine Pollution by Dumping of Wastes and Other Matter" that became effective on August 30, 1975. At present, substances internationally prohibited from being ocean dumped include organohalogens, mercury, cadmium and their compounds, persistent plastics, high-level radioactive wastes, and biological and chemical warfare agents.

1169 - Parker, C. A. (editor). 1978. Annual informal workshop on physical oceanography and meteorology of the Middle Atlantic and New York Bights. NOAA. TM ERL MESA-27. Boulder, CO. 56p. NTIS No: PB 286732.

1170 - Parker, J. G., and F. Wilson. 1975. Incidence of polychlorinated biphenyls in Clyde seaweed. Mar. Pollut. Bull. 6(3):46-47.

1171 - Parker, J. H., I. W. Duedall, H. B. O'Conner, Jr., and R. E. Wilson. 1976. Raritan Bay as a source of ammonium and chlorophyll *a* for the New York Bight Apex. In: M. G. Gross (editor), Amer. Soc. Limnol. Oceanogr. Spec. Symp. 2:212-219.

Abstract: Measurements in June 1974 and 1975 in Raritan, Sandy Hook, and Lower New York Bays showed that water nearest Sandy Hook had low salinity and high Chl *a* (10-60 mg/m^3) and NH_4^+ concentrations (10-40 μM). Sandy Hook Bay had the highest Chl *a* values (45-90 mg/m^3) and low NH^4 concentrations (5-8 μM). High Chl *a* concentrations (approx. 50 mg/m^3) were also found at the center of Raritan Bay accompanied by high NH_4^+ concentrations (60-70 μM). Near Rockaway Inlet, NH_4^+ concentrations were 0-40 mμ while Chl a values were minimal (3-4 mg/m^3). The Narrows also showed high NH_4^+ (15-40 mμ) and low Chl a concentrations (0-10 mg/m^3).

By comparing the salinity, ammonium, and chlorophyll a distributions at the Sandy Hook-Rockaway Point transect with the characteristics of the different sources of water flowing out to the Bight, an identification of water masses at the transect is possible.

1172 - Patchen, Richard C., E. E. Long and B. B. Parker. 1976. Analysis of current meter observations in the New York Bight Apex August 1973-June 1974. NOAA TR ERL 368-MESA-5. Boulder, CO. 24p. NTIS No. PB 263928/AS.

Abstract: Primary tidal constituents are M2 and K1 constituents. M2 direction of progress is west along NY coast, and north along the NJ coast. Frictional bottom boundary layer extends 7 m above bottom in the fall season with a clockwise circulation prevailing but with 30% of the observations indicating counterclockwise circulation. In spring stratification and thermocline depth determine the mean circulation. When stratification is weak there is a strong response of mean current to the wind.

1173 - Patell, James. 1970. Economic aspects of solid waste disposal at sea. Massachusetts Institute of Technology. Cambridge, MA. Sea Grant. Report No. 71-2. 128p.

1174 - Pauls, H. 1975. Waste disposal in offshore areas. Deutsche Gewasserkundliche Mitteilungen. 19(Sonderheft):55.

1175 - Pearce, J. B. 1972. The effects of solid waste disposal on benthic communities in the New York Bight. In: M. Ruivo (editor), Marine Pollution and Sea Life. FAO. Fishing News (Books). London. pp. 404-411.

1176 - Pearce, J. B. 1974. Benthic assemblages in the deeper continental shelf waters of the Middle Atlantic Bight. In: Proc. Estuarine Research Federation Outer Continental Shelf Conference and Workshop, Univ. of Maryland. College Park, MD. pp. 297 317.

Abstract: Many of the taxa (Amphipods and polychaetes) identified from deeper shelf waters are forms important to the diet of demersal finfish. Petroleum based contaminants may eliminate important, sensitive species which could be replaced by less desirous fish. The impact of any contaminant on deep water assemblages are much greater than on similar shallow assemblages because deep water assemblages exist in stable and predictable environments.

1177 - Pearce, J. B., and J. S. Young. 1975. Shell disease in crabs and lobsters from New York Bight. Mar. Pollut Bull. 6(7):101-105.

Abstract: Dumping grounds in New York Bight receive very large quantities of sewage sludge. Lobsters and rock crabs collected in or near the dumping grounds sometimes show various pathological conditions of the shell and gills. In this study the histopathology of 'shell disease', the causative agents, and its effects on respiration are discussed in connection with a possible association with the disposal of solid wastes into the ocean.

1178 - Pearce, J. B., J. P. Thomas, and R. Greig. 1975. Preliminary investigation of benthic resources at deepwater dumpsite 106. In: May 1974 Baseline Investigation of Deepwater Dumpsite 106. NOAA Dumpsite Evaluation Report 75-1. Rockville, MD. pp. 217-228.

1179 - Pearce, John B., Janice V. Caracciolo, Martha B. Halsey, and Leslie H. Rogers. 1976a. Temporal and spatial distributions of benthic macroinvertebrates in the New York Bight. In: M.G. Gross (editor), Amer. Soc. Limnol. Oceanogr. Spec. Symp. 2:394-403.

Abstract: Benthic biota in the New York Bight Apex is characterized by high spatial and temporal variability. A considerable reduction in numbers of individuals per sample was observed between August 1973 and Ausust 1974; on an average, numbers of individuals per station decreased from 417 to 174. Low species diversity (<0.500) was observed at some stations inside the Christiaensen Basin, an area characterized by high organic carbon values in sediment. Certain species, apparently tolerant of carbon-rich deposits of the Christiaensen Basin, were extremely abundant at some stations located in the basin. These included the anemone *Cerianthus*, four species of polychaete, and the bivalve *Nucula*.

1180 - Pearce, J. B., J. Caracciolo, A. Frame,. L. Rogers, M. Halsey, and J. P. Thomas. 1976b. Distribution and abundance of benthic organisms in the New York Bight, August 1968 - December 1971. NOAA DR ERL MESA-7. Boulder, CO. 114p. NTIS No. PB 271343/AS.

Abstract: Presents machine-listed data and statistical calculations (diversity and equitability) on the distribution and abundance of benthic invertebrates at 26 stations in the New York Bight Apex. Data are listed by latitude and longitude as well as by station number. Benthic samples were collected at various times of the year, therefore some seasonal data are available.

1181 - Pearce, J., J. P. Thomas, J. Caracciolo, M. Halsey, and L. Rogers. 1976c. Distribution and abundance of benthic organisms in the New York Bight Apex, ^-6 August 1973. NOAA DR ERL MESA-8. Boulder, CO. 131p. NTIS No. PB 271344/AS.

Abstract: During this period, a series of benthic grab samples were collected at standard sampling stations located on the NY Bight Apex MESA grid. In addition to benthic microfauna, samples were collected for: sediment grain size analyses and sediment heavy metal burdens; benthic meiofaunal studies seabed oxygen consumption rates. This report provides lists of data and certin statistical calculations (diversity and equitability) concerned with the seasonal distribution and abundance of benthic invertebrates.

1182 - Pearce, J., J. P. Thomas, J. Caracciolo, M. Halsey, and L. Rogers, 1976d. Distribution of abundance of benthic organisms in the New York Bight Apex, 26 August - 6 September 1974. NOAA DR ERL MESA-9. Boulder, CO. 88p.

Abstract: During this period, a series of benthic grab samples were collected at standard sampling stations located on the NY Bight Apex MESA grid. In addition to benthic microfauna, samples were collected for: sediment grain size analyses and sediment heavy metal burdens; benthic meiofaunal studies; seabed oxygen consumption rates. This report provides lists of data and certin statistical calculations (diversity and equitability) concerned with the seasonal distribution and abundance of benthic invertebrates.

1183 - Pearce, J., L. Rogers, J. P. Thomas, J. Caracciolo, M. Halsey, and K. McNulty. 1976e. Distribution and abundance of benthic organisms in the outer New York Bight and proposed alternate disposal sites, June 1974 and February 1975. NOAA DR ERL MESA-10. Boulder, CO. 68p.

Abstract: During this period, a series of benthic grab samples were collected at standard sampling stations located on the NY Bight Apex MESA grid. In addition to benthic microfauna, samples were collected for sediment grain size analyses and sediment heavy metal burdens; benthic meiofaunal studies; seabed oxygen consumption rates. This report provides lists of data and certain statistical calculations (diversity and equitability) concerned with the seasonal distribution and abundance of benthic invertebrates.

1184 - Pearce, J., J. Caracciolo, M. Halsey, and L. Rogers. 1977a. Distribution and abundance of benthic organisms in the New York-New Jersey outer continental shelf. NOAA DR ERL MESA-30. Boulder, CO. 75p.

1185 - Pearce, J., L. Rogers, J. Caracciolo, and M. Halsey. 1977b. Distribution and abundance of benthic organisms in the New York Bight Apex, five Seasonal Cruises, August 1973 - September 1974. NOAA DR ERL MESA-32. Boulder, CO. 803p. NTIS No. PB 282679/AS.

Abstract: Between August 1973 and September 1974, personnel of the Division of Environmental Assessment (DEA), Northeast Fisheries Center, conducted a series of five seasonal cruises to stations within the New York Bight Apex in order to assess the impacts of contaminants in sediments on the ecosystem in this area. The data are being made available in the present form for timely assistance to user groups.

This set of data is being compared with the historical data collected by Pearce, et al, from the same area between 1968 and 1971. In addition, the information provided by these samples will be especially useful in determining if seasonal patterns exist within benthic communities in the Bight Apex.

1186 - Pearce, J., J. Caracciolo, M. Halsey, and L. Rogers. 1977c. Distribution and abundance of benthic macrofauna in the sewage sludge disposal area, New York Bight Apex, February 1975. NOAA DR ERL MESA-36. Boulder, CO. 34p. NTIS No. PB 283867/AS.

1187 - Pearce, J. B., J. Caracciolo, F. W. Steimle, Jr. 1977d. Final Report on benthic infauna of Deepwater Dumpsite 106 and adjacent areas. In: Baseline Report of Environmental Conditions in Deepwater Dumpsite 106. NOAA/ODP Dumpsite Evaluation Report 77-1. Rockville, MD. Vol. II. pp. 465-480.

Abstract: Our data show no significant differences in numbers of individuals, numbers of species, or diversity in comparisons of stations within DWD 106 with control stations. Bottom dissolved oxygen measurements are not depressed and are similar within the dumpsite and in control areas. Since stressed environments are often characterized both by low diversity and low concentrations of dissolved oxygen, there is, at this time, no reason to believe that toxic wastes disposed of at site 106 have impinged upon the fauna collected at our sampling sites. However, it is known that degradation of certain categories of organic material is very slow in great depths of water, and the same may be true of toxins. Therefore, with continued dumping and degradation of this toxic material, there is a possibility of development of a stress situation in the future. Because of this, it is suggested that minitoring of this area be continued in order to detect any such changes.

1188 - Pearce, John, Clyde MacKensie, Janice Caracciolo, and Leslie Rogers. 1978. Reconnaissance survey of the distribution and abundance of benthic organisms in the New York Bight Apex, 5-14 June 1973. NOAA DR ERL MESA-41. Boulder, CO. 9p.

Abstract: Between 5 and 14 June 1973, personnel of the Division of Environmental Assessment (DEA), Northeast Fisheries Center, conducted a reconnaissance cruise to stations within the New York Bight Apex. The overall purpose of the investigation was to provide a baseline to be used in assessing the impacts of contaminants from all sources, particularly ocean dumping, on the sediments in this area. The reconnaissance cruise was the first of four phases of our study of the New York Bight Apex which include: 1) reconnaissance, 2) seasonal sampling (five quarterly cruises, August 1973 through September 1974, 3) intensive grid sampling in areas known to be affected by sewage sludge and dredge spoil disposal, and 4) monitoring of selected stations in the Apex.

This data report was prepared to provide machine-listed data and certain statistical calculations, diversity, and equitability, concerned with the distribution and abundance of benthic invertebrate organisms found in the New York Bight Apex.

1189 - Pearce, John B. 1979. Ocean Pulse - A coastal environmental assessment and monitoring program In: Bulletin - The Coastal Society. J. Zinn and R. Payne (editors). 3(2):9-12.

1190 - Pearce, John; J. Caracciolo, R. Greig, D. Wenzloff, and F. W. Steimle, Jr. 1979. Benthic fauna and heavy metal burdens in marine organisms and sediments of a continental slope dumpsite off the northeast coast of the United States (Deepwater Dumpsite 106). In: E. Dahl (editor), The Deep Sea-Ecology and Exploitation. Ambio Spec. Rep. G. pp. 109-114.

Abstract: This paper examines the impact of six to eight years of industrial dumping at a deepwater (2500 m) dumpsite off the east coast of the United States. Data from Smith-McIntyre grab samples (sediment and macrofauna), water and nekton samples within and outside the dumpsite were examined. It is concluded that past dumping activities at Deepwater Dumpsite 106, located 222 km southeast of New York City, have not produced any obvious impact on the community structure or heavy metal body burdens in benthic macrofauna or nekton.

1191 - Pearce, J., C. Mackenzie, and J. Caracciolo. 1979b. Reconnaissance survey of the distribution and abundance of benthic orgnisms in the New York Bight Apex: 5-14 June 1973. U.S. National Oceanic and Atmospheric Administration. NOAA DR ERL MESA-41. Boulder, CO. 12p.

1192 - Pearce, J. B. 1980. The effects of pollution and the need for long-term monitoring. Helgolander Meeresuntersuchungen. 34:207-220.

1193 - Pearce, K. W., and J. D. Vincent. 1963. Investigation into the effects of deep sea pressures on waste materials and disposal containers. Harwell: A.E.R.E. Engineering Division. AERE-M-1254.

1194 - Pearce, K. 1977. Konvention über die tiefseeablagerung von abfallstoffen. [Deep sea disposal of wastes.] Mar. Technol. 8:169-172.

1195 - Pearson, C. R. 1968. Quelques aspects techniques de la decharge des effluents industriels dans les eaux de mer. (Some technical aspects of the discharge of industrial effluents into seawater.) La Tribune du Cebedeau. 21:416-418.

1196 - Pearson, E. A. (editor). 1960. Proceedings of the first International Conference on waste disposal in the marine environment, University of California, Berkeley, July 22-25, 1959. Pergamon Press. New York, NY. 569p. Bound in two parts - Pt. 1, p.1-278: Pt. 2, p. 279-569.

1197 - Pearson, E. A. 1966. Some developments in marine waste disposal. Jour. Proc. Inst. Sew. Purif. pp. 223-234.

1198 - Pearson, E. A., P. N. Storrs, and R. E. Selleck. 1967. Some physical parameters and their significance in marine waste Disposal. In: T. A. Olson and F. J. Burgess (editors), Pollution and Marine Ecology. Proceedings of a conference held at Texas A&M University. March, 1966. Interscience Publishers. New York, NY. pp. 297-315.

1199 - Pearson, E. A. 1972. Marine waste disposal systems. In: B. L. Edge (editor), Coastal Zone Pollution Management. Symposium Proceedings. Clemson Univ. Clemson, SC. pp. 131-154.

1200 - Pearson, E. A. 1975a. Conceptual design of marine waste disposal systems. In: A. L. H. Gameson (editor), Discharge of Sewage from Sea Outfalls. Proceedings of an Inter-National Symposium held at Church House, London. 27 August-2 September, 1974. [Suppelement to Progress in Water Technology.] Pergamon Press. New York, NY. pp. 403-413.

1201 - Pearson, E. A. 1975b. Conceptual design of marine waste disposal systems. In: E. A. Pearson and E. de Fraja Frangipane (editors), Marine Pollution and Marine Waste Disposal. (Supplement to Progress Water Technology.) Proceedings of the 2nd International Congress. San Remo. 17-21 December, 1973. Pergamon Press. New York, NY. pp. 351-358.

1202 - Pearson, E. A., and E. de Fraja Frangipane (editors). 1975. Marine Pollution and Marine Waste Disposal. Proceedings of the 2nd International Congress, San Remo, 17-21 December, 1973. (Supplement to Progress in Water Technology). Pergamon Press. New York, NY. 487p.

1203 - Pearson, E. A. 1975c. Criteria for marine waste disposal in California. In: E. A. Pearson and E. de Fraja Frangipane (editors), Marine Pollution and Marine Waste Disposal. (Supplement to Progress in Water Technology.) Proceedings of the 2nd International Congress. San Remo. 17-21 December, 1973. Pergamon Press. New York, NY. pp. 85-88.

1204 - Pearson, T. H., and S. D. Stanley. 1980. Final report on the monitoring survey carried out on 8th-12th October 1979, Garroch Head sludge dumping ground survey. Scottish Marine Biological Association. Dunstaffnage Marine Research Laboratory. Oban, Argyll. 74p.

1205 - Pearson, T. H. 1981. Final report, Garroch head sludge dumping ground survey. (May & Oct, 1980 Surveys). Scottish Marine Biological Association. Dunstaffnage Marine Research Laboratory. Oban, Argyll. 21p. 28 Figs & Appendices.

1206 - Peddicord, R. K. 1980. Impacts of open-water dredged material discharge. Conference Paper. U.S. Army Corps of Engineers. Waterways Experiment Station. Environmental Lab. Vicksburg, MS. pp. 24-48.

1207 - Peele, E. R., F. L. Singleton, J. W. Deming, B. Cavari, and R. R. Colwell. 1981. Effects of pharmaceutical wastes on microbial populations in surface waters at the Puerto Rico dumpsite in the Atlantic Ocean. Appl. Envir. Microbiol. 41:873-879.

1208 - Pentelow, F. T. K. 1961. Fish, sewage and sea outfalls. Paper presented at a meeting of Institution of Public Health Engineers and the Royal Society of Health. Blackpool. 7p. Also: Sport Fish. Abstr. #5612.

1209 - Pentreath, R. J., D. F. Jefferies, M. B. Lovett, and D. M. Nelson. 1980. The behaviour of transuranic and other long-lived radionuclides in the Irish Sea and its relevance to the deep sea disposal of radioactive wastes. Conference Paper. MAFF. Fisheries Radiobiological Lab. pp. 203-226.

1210 - Pequegnat, W. E., and T. A. Wastler. 1980. Field bioassays for early detection of chronic impacts of chemical wastes upon marine organisms. Helgolaender Meeresuntersuchungen. 33(1-4):531-545.

1211 - Perlmutter, A. 1947. The blackback flounder and its fishery in New England and New York. Bull. Bingham Oceano. Coll. 11-(2):1-92.

1212 - Pesch, G., B. Reynolds, and P. Rogerson. 1977. Trace metals in scallops from within and around two ocean disposal sites. Mar. Pollut. Bull. 8(10):224-228.

1213 - Peschiera, L., and F. H. Freiherr. 1968. Disposal of titanium pigment processing wastes. Jour. Water Poll. Control Fed. 40(1):127-131.

1214 - Peters, L. S., and J. M. O'Connor. (In Press). Factors affecting short-term PCB and DDT accumulation by zooplankton and fish from the Hudson Estuary. In: G. F. Mayer, A. Calabrese, F. A. Cross, D. C. Malins, D. W. Menzel, J. S. O'Connor, and F. J. Vernberg (editors), Ecological Stress in the New York Bight: Science and Management. Estuarine Research Federation. Columbia, SC.

Abstract: Amphipods (*Gammarus daiberi*), mysid shrimp (*Neomysis americana*), and striped bass (*Morone saxatilis*) were exposed to ^{14}C-labeled PCB and DDT in food and in water to determine contaminant accumulations under different conditions. These organisms accumulated both contaminants rapidly from the water. Levels in organisms of approximately 3-5 µg/g were observed during the first 48 hours of exposure in labeled water. PCB uptake from food and in the presence of suspended particulates was lower than that determined for direct accumulation of these compounds from water. However, the means by which foraging organisms serve to alter the transfer of organic contaminants is complex.

Bioaccumulation of chlorinated organics by estuarine organisms is dependent upon a variety of factors, including adsorption to particulates in the water column, amount of food and particulates present, and route of entry into the organism (i.e., via direct water uptake or with food). Each of these factors is discussed with regard to contaminant levels common in the Hudson estuary and the coastal marine environment of the New York metropolitan region.

1215 - Peters, N., and B. Watermann. 1979. Three types of skin papillomas of flatfishes and their causes. Marine Ecology Progress Series. 1(4):269-276.

1216 - Petersen, Harold. 1975. Micronutrient analysis of seawater samples taken at Deepwater Dumpsite 106-- May 1974. In: May 1974 Baseline Investigation of Deepwater Dumpsite 106. NOAA/ODP Dumpsite Evaluation Report 75-1. Rockville, MD. pp. 189-228.

1217 - Peterson, Bruce, and W. S. Overton. 1978. Analysis of data from New York Bight sampling program. 23p.

Abstract: Thirty sites were sampled for biological and environmental data. Data indicate there is an association between levels of metallic and hydrocarbon pollutants and measures of embryo health.

1218 - Peterson, S. A., and K. K. Randolph (editors). 1978. Management of bottom sediments containing toxic substances. Proceedings of the Third US - Japan Experts' Meeting November 1977, Easton, Maryland. US Environmental Protection Agency. Corvallis, OR. EPA-600/3-78-084. 311p.

1219 - Petrilli, F. L., G. P. De Renzi, R. P. Morelli, and S. de Flora. 1979. Survey of the pollution in a coastal area of the Tyrrhenian Sea. Aerial photography. Physico-chemical and microbiological investigations and mutagenic monitoring. Water Res. 13(9):895-904.

1220 - Philpot, W., and V. Klemas. 1979a. Detection of ocean waste in the New York Bight. University of Delaware. College of Marine Studies. Newark, DE. 43p.

Abstract: The application of remote sensing to detection and monitoring of ocean waste disposal in the New York Bight is discussed. Attention is focused on the two major pollutants in this area--sewage sludge and iron-acid waste--and on detecting and identifying these pollutants. The emphasis is on the use of LANDSAT multispectral data in identifying these pollutants and distinguishing them from other substances. The analysis technique applied to the LANDSAT data is the eigenvector. This approach proved to be quite successful in detecting iron-acid waste off the coast of Delaware.

1221 - Philpot, W. D., and V. Klemas. 1979b. Remote detection of ocean waste. SPIE. Vol. 208. Ocean Optics VI. pp. 198.

Abstract: An application of characteristic vector analysis to ocean color spectra for the detection of pollutants in water is described. The technique can be useful for detecting and distinguishing among a variety of substances in the water. The analysis also illustrates the variability of the water color spectra with both concentration and depth distribution. Both concentration and depth information may be extractable from passive multispectral data under certain circumstances.

1222 - Pike, E. B., and A. L. H. Gameson. 1970. Effects of marine sewage disposal. Water Pollut. Control London. 69(4):355-377.

1223 - Pindzola, D., C. T. Davey, and R. A. Erb. 1970. Industrial wasteline study - a system for controlled ocean disposal. Franklin Institute Research Laboratories. Philadelphia, PA. Final Rept. F-C2577. 50p.

1224 - Pindzola, D. 1971. Industrial wasteline study - A study for controlled ocean disposal. EPA 16070 EOI 08/71.

1225 - Pingree, R. D., and L. Maddock. 1977. Tidal eddies and coastal discharge. Jour. Mar. Biol. Ass. (U.K.). 57(3):869-875.

1226 - Pneumo Dynamics Corportion. 1961a. Survey of radioactive waste disposal sites. Technical Report ASD 4634-F prepared under Contract AT(04-3)-331 for AEC. 131p.

1227 - Pneumo Dynamics Corporation. 1961b. Sea disposal container test and evaluation. Technical Report under AEC contract AT(04-3)-367. 132p.

1228 - Pohlman, J. R., and M. M. Lemcoe. 1961. Strength test of protype sea disposal containers for radioative wastes. Final report. U. S. Atomic Energy Commission. San Antonio, TX. (TID-13942.)

1229 - Pohlman, J. R., and A. G. Pickett. 1962. Study of design requirements for sea disposal containers. U.S. Atomic Energy Commission. Div. of Technical Information. TID 15431.

1230 - Pojasek, R. B. 1979. Disposing of hazardous chemical wastes. Environm. Technol. 13:810-814.

1231 - Porch, D., R. W. Bayley, and A. M. Bruce. 1977. Economic aspects of sewage-sludge disposal. Stevenage Water Research Center. Technical Report No. 42. 32p.

1232 - Portmann, J. E., and J. D. Yardley. 1972. The distribution of some metals in shellfish taken from the coasts of England and Wales. ICES. C.M. 72/K:12.

1233 - Portmann, J. E., and K. W. Wilson. 1973. Disposal of wastes at sea. I. Factors to be considered in assessing the suitability of wastes for disposal. ICES. C.M. 73/E:13. 9p.

1234 - Portmann, J. E. 1974a. Working Group for the International Study of the Pollution of the North Sea and its Effects on Living Resources and their Exploitation. - A summary of the results of sewage sludge and dredging spoil analysis in North Europe. ICES. C.M. 74/E:20.

1235 - Portmann, J. E. 1974b. Disposal of sewage sludge to sea; United Kingdom experience and practice. In: A. Freiberger (editor), Pretreatment and Ultimate Disposal of Wastewater Solids. Proceedings of a Research Symposium. May 21 & 22, 1974. Rutgers University. U.S. Environmental Protection Agency. Region III. Philadelphia, PA. EPA/902/9 74 002. pp. 331-348.

1236 - Portmann, J. E. 1977. International marine pollution controls. Mar. Pollut. Bull. 8(6):126-132.

1237 - Portmann, J. E., and A. Preston. 1980. Sources and assessment of pollution. Coastal Discharges. Institution of Civil Engineers. London. pp. 17-22.

Abstract: In assessing the effect or impact of a pollutant it is necessary to have a clear objective, i.e. to know the target it is desired to protect and the environmental concentration of the pollutant that can be tolerated by that target. To some extent this can be done for most currently recognized pollutants, albeit with varying degrees of precision. However, in order to bring about cost-effective controls, it is necessary to know the major sources of each pollutant contributing to target exposure and to establish the necessary degree of control over these inputs in practical and economic terms. Information on sources of pollutants is often inadequate. Suggestions are made on where information is lacking in both source and impact assessment terms and on how priorities can be established for filling the gaps.

1238 - Portmann, J. E., and M. G. Norton. 1981. Disposal to sea of toxic materials. Chem. Ind. No. 8. pp. 285-290.

1239 - Pouch Terminal, Inc. and Carver-Greenfield Corp. 1976. An alternate to the ocean disposal of New York and New Jersey - sewage sludges - industrial wastes - oily wastes. Pouch Terminal, Inc. Staten Island, NY. and Carver--Greenfield Corp. NJ.

Summary: The Carver-Greenfield process located at the Pouch Terminal site can provide a viable economic alternative to ocean disposal barging costs. Substantial reductions in costs/dry ton can be realized with government subsidies of construction costs. Further reductions can be achieved in both barge transportation costs and processing costs if the municipal sewage sludges are prethickened by the addition of waste oils.

It is estimated that if the sewage sludges were prethickened to 8% that the 4 Carver-Greenfield process units would yield approximately $7,500,000/year in surplus energy based on oil costs of $3.00/million B.T.U.'s (steam).

The Pilot Plant (Semi-Works) has been designed to handle all of the waste materials that can be expected to be processed in the production units. Further, by allocating approximately 60% of its process time to handling oily wastes (as noted before), it is expected that the entire capital investment can be paid down within 10 years thru the sale of oil.

1240 - Powers, C. D., R. G. Rowland, and C. F. Wurster. 1976. Dialysis membrane chambers as a device for evaluating impacts of pollutants on plankton under natural conditions. Water Res. 10:991-994.

1241 - Powers, C. D., R. D. Rowland, and C. F. Wurster. 1977a. Dieldrin-induced destruction of marine algal cells with concomitant decrease in size of survivors and their progeny. Environ. Pollut. 12:17-25.

1242 - Powers, C. D., R. G. Rowland, H. B. O'Connors, Jr., and C. F. Wurster. 1977b. Response to PCBs of marine phytoplankton isolates cultured under natural conditions. Applied and Environ. Micro. 34(6):760-764.

1243 - Powers, Charles F., and Richard H. Backus. 1951. The distribution of temperature in New York Harbor and its approaches. In: John C. Ayers, et al., The Hydrography of the New York Area. Status Report No. 10. (Contract N6 DNR 264, task 15). Cornell Univ. Ithaca, NY.

1244 - Powers, Charles F., Earl E. Deubler, Jr., and Richard H. Backus. 1952. The transparency by black and white Secchi disc of the waters of New York and Newport Bights. Cruise Stirni I, July-September, 1951. In: John C. Ayers, et al., The Hydrography of the New York Area. Status Report No. 15. (Contract N6 DNR 264, task 15). Cornell Univ. Ithaca, NY.

1245 - Pratt, S. D., S. B. Saila, A. C. Gaines, Jr., and J. E. Krout. 1973. Biological effects of ocean disposal of solid waste. University of Rhode Island. Graduate School of Oceanography. Kingston, Rhode Island. Technical Report No. 9. 53p.

1246 - Presley, B. J., J. Scott Schofield, and John H. Trefry. 1981. Waste material behavior and inorganic geochemistry at the Puerto Rico waste dumpsite. In: B. H. Ketchum, D. R. Kester, and P. K. Park (editors), Ocean Dumping of Industrial Wastes. Plenum Press. New York, NY. pp. 233-246.

Abstract: Samples of a pharmaceutical waste were taken prior to dumping at a site over the Puerto Rico Trench and analyzed for the cations Ag, Al, As, Ca, Cd, Co, Cu, Fe, Hg, K, Li, Mg, Mn, Mo, Na, Ni, Pb, Sb, Sr, and Sn. Iron and Mn were found at concentrations of 170 ppm and 5.2 ppm respectively with the other heavy metals present at below 1 ppm. Of the remaining cations, the most abundant were: Ca at 290 ppm, K at 340 ppm, Mg at 33 ppm, and Na at 9,900 ppm.

The major anions were Cl^- at 8,000 ppm and SO_4^{2-} at 13,500 ppm. Total phosphorus was 105 ppm and PO_4^{3-}-P was present at 53 ppm. The quantity of organic material, pH, and physical characteristics were also determined. Analyses of surface water samples, after dumping, for Cd, Cu, Fe, Hg, and Ni indicated only Fe levels were distinctly affected by dumping. The highest value noted in the study was 30 ppb, 30 minutes after a dump. The results suggest rapid depletion of surface Fe and indicate a plume surface half-life of about 2.4 minutes.

Suspected dump-induced iron contamination of zooplankton was seen five hours after one dump but no effects were noted after 24 hours. The toxic trace metals were thus generally present in low concentrations initially and rapidly dispersed in the water after dumping.

1247 - Preston, A., and D. F. Jefferies. 1967. Assessment of the principal public radiation exposure from, the resulting control of, discharges of aqueous radioactive waste from the United Kindgom Atomic Energy Authority Factory at Windscale, Cumberland. Health Physics. 13(5):477-485.

1248 - Preston, A., D. F. Jefferies, J. W. R. Dutton, B. R. Harvey, and A. K. Steele. 1972. British Isles coastal waters: the concentrations of selected heavy metals in sea water, suspended matter and biological indicators-a pilot survey. Environ. Pollut. 3:69.

1249 - Preston, A. 1977. The study and control of environmental radioactivity and its relevance to the control of other environmental contaminants. Atomic Energy Review. 15(3):374-405.

1250 - Priano, Michael Philip, Jr. 1971. Distribution and numerical taxonomy of heterotrophic bacteria in the New York Bight, April and August, 1971. Masters Thesis. Adelphi University. Garden City, NY. 70p.

Abstract: Heterotrophic bacteria's most significant role in any ecosystem is mineralization or the conversion of organic matter to the inorganic state. The accumulation of organic wastes and enteric bacteria in the water column in the Bight might be due to changes in biochemical and mineralization activities due to changes in aerobic heterotroph population.

Numerical taxonomy was used to show the distribution of aerobic heterotrophs in the New York Bight. The data obtained is intended to add to the information of the Bight and is too limited in scope to show changes in the Bight's micro-ecology.

1251 - Pritchard, E. W., A. Okubo, and E. Mehr. 1962. A study of movement and diffusion of an introduced contaminant in New York Harbor waters, Chesapeake Bay Institute. Tech. Report 31. 89p.

1252 - Proni, J. R., F. C. Newman, R. L. Sellers, and C. Parker. 1976. Acoustic tracking of ocean-dumped sewage sludge. Science. 193:1005-1007.

Abstract: With a modified 200-kilohertz acoustic echo sounder, it has been possible to detect and map sewage dumped into the ocean over several hours. The three-dimensional distribution of suspended material and its rate of diffusion are indicated after digital processing of the data.

1253 - Proni, J. R., and Donald V. Hansen. 1981. Dispersion of particulates in the ocean studied acoustically. The importance of gradient surfaces in the ocean. In: B.H. Ketchum, D.R. Kester, and P.K. Park (editors), Ocean Dumping of Industrial Wastes. Plenum Press. New York, NY. pp. 161-173.

Abstract: Knowledge of the rate of dispersion of particulate matter dumped in the upper ocean and of the processes governing that dispersion is fundamental to an understanding of the effects of dumping on the ocean environment. Accurate chemical dilution rates of dumped material cannot be determined without a knowledge of the spacial distribution of ocean dumped material and the way in which that distribution changes with time. Acoustical methods offer one approach to studying ocean dispersion of dumped material. Acoustical results presented herein show some of the complicated effects which gradients in oceanic parameters, such as temperature and density, produce in the dispersion of particulate matter dumped in the ocean.

1254 - Psuty, N. P., K. F. Nordstrom, R. W. Hastings, and S. Bonsall. 1974. Dredged spoil disposal on the New Jersey wetlands: The problem of environmental impact assessment. Shore and Beach. 42(1):25-30.

Abstract: The advantages and disadvantages of maintenance dredging along parts of the New Jersey Intracoastal Waterway are contrasted and discussed. Continued maintenance dredging cannot be accomplished without some evironmental destruction. The major environmental problems are associated with the disposal of dredged spoil, the destruction of benthic organisms caught in the dredging operation, changes in the quality of water and bottom sediments, and the burial of organisms by the dumping of dredged spoil. Continued dredging practices should be based on a restructured cost-benefit ratio. Recreational, commercial, and environmental variables also should be considered. Alternatives to dredge spoil disposal on the marsh include creation of tidal marshlands, development of shoals for shellfish, beach nourishment, and an improvement of the sandy soils of the adjacent coastal uplands. Recommended studies include analyses of of the long-term effects of subaqueous disposal on benthic organisms; the recolonization and succession of spoils including beach fill, soil enhancement, and construction; and environmental impacts related to aesthetics.

1255 - Public Hearing on Relocating Sewage Sludge Ocean Dumping Sites. 1977. Report of the Hearing Officer. Toms River, NJ. 103p.

Abstract: Major topics covered by the hearing were: a) wether or not sludge dumping at the present New York Bight Site was the cause of fish kills in 1976. b) the potential and present impacts of sludge dumping at the present sites. c) benefits that would ensue from cessation of dumping. d) adverse effects of dumping on new sites. e) the balance between the cost of using more distant sites and the benefits that would result. f) the need for an EIS to support any relocation of sludge dumping sites. g) the statutory requirement that sites off the continental shelf shall be utilized where feasible.

1256 - Rachor, E. 1972. On the influence of industrial waste containing H_2SO_4 and $FeSO_4$ on the bottom fauna off Hegoland (German Bight). In: M. Ruivo (editor), Marine Pollution and Sea Life. Fishing news (Books) Ltd. London. pp. 390-392.

1257 - Rachor, E., and V. Dethlefsen. 1974. Effects of acid-iron waste disposal in the marine environment. ICES. C.M. 74/E:27.

1258 - Rachor, E. 1977. Faunenverarmung in einem schlickgebiet in der nahe Helgolands. Helgolander wiss. Meeresunters. 30:633-651.

1259 - Rachor, E. and S. A. Gerlach. 1978. Changes of macrobenthos in a sublittoral sand area of the German Bight, 1967 to 1975. Rapp. P.-v. Reun. Cons. Int. Explor. Mer. 172:418-431.

1260 - Rachor, E. 1979. Probleme der koexistenz von fischerei und anderen wirtschafszweigen bei der nutzung des deutschen festlandsockels der Nordsee. Arbeiten des Deutschen Fischerei-Verbandes. H. 27:135-145.

1261 - Rachor, E. 1980. The inner German Bight - an ecologically sensitive area as indicated by the bottom fauna. Helgolander Meeresunters. 33:522-530.

1262 - Rawn, A. M. and F. R. Bowerman. 1954. Disposal of digested sludge by dilution. Sewage and Industrial Wastes. 26:1309-1314.

1263 - Redfield, Alfred C., and Lionel A. Walford. 1951. A study of the disposal of chemical waste at sea. Report of the committee for investigation of waste disposal. Nat. Acad. Sci. Nat. Res. Council. Pub. No. 201. Washington, DC. 49p.

1264 - Reed, A. W. 1975. Ocean waste disposal practices. Noyes Data Corporation. Park Ridge, NJ. 336p.

1265 - Reed, L. 1972a. An ocean of waste. Conservative Political Centre. London CPC No. 500. 74p. (ISBN 0 85070 496 0).

1266 - Reed, L. 1972b. Waste management in the oceans. Fish. Ind. Rev. 2(3):2-4.

1267 - Rees, E. I. S., A. J. M. Walker, and A. R. Ward. 1972. Benthic fauna in relation to sludge disposal. U.K. Department of the Environment. Out of sight - out of mind. Vol. 2. pp. 297-343.

1268 - Reid, P. C. 1975. Large scale changes in North Sea phytoplankton. Nature. 257:217-219.

1269 - Reijnders, P. J. H. 1980. Organochlorine and heavy metal residues in Harbour seals from the Wadden Sea and their possible effects on reproduction. Netherlands Jour. of Sea Res. 14(1):30-65.

1270 - Renzoni, A. 1973. Trouble for the Mediterranean. Mar. Pollut. Bull. 4:19.

1271 - Reutter, John G. Associates-Consulting Engineers. 1979. Report to the Atlantic County Sewerage Authority on the ocean monitoring program phase I May, 1979. John G. Reuther Associates. Camden, NJ. 175p.

1272 - Rice, E. K., and J. W. Johnson. 1954. Waves as a factor in effluent disposal. How wind mixes and moves the surface layers of large bodies of water. Water Sewage Works. 101. (4):172-175.

1273 - Richards, F. A. 1969. Some chemical and geochemical processes which interact and influence the distribution of wastes introduced into the marine environment. U.S. National Academy of Sciences. Background papers on coastal wastes management. Washington, DC. pp. XI-1 to XI-25.

1274 - Riley, B. T. Jr. 1975. Summation of conditions and investigations for the complete combustion of organic pesticides. Municipal Environmental Research Lab. Cincinnati, OH. EPA-600/2-75-044. 69p.

1275 - Roethel, F. J., J. H. Parker, I. W. Duedall. 1977. Coal waste disposal in the ocean: II. construction of an artificial reef from stabilized scrubber sludge and fly ash. Trans. Amer. Geophys. Union, Abstr. Publ. 58(12):1176.

Abstract: Nine blocks of stabilized scrubber sludge, one cubic foot in volume, were placed in approximately twenty feet of water (mean high tide) in an area known as Conscience Bay in Long Island Sound. These blocks were arranged to maximize the surface area in contact with the seawater and also produce little holes and crevices. A duplicate habitat was constructed of the same size and shape concrete blocks. Within three weeks after placement of these blocks, a heavy crop of attached plants and animals were observed growing on both the concrete and scrubber sludge blocks. Measurement of community respiration was conducted bi-weekly and samples of the attached organisms were assayed for lead, zinc, cadmium, mercury, selenium, silver, arsenic and copper. Detailed photographic surveys reveal no movement of the scrubber blocks, however, some slight burial of these blocks due to the strong tidal currents in the bay appears evident. This continuing research is the first phase in determining whether disposal of stabilized scrubber sludge in the marine environment is an acceptable alternative to land disposal.

1276 - Rogers, B. A. 1972. Toxicity of incinerator residues for marine organisms. Toxicity of suspended mineral solids to estuarine fish. In: M. W. First (editor), Municipal Waste Disposal by Shipborne Incineration and Sea Disposal of Residues. Report. Harvard University School of Public Health. Boston, MA. pp. 4:71-4:105.

1277 - Rohatgi, N. K., and K. Y. Chen. 1976. Fate of metals in wastewater discharge to ocean. Jour. Environ. Eng. Div. Amer. Soc. Civil Eng. 102(EE3). pp. 75-85.

1278 - Rolfe, M. S. 1973. Disposal of wastes at sea. II. Factors to be considered in the selection of a dumping ground. ICES. C.M. 73/E: 14. 9p.

1279 - Ropes, J. W., A. S. Merril, S. A. Murawski, S. Chang, and C. L. Mackenzie, Jr. 1979. Impact on clams and scallops. Part 1. Field survey assessment. In: R.L. Swanson and C.J. Sinderman (editors), Oxygen Depletion and Associated Benthic Mortalities in New York Bight, 1976. NOAA Professional Paper No. 11. U.S. Department of Commerce. Rockville, MD. pp. 263-275.

Abstract: For surf clams, a 6,750-km^2 area of mortality was delimited off New Jersey. The area extended from immediately north of Manasquam Inlet to immediately south of Atlantic City, and seaward to about 37 mi. An almost complete kill (92%) took place in the mortality area, but was least intense in the 3- to 15 km-wide beach zone. An estimated 61.5 percent of the total surf clam biomass was lost off New Jersey. Surf clam landings were substantially lower (31%) in New Jersey during 1976 than during 1975.

The principal ocean quahog resource occurs deeper than 37 m and thus only the shoreward margin of the population was affected. The mortality area was 9,105 km^2, and 25.4 percent of the quahog biomass within it was lost. Of the entire New Jersey ocean quahog resource, 6.3 percent of the biomass was lost. New Jersey vessels began fishing ocean quahogs in 1976 and landed 71.7 percent of the U.S. total.

The principal sea scallop resource also occurs deeper than 37 m, and only the shoreward margin of the population was affected. From 8.8 to 12.9 percent of the entire New Jersey scallop resource was killed. Scallop landings increased four-fold in 1976.

1280 - Rosen, Jerome M. 1972. Analysis of explosives in sea water. Bull. of Environ. Contam. and Toxicol. 7(2/3):177-181.

1281 - Rosenfeld, A. 1976. Infectious diseases in commercial shellfish on the Middle Atlantic coast. In: M.G. Gross (editor), Amer. Soc. Limnol. and Oceanogr. Spec. Symp. 2:414-423.

Abstract: Several important molluscan and crustacean food organisms of the middle Atlantic coast of the United States are subjected to natural and man-caused perturbations and to infectious deseases.

Among the naturally occurring diseases, the protozoan oyster parasites *Minchinia nelsoni* and *Minchinia costalis* cause devastating oyster mortalities, particularly in Delaware and lower Chesapeake Bays. *Minchinia nelsoni* infections continue to be prevalent in Delaware Bay, with apparent resistance to the disease having developed there.

Several other infectious diseases agents cause pathology or damage to these bivalve molluses including viruses, bacteria, other protozoans, fungi, helminths, and crustacean parasites. Except for *Minchinia* and *Labyrinthomyxa marina* (=*Dermocystidium marinum*), none of these organisms is known to cause massive mortalities of adult bivalve molluses. However, expression of herpes-type inclusions in hemocytes may be enhanced in oysters exposed to thermal effluents.

Proliferative cell conditions in marine shellfish have been reported with increased frequency recently and neoplasias of molluscs in exizootic proportions have been reported from both U.S. coasts. Gonadal neoplasms in soft-shell clams (*Mya arenaria*) have been found in areas associated with oil spills. Studies of benthic crustaceans from Sandy Hook Bay and the New York Bight show that the parasitic ameoba *Paramoeba perniciosa* occurs in rock crabs *Cancer irroratus* and American lobsters *Homarus americanus*.

Lobsters, shrimp, and rock crabs near sludge and spoil disposal grounds in the New York Bight had shell erosion and ulcers. Naturally occurring bacteremias in lobsters and blue crabs have been reported and may play a role in crustacean eqizootics.

Viruslike infections in blue crabs from Chincoteague Bay have recently been noted. Polychlorinated biphenyls may activate production of a *Baculovirus* in shrimp exposed to sublethal levels of these compounds. Viruses may be latent in shellfish and activated by the presence of certain chemicals.

1282 - Rosset, F. 1962. Experience de dispersion d'un effluent dans la mer. Cah. Oceanogr. 14:103-131.

1283 - Rothmann, H., and D. Bartlett. 1977. Sewage - pollutant or food? An assessment of alternative ways of disposing of sewage. ISTA Doc. Ser. 1977. No. 1. pp. 161-167.

1284 - Rowe, G. T. 1971. The effects of pollution on the dynamics of the benthos of New York Bight. Thalassia Jugoslavia. 7(1):353-359.

1285 - Rowe, Gilbert T., Kenneth L. Smith, Jr., and C. Hovey Clifford. 1976. Benthic-pelagic coupling in the New York Bight. In: M. G. Gross (editor), Amer. Soc. Limnol. Oceanogr. Spec. Symp. 2:370-376.

Abstract: Bottom sediment oxygen consumption can be used to estimate how much of the energy incorporated by the phytoplankton is utilized by the benthos, and such rates are directly related to temperature, organic matter in the sediment, availability of dissolved oxygen, and primary production in the water column. Nutrient flux out of sediments has been measured directly by incubating areas of bottom under bell jar-like chambers. Fluxes of major inorganic plant nutrients are often high, indicating that in most mearshore environments most regeneration occurs on the bottom.

In the New York Bight, oxygen consumption by the bottom was measured in four different seasons; we concluded it was high enough to oxidize a large fraction of the daily input of sludge. Bottom water ammonia gradients suggested too that remineralization rates were high on the bottom in the Bight. Samples taken in August 1975, in Christiaensen Basin, along with measurements in situ of ammonia flux from the bottom, confirmed that the sediments enriched by sewage sludge are regenerating nutrients but at that time not at rates as high as our earlier predicted rates for the mid-Atlantic Bight.

1286 - Rowe, G. T., R. L. Haedrich, P. T. Polloni and C. H. Clifford. 1977. Epifaunal megabenthos in DWD-106. In: Baseline report of Environmental Conditions in Deepwater Dumpsite 106. NOAA/ODP Dumpsite Evaluation Report 77-1. Rockville, MD. Vol. II. pp. 459-464.

1287 - RWS North Sea Directorate. 1977. Disposal of chemicals used in offshore industry. ICES. C.M. 77/B:14.

1288 - Ryther, J. H. 1954. The ecology of phytoplankton blooms in Moriches Bay and Great South Bay, Long Island, New York. Biol. Bull. 106:198-209.

1289 - Roper, M. M., K. C. Marshall. 1978. Biological control agents of sewage bacteria in marine habitats. Aust. Jour. Mar. Freshwater Res. 29:335-343.

1290 - Sabo, J. J., and P. H. Beckosian. (editors). 1963. Studies of the fate of certain radionuclides in estuarine and other aquatic environments. PHS Publication. No. 999-R-3.

1291 - SACSA (OSLO). 1975. The incineration of industrial waste at sea. SACSA(75)9. Paper presented by the French Delegation to the OSLO Commission. 22p. (Translated to English).

1292 - SACSA (OSLO). 1976. Incineration at sea of industrial wastes containing chlorine. Report on the experiments in connection with the activities of the incinerator ship Matthias II used by the Incimer Company. Meeting of the Working Party on Incineration at Sea. The Hagne, 24th May. SACSA(76)3 France. 44p. (Translated to English).

1293 - Saila, S. B., T. T. Polgar, and B. A. Rogers. 1968. Results of studies related to dredge sediment dumping in Rhode Island Sound. Proc. Annual New England Anti-Pollution Conference. pp. 71-80.

1294 - Saila, S. B. 1971. Container dumping in the North Sea. Mar. Pollut. Bull. 2:125-127.

1295 - Saila, Saul B., Robert A. Pikanowski, and Douglas S. Vaughan. 1976. Optimum allocation strategies for sampling benthos in the New York Bight. Estuar. and Coastal Mar. Sci. (4):119-128.

Abstract: Sampling plans based on historical data available from the Marine Ecosystems Analysis (MESA) New York Bight Project are described and illustrated. These plans are efficient in the sense that they provide the requisite levels of precision for the minimum expenditure of time and effort. Sampling for several variables simultaneously in a two-stage scheme can also be optimized. The optimization procedure is explained and illustrated in detail and the results for a realistic case are given.

Relatively few replicates (approximately five) are shown to be required to test reasonable defined hypothesis concerning differences in trace elements between two stations. It is also shown that an optimum two-state sampling plan for *Cerianthus americanus* involved relative few replicates within a station (3) and a fairly small number of stations (7) in order to satisfy derived variance and cost constraints for the New York Bight with an acceptable level of confidence (90%) for 50% differences in abundance of transformed mean values. Sample sizes for six other species were also determined and a compromising allocation is discussed.

1296 - Saila, S. B., E. L. Anderson, and H. A. Walker. 1978. Sampling design for some trace elemental distributions in New York Bight sediments. In: K.L. Dickson, J. Cairns, Jr. and R.J. Livingston (editors), Biological Data in Water Pollution Assessment: Quantitative and Statistical Analyses. American Society for Testing and Materials. ASTM STP 652. Washington, DC. pp. 166-177.

Abstract: The paper outlines a model to describe the spatial distribution of selected trace elements (chromium, copper, nickel, lead, and zinc) in New York Bight sediments. Empirical tests of the fit of existing data by means of cumulative probability plots on log-normal probability paper and goodness-of-fit tests demonstrate conformity to the log-normal distribution predicted by the model. An adaptive sampling strategy for describing the spatial distribution of the trace elements is developed, based on a knowledge of the sampling distributions of the elements. The spatial distribution of the sampling stations conforms to the expected nonuniformity of the trace metal concentrations of interest, providing for a more efficient allocation of sampling effort.

1297 - Samuels, William Benjamin. 1979. Bioassay of New York harbor waters using the marine diatom *Skeletonema costatum* (Greville) Cleve as test organism. Ph.D Dissertation. Fordham University. New York, NY. 161p.

Abstract: Field measurements of *Skeletonema costatum* standing crop and laboratory bioassays of growth rate as a function of harbor water concentration and treatment were made for a fifteen month period (January, 1977 to March, 1978) at one station in New York Harbor. The seasonal growth cycle of *S. costatum* was: maximum standing crop in the late winter-early spring, a decline in the late spring to a minimum during the summer, and a subsequent recovery of the crop with a recurrent maximum in the late winter and early spring. Growth rates followed the same pattern. Laboratory analysis showed nutrient saturation throughout the year.

1298 - Sandy Hook Marine Laboratory. 1969. Biological factors to be considered in disposal of obsolete gas munitions in the sea. Report to the Dept. of the Interior Working Group on Possible Effects of Disposal of Chemical Warfare Munitons in the Sea. June 27. 21p. (mimeo).

1299 - Sandy Hook Laboratory. 1971. The effects of waste disposal in the New York Bight. National Marine Fisheries Service, Middle Atlantic Coastal Fisheries Center. Highlands, NJ. Technical Report. Vol. 9. 749p. NTIS No. AD 7353 through AD 739539.

1300 - Sandy Hook Laboratory. 1972a. The effects of waste disposal in the New York Bight. Final Report submitted to the U.S. Army Corps of Engineers, Coastal Engineering Research Center by the NOAA/NMFS Middle Atlantic Coastal Fisheries Center. Highlands, NJ.

1301 - Sandy Hook Laboratory. 1972b. The effects of waste disposal in the New York Bight. Summary Final Report, Middle Atlantic Coastal Fisheries Center, National Marine Fisheries Service. Highlands, NJ. 70p. NTIS No. AD 743936.

1302 - Sar, Uhashi-Katsuko. 1981. Ocean dumping of radioactive waste. Kogai Kenk Yu. 10. pp. 11-21. Review of ocean dumping of radioactive wastes.

1303 - Saruhashi, K. 1976. Disposal of radioactive waste into deep seas. Jour. of Radiation Res. 17(1):42-43.

1304 - Sawyer, T. K., R. Cox, and M. Higginbottom. 1970. Hemocyte values in healthy blue crabs, *Callinectes sapidus*, and crabs infected with the amoeba, *Paramoeba perniciosa*. J. Invertebr. Pathol. 15:440-446.

1305 - Sawyer, T. K., S. A. Maclean, and J. Ziskowski. 1976. A report on *Ephelota* sp. (Ciliata, Suctorida) as an epibiont on the gills of decapod crustaceans. Trans. Amer. Micros. Soc. 95(4):712-717.

1306 - Sawyer, T. K., G. S. Visvesvara, and B. A. Harke. 1977. Pathogenic amoebae (Acanthamoeba) from brackish and ocean sediments, with a description of *Acanthamoeba hatchetti*, N.Sp. Science. 196:1324-1325.

1307 - Sawyer, T. K., S. A. MacLean, J. E. Bodammer and B. A. Harke. 1979. Gross and microscopical observations on gills of rock crabs *Cancer irroratus* and lobsters *Homarus americanus*) from nearshore waters of the Eastern United States. In: D.H. Lewis, and J.K. Leong (editors), Proceedings of the Second Biennial Crustacean Health Workshop. Galveston, TX. TAMU-SG-79-114. pp. 68-91.

Abstract: Rock crabs, *Cancer irroratus*, were collected from Sandy Hook and Raritan Bays, New Jersey, the New York Bight Apex near Ambrose Light, and from coastal waters of the eastern United States, ranging from Maine to North Carolina. American lobsters, *Homarus americanus*, were collected only from the New Jersey-New York stations. Visual examinations of gills were made to record their condition as clean, discolored, or black, and histological studies were made to study gill-fouling microorganisms and internal parasites. Fouling organisms included bacteria, diatoms, peritrich and suctorian sessile ciliates, amoebae, and copepods. Internal parasites included amoebae, dinoflagellates, probable microsporidans, gregarines, acanthocephalans, and larval cestodes, trematodes, and nematodes. Evidence of bacterial phagocytosis and cellular nodule formation with melanization was extensive in gill, hepatopancreas, and hindgut hemolymph channels. Gross observations on over 3,000 specimens and histological findings in over 1,000 of them are summarized.

1308 - Sawyer, T. K. 1979. Species diversity among marine protozoa in sediment from a sewage disposal site. Jour. of Protozool. Vol. 26. No. 3. Pt. 1. p. 22A. No. 60.

Abstract: Sea-bottom sediments were collected from 10 stations located in or near a New York sewage disposal site. Water depth ranged from 31 to 107 feet, and sediments ranged from clean sand to foul black silty mud. Most of the study area has been closed to shellfishing and is known to have persistently high numbers of fecal and total coliform contaminants. Each sample was cultured in triplicate (3 subsamples) on agar prepared in distilled-water, brackish-water, or seawater to which bacteria were added as a food source. Five of the stations yielded one or more species of _Acanthamoeba_ (_castellanii_, _rhysodes_, _polyphaga_, _culbertsoni_, _hatchetti_). Upon subculture all 5 were found to grow throughout the range of distilled-water to seawater. _Acanthamoeba_ usually did not appear on parent cultures made with seawater because marine species usually cleared bacterial food before slower growing species became established. Since the _Acanthamoeba_ species all are well-known soil or freshwater amoebae, it is possible that they are introduced in the disposal wastes or are native species which flourish wherever bacterial food is abundant. Full-strength seawater-agar plates yielded: _Paramoeba pemaquidensis_, _P. aesturina_, _Vexillifera ottoi_, _Flabellula sp._, _Clydonella vivax_, _C. wardi_, _C. rosenfield_, _Platyamoeba flabellata_, _P. langae_, _P. murchelanoi_, _Lingulamoeba leei_, _Hyalodiscus angelovica_, _Stygamoeba polymorpha_, and unidentified small limax amoebae. All of the marine species previously were identified from surface waters of Chincoteague Bay, Virginia. Other incompletely identified species included a minute heliozoan, a small testacean of about 30 µm in diameter, a colorless _Labyrinthula_ with distinct slime tracks, and a species with reticulate filose pseudopodia which resembled _Lieberkuhnia_ and a large plasmodial form which resembled proteomyxans belonging to the order Leptomyxida. The last two forms were isolated only from black sewage-sludge sediment.

1309 - Sawyer, T. K., M. Light, B. A. Harke, and E. J. Lewis. 1979. Marine amoebae (Protozoa: sarcodina) as potential indicators of oiled bottom sediments in the Gulf of Mexico. Trans. Amer. Microsc. Soc. 98:153.

1310 - Sawyer, Thomas K. 1980. Marine amebae from clean and stressed bottom sediments of the Atlantic Ocean and Gulf of Mexico. Jour. of Protozool. 27(1):13-32.

1311 - Sawyer, Thomas K. (In Press). Distribution and seasonal incidence of "Black Gill" in the rock crab, *Cancer irroratus*. In: G. F. Mayer, A. Calabrese, F. A. Cross, D.C. Malins, D. W. Menzel, J. S. O'Connor and F. J. Vernberg (editors), Ecological stress in the New York Bight: Science and Management. Estuarine Research Federation. Columbia, SC.

Abstract: Rock crabs, *Cancer irroratus*, from coastal sewage dumpsites near New York, New Jersey, Delaware, and Maryland were examined for evidence of stress of disease attributable to the effects of ocean dumping. Excessive gill fouling by bacteria, debris, diatoms, copepods, amoebae, and stalked ciliates frequently was noted in microscopical studies but could not be used as evidence for environmental stress. Gills that were black over 50% or more of their tissue mass were observed in crabs collected from ocean disposal sites and appeared to be useful indicators of environmental degradation. The "black gill" condition is believed to be caused by the accumulation of black sludge sediment between the gill lamellae. Summarized data showed that blackened gills were present in 4.5% (134/2,920) of the crabs on a year-round basis, or in up to 10% when only intermolt animals were considered. Blackened gills were not observed within a second group of 268 crabs collected from control locations (Georges Bank; Montauk Point, New York; Sheepscott River, Maine). An additional 846 specimens were collected at stations from Cape Cod, Massachusetts to Cape Hatteras, North Carolina during surf clam assessment and scallop cruises. Eight percent (17/217) of the crabs from the surf clam stations and less than 1% (3/629) of the crabs from the scallop cruise had black gills. Station data from the two cruises showed that affected crabs usually were collected from locations in or near sewage disposal sites. Observations on the incidence of black gills in *C. irroratus* provided evidence that the condition occurs primarily in areas where bottom sediments are altered already by the accumulation of sewage sludge deposits.

1312 - Sawyer, Thomas K. and Shawn M. Bodammer. (In Press). Marine amoebae, protozoa: Sarcodina, as indicators of healthy or impacted sediments in the New York Bight Apex. In: I. W. Duedall, B. H. Ketchum, P. K. Park and D. R. Kester (editors), Wastes in the Ocean. Vol. I: Industrial and Sewage Wastes in the Ocean. John Wiley and Sons, Inc. New York, NY.

Abstract: Thirty-two species of marine amoebae were identified from the Northwest Atlantic Ocean in the New York Bight Apex near active sewage, dredge, and acid waste dumpsite. Twenty-seven of the 32 species were present in surface waters, 25 in bottom waters, and 20 in sediment core samples. Bottom cores from the sewage site yielded 11 of the 32 species, acid waste cores 11, dredge spoil cores 9, and control station cores 13. Only 3 of the 12 species were present in cores from all of the collection sites - *Paramoeba pemaquidensis*, *Clydonella vivax*, and *Platyamoeba langae*. The recovery of the three species from all collection sites indicated that they might be useful indicators for monitoring ocean sediments for effects more serious than those brought about by existing disposal practices. The presence of 11 amoeba species from the New York sewage dumpsite, already depleted of most macrofaunal species, indicated that the polluted sediments still support microfaunal species which contribute to organic decay and nutrient regeneration. The 20 species present in sediments are bactivorous and thrive at the lowest level of the food web. Environmental modifications which might disrupt protozoan/bacterial interactions in surficial sediments can be measured qualitatively by monitoring the seabottom for changes in protozoan species diversity. Preliminary studies on species diversity showed that, when seabottom sediments were cultured on distilled-water agar to test for physiologically adaptable species, well-known freshwater or soil species of *Acanthammoeba* were common at stations harboring sewage-associated bateria. Sewage sites yielded species of *Acanthamoeba* from 20 to 80% of the sediment samples, and sites that were not affected by sewage sludge yielded the amoebae from 0 to 5% of the samples.

1313 - Sayler, G. S., R. Thomas, and R. R. Colwell. 1978. Polychlorinated biphenyl (PCB)-degrading bacteria and PCB in estuarine and marine environments. Estuar. and Coast. Mar. Sci. (6):553-567.

Abstract: Surface water and sediment samples collected throughout the Chesapeake Bay and along the southeast Atlantic Coast, extending from Miami, Florida to Cape Hatteras, North Carolina, were analyzed for the presence of polychlorinated biphenyl (PCB)-degrading bacteria and PCB. PCB-degrading bacteria were recovered from all samples for which enumeration was attempted. In addition, the occurrence of PCB was detected in all samples analyzed for PCB. Both PCB-degrading bacteria and PCB were found in higher concentrations in estuarine waters and sediment, compared with marine samples. The data obtained indicated a greater correlation between PCB concentrations and numbers of PCB degrading bacteria in areas of urbanization as opposed to areas distant from land. These results suggest assessment of existing PCB-degrading bacteria provides a potential indicator of PCB contamination. Because PCB-degrading bacterial populations were much more highly correlated with nutritional factors governing microbial growth, the extent of microbial PCB degradation by natural microbial populations remains undetermined.

1314 - Schaeffer, A. A. and S. R. Das. 1972. Occurrence of pathogenic *Naegleria aerobia*, *Hartmannella culbertsoni* and *H. rhysodes* in sewage sludge samples of Lucknow. Curr. Sci. 41. pp. 277-281.

1315 - Schaeffer, Oliver A. Date unknown. Anthropogenic fluxes of carbon into the sediments of the New York Bight. Final Report to NOAA, MESA New York Bight Project. Dept. Earth & Space Sciences. State Univ. of New York. Stony Brook, NY. 16 p.

Abstract: In order to evaluate the influence of human activity and the organic carbon in the New York Bight, a study of isotopic carbon variations in surface sediments and one 63 cm long sediment core has been completed. On the basis of our $^{13}C/^{12}C$ measurements, it appears that: (1) sewage is depleted in ^{13}C as compared to average organic sedimentary carbon, and (2) ^{13}C values from sediments of the New York Bight show systematic variations and are lowest near the sewage dump site.

1316 - Schafer, H. A., and W. Bascom. 1976. Sludge in Santa Monica bay. Ann. Rep. Coastal Water Res. Proj. pp. 77 - 82.

1317 - Scheraga, M., M. Meskill, and C. D. Litchfield. 1979. Analysis of methods for the quantitative recovery of bacteria sorbed onto marine sediments. Amer. Soc. for Testing and Materials. Washington, DC. STP 673. 18p.

Abstract: Adsorption and desorption studies using both motile and nonmotile bacterial strains were performed with sedimentary sludge, clay, and sand in order to determine the most reproducible means for the recovery of viable bacteria from particles. The organisms used were _Flavorbacterium oceanosedimentum_, _Aeromonas proteolytica_, and _Escherichia coli_, EPA 104. Viable cell numbers were determined using surface spread plate counts, direct counts via epifluorescence microscopy, and adenosine triphosphate analysis (ATP). Adsorption of the organisms onto sedimentary materials obtained from the New York Bight followed the Langmuir adsorption isotherm regardless of the length of time the cultures were in contact with the sediment. This implies that the initial events in the attachment of both nonmotile and motile bacteria to these types of sediment are primarily due to physical chemical forces. Desorption studies employed acidic, neutral, and basic surface-active agents; acid and basic pH treatments; osmotic shock; and peptone-water treatments. Physical manipulation of the sediment via shaking or agitation in the presence of 0.00001 percent cetyl trimethylammonium bromide appears to be the most effective and least toxic procedure for the recovery of viable microorganisms.

1318 - Schmidt, D. 1976. Distribution of seven trace metals in sea water of the Inner German Bight. ICES. C.M. 76/C:10.

1319 - Schmidt, D. 1977. Use of aircraft remote sensing in marine chemistry and pollution research with particular reference to the dumping of TiO_2 waste solutions. ICES. C.M. 77/C:10.

1320 - Schmidt, D. 1978. Fernerkundung von meeresverschmutzung mit flugzeugen und satelliten. Sonderdruck aus Meerestechnik. Nr. 1. pp. 27-28.

1321 - Schubel, J. R., W. M. Wise, and J. Schoof (editors). 1979. Questions about dredging and dredged material disposal in Long Island Sound. Marine Sciences Research Center. State Univ. of New York. Stony Brook, NY. Special Report 28 Ref. 79-11. 136p.

Abstract: This booklet was designed to provide, when read from start to finish, an over-view of the history of dredging and dredged material disposal in the Long Island Sound, an assessment of how these activities have affected the Sound and its biota, and examination of alternative modes of disposal, and a general discussion of research priorities. This booklet was also designed to provide answers to specific questions you may have about these topics without having to read the entire volume.

The questions were compiled at a series of workshops in which scientists, environmental decision makers, and lay people participated.

1322 - Schwab, Claude Raymond. 1980. The effects of ocean-dumped wastes on marine phytoplankton. Masters Thesis. Texas A&M University. College Station, TX. 104p.

Abstract: The effects of ocean-dumped wastes on the marine phytoplankton *Skeletonema costatum* were studied by monitoring the ^{14}C uptake and the adenosine triphopsphate (ATP) levels of laboratory cultures exposed to varying levels of waste. Measurements of ^{14}C uptake and ATP levels were found to be an effective way to monitor the physiological state and function of the cultures. These parameters provided sensitive and rapidly responding data that, when compared with each other, gave an insight into the types of chemicals which cause the observed effects and their mode of action on the phytoplankton.

Wastes from a Gulf of Mexico and a Puerto Rico dumpsite was studied. The Gulf of Mexico waste was a biosludge originating in the chemical plant biotreater of the Shell Chemical Company Deer Park Manufacturing Complex at Deer Park, Texas. At levels less than approximately 5%, this waste caused significant inhibition of both ^{14}C uptake and ATP compared to controls, whereas higher levels of most forms of the waste were toxic to phytoplankton. At the high dilutions normal in ocean dumping from a barge, the only probably acute effect would be slight stimulation of ^{14}C uptake.

The Puerto Rico dumpsite waste studied was a variable combination of batch and process wastes from several pharmaceutical manufacturing companies. A composite waste as well as samples from individual companies were tested. Although their effects on ATP were somewhat variable in general, they caused severe reductions in ATP levels. All caused drastic inhibition of ^{14}C uptake even at very low levels of waste, and therefore would be expected to affect phytoplankton photosynthesis in the dumpsite area even at high dilutions.

1323 - Schwab, Claude R., Theodor C. Sauer, Jr., Guinn F. Hubbard, Hussein Abdel-Reheim and James M. Brooks. 1981. Chemical and biological aspects of ocean dumping at the Puerto Rico dumpsite. In: B. H. Ketchum, D. R. Kester and P. K. Park (editors), B. H. Ketchum, D. R. Kester and P. K. Park (editors). Ocean Dumping of Industrial Wastes. Plenum Press. New York, NY. pp. 247-273.

Abstract: Chemical and biological studies at the Puerto Rico Ocean Dumpsite during February 1978 are reported. Dye and various chemical and biological parameters were measured with time within the waste plume after the dumps. Several parameters (e.g., ammonia, POC, DOC, volatile organics, and transmissometry traces) were elevated in the waste plume compared to the baseline values and were, therefore, very useful in tracking the dispersion of the waste.

The volatile organics consitute the major non-aqueous fraction of the waste. Butanol and dimethyl aniline constitute about 70% of this fraction. Dimethyl aniline was very persistent in the dumpsite having a concentration of 0.1 $\mu g\ l^{-1}$ three days after dumping. Evaporation studies indicate that benzene and toluene concentrations decrease to background levels within 12 hrs following dumping. Many hydrocarbons, halocarbons, alcohols, esters, ketones and nitrogen compounds were identified in the waste and surface waters of the dumpsite.

Phytoplankton studies indicate no significant differences existed between samples taken from similar depths inside or outside of the dumpsite. Laboratory toxicity tests using ^{14}C uptake and ATP indicate that 0.14% of the composite waste causes a 76% decrease in growth of the phytoplankton Skeletonema costatum.

1324 - Scoullos, M., F. Oldfield, and R. Thompson. 1979. Magnetic monitoring of marine particulate pollution in the Elefsis Gulf, Greece. Mar. Pollut. Bull. 10:287-291.

Abstract: Measurements of magnetic susceptibility and saturation isothermal remanent magnetization (SIRM) on both filtered particulates and sediment samples provide a cheap, rapid and non-destructive method of particulate pollution monitoring in marine environments where ferrimagnetic iron oxides are important components of the discharge from industrial or urban complexes. The value of this approach is illustrated by reference to results obtained from the Elefsis Gulf, Greece, where the major source of particulate metal pollution is an iron and steel works discharging waste into a settling lagoon connected to the waters of the Gulf.

1325 - Searl, Thomas D., H. L. Huffman Jr., and J. P. Thomas. 1977. Extractable organics and nonvolatile hydrocarbons in New York Harbor waters. In: Procedings, Fifth Conference on the Prevention, Behavior, Control and Clean-up of Oil Pollution. 8-10 March 1977. New Orleans, La. Amer. Petrol. Inst. Washington, DC. 13p.

Abstract: In conjunction with the Marine Ecosystem Analysis program in the New York Bight, selected stations in New York Harbor were sampled during November 1974, and March and June 1975, for carbon tetrachloride extractable organics and nonvolatile hydrocarbons. The extractable organics ranged from 64 to 840 µg/l with a mean value for the Harbor of 158 µg/l. The standard deviation, 2 sigma , of the test method, based on duplicate samples, was 26 µg/l. The nonvolatile hydrocarbons ranged from 14 to 270 µg/l with a mean of 39 µg/l. The standard deviation, 2 sigma , was 5 µg/l. The highest concentration of organics and hydrocarbons was found in Newtown Creek; the lowest at the entrance to Ambrose Channel. The average concentrations are about a factor of ten above those reported for the open ocean. The concentrations obtained on the three sampling dates did not differ widely.

To help establish distribution patterns, ammonia, salinity, turbidity, and other measurements were made. In Arthur Kill, ammonia was constant at 50 µg-atoms N/l while the ratio of hydrocarbons to organics dropped from

0.29 at Newark Bay to 0.12 at Raritan Bay. This suggests that oxidation and/or adsorption of hydrocarbons onto particulates with subsequent settling took place.

1326 - Segar, D. A., and R. E. Pellenbarg. 1973. Trace metals on carbonate/organic rich sediments: Effects of industrialization and urbanization. Mar. Pollut. Bull. 4(9):138-142.

1327 - Segar, Douglas A. 1974. Evaluation of chemical characteristics of designated alternate dump sites. NOAA Atlantic Oceanographic and Meteorological Laboratories. Miami, FL. 8p.

Abstract: Nutrient chemistry of water at alternate dumpsites is less affected by coastal runoff than at the present dumpsites. Alternate dumpsites have lower dissolved and suspended trace metal and organic concentrations. Coarse grained sediments at the bottom of alternate dump sites indicate that low density sludge will not remain at its deposition site but will be resuspended and further dispersed.

1328 - Segar, D. A., P. G. Hatcher, G. A. Berberian, L. E. Keister, and M. A. Weiselberg. 1974. The chemical and geochemical oceanography of the New York Bight Apex region as it pertains to the problem of sewage sludge disposal. NOAA Atlantic Oceanographic and Meteorological Laboratories. Miami, FL. Special Report No. 1. Part III. 33p.

Abstract: Large quantities of contaminants are being introduced into the New York Bight from sources such as the Hudson River, acid-iron waste dumping, dredge spoil dumping, sewage-sludge dumping, and the various outfalls along the New Jersey and Long Island shorelines. This report is primarily concerned with the sewage-sludge dumping as it affects the chemical qualities of the New York Bight ecosystem. Ongoing studies to identify and trace the various sewage contaminants are at the initial stage of investigation. However at this time it appears that the carbohydrate/total organic carbon (TOC) ratio in sediments together with the TOC value itself may provide an indication of the geographical and quantitative extent of sewage derived contaminants.

1329 - Segar, D. A., and A. Y. Cantillo. 1976. Trace metals in the New York Bight. In: M.G. Gross (editor), Amer. Soc. Limnol. and Oceanogr. Spec. Symp. 2:171-198.

Abstract: Large quantities of trace metals are introduced to the New York Bight Apex from many sources. Distributions of dissolved Mn, Fe, Cd, Cu, and Zn are extremely nonuniform in the waters of the Apex due to the many sources and complex reactions taking place. Estuarine discharge and dredge spoil dumping are major sources, while sewage sludge and acid waste dumping are minor sources for most elements studied. Much of the dissolved Cu and Fe occurs in a chemical form that is not extractable by chelation/solvent extraction, even after acidification. The quantity of this metal fraction increases with distance from the Hudson-Raritan estuary. Loss from solution of some elements, notably Mn, occurs when estuarine water mixes with oceanic water. Metals, particularly Zn, are released to solution during ocean dumping of sewage sludge and other materials. Concentrations of dissolved metals in the Apex are higher than on the open shelf and higher in summer than in spring and fall. This suggests that the Apex flushes slower in summer, as inputs do not vary significantly with season.

Budget calculations show that contaminant metals, exemplified by Cu and Zn, do not accumulate in the Apex but are rapidly removed either to the estuaries or the surrounding shelf waters. Mean residence times of contaminant metals in the Apex waters are less than 6 months, perhaps considerably less.

1330 - Segar, D. A. and G. A. Berberian. 1976. Oxygen depletion in the New York Bight Apex: Causes and consequences. In: M.G. Gross (editor), Amer. Soc. Limnol. and Oceanorgr. Spec. Symp. 2:220-239.

Abstract: Dissolved oxygen concentrations in water of the New York Bight are near saturation except in summer when a stable thermocline exists and concentrations in the lower layer can drop to 10% of saturation.

Mass balances of oxygen and carbon cycles in the Apex were examined. Photosynthetically produced carbon accounts for most of the oxygen demand, particularly in summer. Oxygen demand due to sewage sludge and dredge spoils is small compared to that from organic carbon produced in situ. Oxygen demand of particualte and dissolved organics in the estuarine discharge may be as great as the sewage sludge and dredge spoils together.

Midsummer primary productivity in the Apex is high due to nutrient inputs, particularly nitrogen. Most nitrogen, supplied to the Apex in forms suitable to support photosynthetic production, comes from the discharge of the Hudson-Raritan-Passaic systems. Most of this nitrogen comes from liquid effluents of sewage treatment plants discharged to the rivers.

Ocean dumping in the Bight Apex does not cause the low oxygen concentrations found in summer. These are caused primarily by nitrogen supply from rivers. Improvement in dissolved oxygen concentrations could be achieved by removing nitrogen from sewage treatment plant effluents.

1331 - Segar, Douglas A. (Unpublished Manuscript). A review of the impact of dredged material disposal in the New York Bight Apex, with emphasis on chemical processes. Draft NOAA TR MESA (1977). MESA, Stony Brook, NY. 115p.

Abstract: The oxygen demand of dredged material does not cause the serious oxygen depletion in the bottom waters but it can aggravate the conditions.

NH_4^+ concentration may reach toxic levels but is diluted rapidly. Manganese and cadmium are released from dredged material. Sulphide ions (highly toxic) are released from anoxic dredged material and is a hazard to the biota.

It appears that contaminant rich fine grained particles are preferentially dispersed and a significant portion of the contaminants are not retained within the dumpsite mound.

1332 - Seki, S. and K. Ouchi. 1976. Hydraulic pressure-resistance test with monolithic cement solids of practical size for the sea disposal. Japan Atomic Energey Research Inst. Tokyo. 14p.

1333 - Selleck, R. E. 1967. Some physical parameters and their significance in marine waste disposal. In: T. A. Olson, and F. J. Burgess (editors). Proceedings of the Conference on the Status of Knowledge, Critical Research Needs, and Potential Research Facilities Relating to Ecology and Pollution Problems in the Marine Environment. Proceedings. Held in Galveston, Texas. 24-26 March 1966. pp. 297-315.

1334 - Selleck, R. E. 1975. The significance of surface pollution in coastal waters. Progr. Water Technol. Suppl. 1975. Pap. No. 15:143-153.

1335 - Sellers, J. W. 1973. Solid waste disposal and ocean dumping. Marine Affairs Jour. No. 1:52-77.

1336 - Seward, Michael 1977. The suffocating sea - sludge accumulation in New Jersey Bight. Oceans. 10(3):3.

Abstract: Fouled water in a 3000 square mile area of New Jersey is plagued by fish kills, depleted oxygen (0-1 ppm instead of normal 5-6 ppm) and algal blooms.

1337 - Shapiro, R. 1956. Pollution survey sampling of New York harbor. Sewage and Ind. Wastes. 28:1504-1506.

1338 - Sharp, Jonathan H. (In Press). In situ phosphate regeneration in Middle Atlantic coastal wasters. In: G. F. Mayer, A. Calabrese, F. A. Cross, D. C. Malins, D. W. Menzel, J. S. O'Connor and F. J. Vernberg (editors), Ecological Stress in the New York Bight: Science and Management. Estuarine Research Federation. Columbia, SC.

Abstract: Concern exists over man's disposal of bioactive nutrients into the marine environment. These nutrients can stimulate microbial plant production that, if not consumed in marine food chains, can cause serious oxygen demands. To understand the impact of man's input of nutrients it is necessary first to better gain an understanding of natural nutrient cycles. This paper attempts to gain a better understanding of natural phosphate regeneration in bottom waters of three regions in Middle Atlantic coastal waters and to evaluate the contributions that in situ nutrient regeneration could make toward supporting microbial production. It is striking that the Apex of the New York Bight with high production receives a much smaller proportion of its phosphate supply from in situ regeneration than adjacent New Jersey and Long Island coastal waters. The approach taken here is the first of its kind for quantifying and comparing coastal nutrient regeneration.

1339 - Shelef, G. 1975. Criteria for marine waste disposal in Israel. In: E. A. Pearson, and E. de Fraja Frangipane (editors), Marine Pollution and Marine Waste Disposal. (Supplement to Progress in Water Technology.) Proceedings of the 2nd International Congress. San Remo 17-21 December 1973. Pergamon Press. New York, NY. pp. 67-73.

1340 - Shelton, R. G. J. 1970. The effects of the dumping of sewage sludge on the fauna of the Outer Thames Estuary. ICES. C.M. 70/E:8. 9p.

1341 - Shelton, R. G. J. 1971. Sludge dumping in the Thames Estauy. Mar. Pollut. Bull. 2:24-27.

1342 - Shenton, Edward D. 1970. Evaluating waste disposal at sea - the critical role of information management. Marine Technology Society. Seventh Annual Meeting. (Held in Wash., D. C., Aug. 16-18, 1971). pp. 358-363.

1343 - Shepherd, J. G. 1978. A simple model for the dispersion of radioactive wastes dumped on the deep-sea bed. Mar. Sci. Communications. 4(4):293-327.

1344 - Sherif, M. A. 1970. Disposal of waste material in tectonic sinks. Nature. 228(5267):154-156.

1345 - Sherman, Kenneth, Donna Busch, and David Bearse. 1977. Deepwater dumpsite 106: Zooplankton studies. In: Baseline Report of Environmental Conditions in Deepwater Dumpsite 106. NOAA/ODP Dumpsites Evaluation Report 77-1. Rockville, MD. Vol. II. pp. 233-303.

1346 - Sherwood, M. J., and R. A. Bendele. 1974. Fin erosion in southern California fishes. Southern California Coastal Water Research Project, 1974 Annual Report. El Segindo, CA. pp. 39-44.

1347 - Sherwood, Marjorie J. (In Press). Fin erosion, liver condition, and trace contaminant exposure in fishes from three coastal regions. In: G. F. Mayer, A. Calabrese, F. A. Cross, D. C. Malins, D. W. Menzel, J. S. O'Connor and F. J. Vernberg (editors), Ecological Stress in the New York Bight: Science and Management. Estuarine Research Federation. Columbia, SC.

Abstract: Fin erosion in demersal fishes is one example of a response common to several regions receiving waste discharges. The regions include southern California, Puget Sound, and New York. Although the precise causes of the disease remain undetermined, fin erosion appears to result from the exposure of susceptible species to contaminated sediments. The present study demonstrates that in these three regions, the levels of total PCB's in muscle, liver, and brain tissues were higher in fishes with fin erosion from contaminated sites than in apparently unaffected speciments from control sites. PCB exposure may contribute to the development of the disease.

A second response common to the three regions involves changes in the size and/or lipid content of the liver. This response also may be related to chlorinated hydrocarbon exposure.

1348 - Shih, H. H. 1971. A literature survey of ocean pollution. Catholic University of America. Institute of Ocean Science and Engineering. Washington, DC. Report No. 71-6. 116p.

1349 - Sieburth, J. McN., and A. S. Dietz. 1974. Biodeterioration in the sea and its inhibition. [incl. dumped organic wastes]. In: R. R. Colwell, and R. Y. Morita (editors), Effect of the Ocean Environment on Microbial Activites. Conference Proceedings. University of Maryland. 25-30 August 1972. University Park Press, Baltimore, MD. pp. 318-326.

1350 - Simensen, T. and H. Munthe-Kaas. 1973. Development of numerical fjord models for the functional planning of a regional waste water disposal system. In: Advances in Water Pollution Research. Proceedings of the 6th International Conference. Jerusalem. 18-23 June 1972. Pergamon Press. New York, NY. pp. 853-863.

1351 - Simpson, D. C., M. F. Devine, C. E. Warsh, E. R. Meyer, T. P. O'Connor, and P. Kilho Park. 1978. Ocean-dumping research and monitoring: strategies and tools. IEEE J. of Oceanic Eng. Vol. OE-3. No. 4:165-171.

Abstract: Ocean dumping studies are designed to establish critical indexes of environmental quality to assess the impacts of ocean dumping.

Strategies emphasized are: 1) source function characterization including determinations of toxicities of original wastes, 2) environmental characterization, 3) waste dispersion studies in the laboratory and at sea, and 4) waste biota interaction studies, both in the short term, within a life cycle of organisms affected, and in the long term, over many life cycles.

Present tools include: 1) quantitative mapping of pollutant distribution by aerial remote sensing and moored and drifting arrays, 2) realtime tracking of waste dispersion in the ocean by 20- and 200-kHz acoustic records, 3) continuous multidepth water sampling to obtain sea truth, 4) in situ bioassay devices, and 5) mathematical dispersion models, some of which are similar to atmospheric chimney plume dispersion models.

1352 - Simpson, Duane C., Thomas P. O'Connor, and P. Kilho Park. 1981. Deep-ocean dumping of industrial wastes. In: R. A. Geyer (editor), Environmental Pollution. 2: Dumping and Mining. Elsevier. New York, NY. pp. 379-402.

Abstract: Summary of the state of knowledge of deep ocean dumping at the 106-mile dumping site (Northern Atlantic) off the east coast U.S. for industrial wastes as pertaining to: National and International legislation, site location and physical characteristics, waste characteristics, waste-seawater interaction, dispersion, dilution, waste plume advection, biolgoical effects (phyto and zoo).

1353 - Sindermann, C. J. 1974. An assessment of human impact on coastal ecosystems and living resources of the New York Bight. ICES. C.M. 74/E:39. 6p.

Abstract: Reduction of species diversity and biomass, antibiotic and heavy metal resistance in pollution bacteria, increased occurrence of "fin rot" and gill disorders, and lower fish biomass indicte degradaion in New York Bight.

1354 - Sindermann, C. J. 1976. Effects of coastal pollution on fish and fisheries -- With particular reference to the middle Atlantic Bight. In: M. G. Gross (editor), Amer. Soc. Limnol. and Oceanogr. Spec. Symp. 2:281-301.

Abstract: Industrial contamination of coastal waters exerts great local impact on fish and shellfish populations. Known offenders include heavy metals halogenated hydrocarbons, and petroleum residues. Even in low concentrations, many industrial chemicals have profound effects on some or all life stages of marine animals; effects may be reflected in mortality, increased occurrence of abnormalties, and physiological disturbances, with resultant slow growth or spawning inhibition.

Many toxic chemicals in industrial effluents and agricultural runoff-pesticides, heavy metals, and petroleum products-are also components of sewage treatment plant effluents or ocean outfalls from domestic sewage systems. Additionally, bacteria and viruses force closure of many areas to shellfish harvesting. Still another effect is increased organic loading of coastal waters which should increase productivity, but unmanaged it may contribute to reduced oxygen levels, to blooms of toxic phytoplankton, or of algal species too small to be utilizied by shellfish, menhaden, and other filter-feeding animals.

Orderly development of marine aquaculture in coastal waters can be severely impeded by contaninant-related public health problems, both chemical and microbial, and by contaminant-induced problems associated with survival, reproduction, and growth of cultivated animals.

Long term effects on sportfish abundance are difficult to quantify. In some instances party boats have shifted from bays and estuaries to open waters and offshore reefs. Localized effects of pollutants on some species, particularly flatfish, have been observed, but increases in abundance of striped bass, weakfish, and bluefish suggest that conclusions about human effects on population changes must be assessed very carefully.

Available evidence suggests that environmental stress may precipitate diseases in fish and shellfish. Public health aspects are presently confined largely to those microorganisms and toxic substances which do not produce obvious disease in fish and shellfish-but which may be accumulated or transmitted passively to humans from aquatic animal vectors (hepatitis, cholera, typhoid, and mecury poisoning). A few pathogens of fish and shellfish (e.g. *Vibrio parahemolyticus*) may be transmissible to humans, but a clear role for pollutant effects has not been demonstrated.

Evidence exists for localized effects on fisheries, but there is as yet little specific evidence of widespread damage to major fisheries resource populations resulting from coastal pollution.

1355 - Sindermann, C. J., and F. W. Steimle, Jr. 1977. Oxygen depletion and mass mortalities of shellfish in the Middle Atlantic Bight of the United States in 1976. ICES. C.M. 77/E:13.

1356 - Sindermann, C. J. 1979a. Use of pathological effects of pollutants in marine environmental monitoring programs. ICES Workshops on Problems of Monitoring Biolgical Effects of Pollution in the Sea. Paper No. 15. 17p.

Conclusions: Stress from pollutants may be expressed in marine animals in a number of ways, including diminished reproductive activities; damage to genetic material of the egg or embryo, with resulting mortality or abnormal development; direct chemical damage to cell membranes or tissues; modification of physiological and biochemical reactions; changes in behavior, often due to chemical damage to sensory equipment; increased infection pressure from facultative microbial pathogens; and reduced resistance to infection. Some of the morphological manifestations of stress may be useful as indicators in pollution monitoring programs.

Among the numerous pathological signs that have been associated with pollution stress, chromosomal anomalies and skeletal abnormalities are amenable to quantification and seem satisfactory as monitoring approaches. Certain neoplasms in molluscs also offer promise. Some of the best qualitative indicators of pollution -- those with the most obvious association with degraded environments (fin rot, ulcers, and shell erosion in Crustacea) -- are more difficult to quantify but could have utility.

Animal tissues repond to trauma in specific circumscribed predictable ways, so it is unlikely that unique pathologies will be found associated with specific contaminants. Also, it is unlikely that a single definitive pathological indicator of pollution will be found; rather several criteria should be selected and utilized, based on different biological disciplines (pathology, physiology, biochemistry, genetics, behavior, ecology), each providing its own independent perspective.

1357 - Sindermann, C. J. 1979b. Pollution associated diseases and abnormalities of fish and shellfish: A review. Fishery Bulletin. 76(4):717-749.

Abstract: The relationship of disease and environmental stress is becoming increasingly well established with time. Human activities-particularly those that result in chemical additions to the coastal estuarine environment - have increased the potential stresses on fish and shellfish inhabiting those areas. Circumstantial evidence for associations of pollutants with certain fish and shellfish diseases and abnormalities is accumulating.

This paper attempts to review and evaluate existing information about associations of diseases and marine environmental degradation. Emphasis has been placed on: diseases caused by contaminant stress and related facultative pathogens; stress-provoked latent infections; environmentally induced abnormalities; genetic abnormalities associated with mutagenic and other properties of contaminants; experimentally induced lesions; contaminant effects on resistance and immune responses; and pollutant-parasite interactions.

There are several diseases particularly fin erosion and ulcers in fish and shell disease in crustaceans, for which a relationship with pollution seems evident and there are a number of other diseases or abnormalities (such as certain neoplasms and skeletal anomalies) for which a relationship with pollution is indicated. Furthermore, there is some evidence that certain latent viral infections may be provoked into patency by environmental stress.

1358 - Sindermann, Carl J., and R. Lawrence Swanson. 1979. Historical and regional perspective. In: R. L. Swanson and C. J. Sinderman (editors), Oxygen Depletion and Associated Benthic Mortalities in New York Bight, 1976. NOAA Professional Paper No. 11. U. S. Dept. of Commerce. Rockville, MD. pp. 1-16.

Summary: In the summer and autumn of 1976, mass mortalities of shellfish and other marine species occurred in the New York Bight, apparently because of extreme oxygen depletion and hydrogen sulfide formation in bottom waters. First reports of a developing problem reached the scientific community during the July 4th weekend. Sport divers, lobstermen, and trawler fisherman observed and reported dead and dying animals of all kinds on fishing reefs and wrecks and on fishing grounds off the northern New Jersey coast. Within a few weeks, mortalities were reported southward some 90 km and seaward some 60 km. Planned surveys and monitoring in the Bight by the National Oceanic and Atmospheric Administration (NOAA) were increased in areal coverage and intensity to determine the extent of the problem and to assess the damage. At its maximum extent, oxygen-deficient bottom water-less than 2.0 ml/l and sometimes with zero values of dissolved oxygen (D.O.)-was found from Sandy Hook to Cape May, a distance of 150 km along the New Jersey coast in a zone or corridor 10 to 100 km off the coast. This environemntal event of major proportions involved an 8,600-km^2 area of the continental shelf off the coast of New Jersey.

1359 - Singleton, F. L., J. W. Deming, E. R. Peele, B. Cavari, B. Gunn, and R. R. Colwell. (In Press). Microbial communities in surface waters at the Puerto Rico dumpsite. In: I. W. Duedall, B. H. Ketchum, P. K. Park and D. R. Kester (editors), Wastes in the Ocean. Vol. II: Industrial and Sewage Wastes in the Ocean. John Wiley and Sons, Inc. New York, NY.

Abstract: A variety of microbiological parameters were determined for surface waters in and surrounding the Puerto Rico dumpsite which is used for disposal of pharmaceutical wastes. Specific activities of microbial populations were derived from comparisons of activity measurements (uptake of radiolabelled substrates or substrate-responsive cell numbers determined by epifluorescent microscopy) and total cell numbers. Highest values were observed in samples from stations in or near the dumpsite. Similarly, largest numbers of colony-forming bacteria, enumerated on marine agar, were obtained in the vicinity of the dumpsite.

Total colony-forming bacteria were enumerated at all stations, with several different culture media, and randomly selected isolates were identified to develop diversity indices for the culturable bacterial community. Bacteria isolated on marine agar were found to be predominantly members of the Vibrio/Aeromonas group, with the more typical marine pseudomonads comprising less than 9% of the community. In the vicinity of the dumpsite, large numbers of Gram-positive bacteria, that is, micrococci, staphylococci, and bacilli, were recovered from water samples plated on marine agar, which selects against bacteria requiring sea salts for growth.

Results obtained, in particular diversity index measurements and the persistence of culturable, waste-specific organisms at the dumpsite, suggest that alterations in the natural microbial populations of surface waters of the Puerto Rico dumpsite and environs have occurred.

1360 - Slanksy, C. M. 1971. Principles for limiting the introduction of radioactive waste into the sea. Atomic Energy Review. Vienna. 9(4):854-867.

1361 - Smith, David D. 1969. Marine disposal of solid wastes. Dillingham Corporation. Applied Oceanography Division. La Jolla, CA. Summary No. AOD-4008. 7p.

1362 - Smith, D. D. and R. P. Brown. 1970. Deep-sea disposal of liquid and solid wastes. Industrial Water Engineering. 7:20-24.

1363 - Smith, D. D. and R. P. Brown. 1971. Ocean disposal of barge-delivered liquid and solid wastes from U.S. coastal cities. U.S. Environmental Protection Agency, Solid Waste Management Series No. SW. 192. 119p.

1364 - Smith David D. 1979. Dredged material, ocean disposal and the regulatory maze. In: H. D. Palmer and M. G. Gross (editors), Ocean Dumping and Marine Pollution. Dowden, Hutchinson and Ross, Inc. Stroudsburg, PA. pp. 241-262.

Abstract: Dredged material disposed of at sea each year totals about 50 to 75 x $10^6 m^3$, which is roughly 15 to 20 percent of the total sediment dredged in the U.S. annually. These volumes are discharged at some 125 designated ocean disposal sites, of which about 60 percent are located within the three mile limit.

Disposal of dredged material requires a permit issued by the Corps of Engineers in accordance with Environmental Protection Agency regulations and criteria. Obtaining this permit requires maneuvering through a complex procedural network (i.e. the regulatory maze) comprised in part by the numerous and varied criteria that govern the review and processing of the permit application. For a dredging project of even moderate size, negotiating this maze and obtaining the permit may require one to two years, including preparation and review of an Environmental Impact Statement.

The regulations governing ocean disposal also prescribe additional sets of criteria for evaluating (a) the dredged material to be disposed of, (b) the site proposed for disposal use and (c) the environmental effects of the proposed disposal activity. Thus, literally dozens of criteria must be considered by the Corps in reaching a decision as to whether or not a permit for ocean disposal should be issued. Unfortunately, the governing regulations contain no recommended framework or standard methodology within which these many criteria can be applied--a major omission that requires resolution. Until such a framework or methodology is established, the decisions reached using the criteria must be recognized as highly subjective and potentially vulnerable to litigation.

1365 - Smith, Ethan T. 1970. Water quality of the New York harbor and Bight area. In: Alan A. Johnson (editor), Water Pollution in the Greater New York Area. Gordon and Breach Science Publishers. New York, NY. pp. 39-52.

1366 - Smith, Richard H. 1969. Deep sea refuse disposal--patent no. 3,456,824. Official Gazette. U.S. Patent Office. Vol. 865. No. 4. p. 1213-1214. July 22.

1367 - Smithsonian Institution, Oceanography, and Limnology Program. 1975. Marine pollution and marine waste disposal. (Supplement to Progress in Water Technology.) Proceedings of the 2nd International Congress. San Remo. 17-21 December 1973. Pergamon Press. New York, NY. pp. 15-18.

1368 - Smyrl, W. H. and L. L. Stephenson. 1977. Behavior of candidate canister material in deep ocean environments. Sandia Labs: Albuquerque, NM. 35p.

1369 - Snook, W. G. G. 1966. Marine disposal of trade wastes. Chemistry and Industry. Nov. 16. pp. 1593-1598.

1370 - Snook, W. G. G. 1975. Design and construction of submarine pipelines and diffusers. Marine Municipal and Industrial Wastewater (SIC) Disposal: Third International Congress: Preprints. XV-1-XV-25.

1371 - Snook, W. G. G. 1979. Design of major regional sewerage systems with disposal at sea. Progr. Water Technol. 4:197-211.

1372 - Sondergutachten. 1980. Umweltprobleme der Nordsee. Der Rat von Sachverstandigen fur Umweltfragen. Verlag W. Kohlhammer GMBH Stuttgart. ISBN 3-17-003214-3. 503p.

1373 - Southern California Coastal Water Research Project (SCCWRP). 1978. The effects of the ocean disposal of municipal waste. Summary Report of the Commission. El Segindo, CA. 25p.

1374 - Southgate, B. A. 1948. Treatment and disposal of industrial waste waters. D.S.I.R. Publ. HMSO. London. 327p.

1375 - Spagnoli, J. J., and C. Skinner. 1977. PCBs in fish from selected waters of New York State. Pest. Monit. Jour. 11(2):69-87.

Abstract: PCB residues in fish from 41 stations throughout NYS were monitored in 1975. Nearly all fish had detectable amounts although the levels of contamination and specific Aroclors differed. The Hudson River contained the highest known PCB concentrations within the U.S., often exceeding 100 ppm. Lake Ontario salmonids, Cayuga Lake trout and Onondaga Lake also had abnormally high levels. Marine water samples generally have levels substantially below 5.0 ppm. As expected, lightly industrialized watersheds have low PCB levels (Chemung, Susquehanna, Delaware).

Hudson River: Contamination is widespread from the Hudson Falls to the mouth, with level of contamination decreasing in lower reaches. Biomagnification transfer model is presented. Levels above 50 ppm are common in the larger oilier fish, with the highest recorded concentration in a large eel as 559.25 ppm. Fort Edward contributes a substantial load of PCB to the river.

1376 - Sperry Rand Corporation. 1971. Ocean waste dumping operations monitoring. NTIS No. AD 735378.

1377 - Spiess, F. N., and S. M. Sanders. 1971. Survey of chase disposal area (NITNATOW). Scripps Institution of Oceanography, Marine Physical Laboratory. San Diego, CA. (SIO Reference 71-33). 20p.

1378 - Squires, D. F. 1981. The bight of the big apple. New York Sea Grant Institute, Albany, NY. 84p.

1379 - Stadler, D. 1977. Chlorinated hydrocarbons in seawater of the German Bight and the Western Baltic. ICES. C.M. 77/E:4. (Also DHI Report 77/17).

1380 - Stainken, D. and J. Rollwagen. 1979. PCB residues in bivalves and sediments of Raritan Bay. Bull. of Envir. Contam. and Toxicol. 23:690-697.

Abstract: Sediments and bivalves (clams, _Mercernaria mercernaria_ and _Mya aremaria_ and oysters _Crassostrea virginica_) in the Raritan Bay-Lower New York Bay area were analyzed to determine the amount and types of PCB compounds present. No relationship was found between sediment silt-clay content and sediment PCB content, sediment silt-clay content and tissue PCB content, or tissue PCB content and sediment PCB content. Tissue PCB content was usually higher than sediment content. Sediments appeared to contain mixtures of various homologs of Aroclors; Aroclors 1016 and 1242 appeared in all samples. Mixtures of Aroclors also occurred in tissue, with 1016 and 1242 the most prevalent. The types and amounts of PCBs found may indicate downstream transport; stations with higher values are toward the center of the bay and may represent concentration and deposition from hydrographic conditions. Higher tissue values of PCB in the bivalves occurred in the center and lower portions of the bay.

1381 - Stalling, D. L. and F. L. Mayer. 1972. Toxicities of PCB's to fish and environmental residues. Environ. Health Perspectives. pp. 159-164.

1382 - Stander, G. J., W. D. Oliff, and D. J. Livingstone. 1967. Problems in using the sea as part of a waste disposal system. Ann. Conf. of Inst. Wat. Pollut. Control, held at Torquay, 20-23 June 1967. Conf. Pap. No. 3. 11p.

1383 - Stander, G. J., W. D. Oliff, and D. J. Livingstone. 1968. Problems in using the sea as part of a waste disposal system. Water Poll. Control. Fed. 67(2):153-158.

1384 - Stanford, Harold M. 1978. What happens to sewage sludge over several hours after it is dumped into the ocean? MESA New York Bight Project. Stony Brook, NY. 10p.

Abstract: A summary of field experiments and sludge characteristics with descriptions of typical transport and dispersion patterns. Plots of salinity, temp., density and extinction coefficient with depth before and after sludge dumping.

1385 - Stanford, H. M., Joel S. O'Connor, an R. Lawrence Swanson. 1981. The effects of ocean dumping on the New York Bight ecosystem. In: B. H. Ketchum, D. R. Kester, and P. K. Park (editors), Ocean Dumping of Industrial Wastes. Plenum Press. New York, NY. pp.53-86.

Abstract: The effects of ocean dumping on a marine ecosystem must be considered in the context of the character and quantity of all contaminants reaching the system. The highly variable nature of natural factors within the ecosystem and of waste inputs to it make the direct determination of cause-effect relationships of anthropogenic materials quite difficult to discern. Indirect methods for such determination are possible, obviating the need for exhaustive study of waste materials and the ecosystem. All of these factors are addressed in a case study of the New York Bight ecosystem.

1386 - Stanley, H. G. and D. W. Kaplanek. 1976. A bibliography on ocean waste disposal. 2nd. Edition. Interstate Electronics Corp. Anaheim, CA. EPA-68-01-0796. 154p.

1387 - Staples, K. D., J. D. Watson. and D. M. Watson. 1974. Investigation for marine diposal of effluents. Water Poll. Control Fed. 73(4):417-424.

1388 - State University of New York, Stony Brook. 1973. Marine Sciences Research Center. Final report of the oceanographic and biology study for Southwest Sewer District No. 3. Vol. I.

1389 - Starr, R. B., G. A. Berberian, and M. A. Weiselberg. 1976. MESA New York Bight Project, Expanded Water Column Characterization Cruise (XWCC-1) on the R/V ADVANCE II, January 1975. NOAA DR ERL MESA-22. Boulder, CO.

1390 - Starr, R. B., Hazelworth, J. B., and G. A. Berberian. 1976. MESA New York Bight Project, Expanded Water Column Characterization Cruise (XWCC-6), NOAA Ship GEORGE B. KELEZ, 29 September - 4 October 1975. NOAA DR ERL MESA-25. Boulder, CO. 132p. NTIS No. PB 271401/AS.

Abstract: During September and October 1975, the NOAA Ship GEORGE B. KELEZ made an expanded water column characterization cruise (XWCC-6) in the New York Bight. The objective of this cruise was to collect data to be incorporated into an analysis of the water characteristics and movements. This report presents the physical and chemical data from this cruise and describes the measurement methods and procedures for reducing the data.

1391 - Starr, R. B., J. B. Hazelworth, S. R. Cummings, and G. A. Berberian. 1977. MESA New York Bight Project Expanded Water Column Characterization Cruise (XWCC-10), NOAA Ship GEORGE B. KELEZ. 28 June-1 July 1976. NOAA DR ERL MESA-28. Boulder, CO. 91p. NTIS No. PB 275442/AS.

Abstract: During the period 28 June - 1 July 1976, an oceanographic cruise was made on the NOAA Ship GEORGE B. KELEZ in the New York Bight. The objective of the cruise was to supply data to provide a base for analysis of the water characteristics and movements in the highly impacted ecosystem. This report presents the corrected physical and chemical (nutrient) data from this cruise, and describes the Neuston net towing operation and results.

1392 - Starr, Robert B., and Frank W. Steimle, Jr. 1979. Temporal development of physical characteristics. In: R. L. Swanson and C. J. Sinderman (editors), Oxygen Depletion and Associated Benthic Mortalities in New York Bight, 1976. NOAA Professional Paper No. 11. U.S. Department of Commerce. Rockville, MD. pp. 17-50.

Summary: Study of the oceanic conditions and events, and their progression in the New York Bight during 1976, indicates the presence of warmer-than-normal bottom waters early in the year. This is in accord with the findings of Hazelworth and Cummings (1979), and suggests a larger amount of offshore water in the Bight than usual. There may be a connection between this presence of offshore water and the large concentration of *Ceratium tripos* in the Middle Atlantic Bight in 1976. By June, however, surface and bottom waters were nearly normal.

The distribution of properties in the Hudson Shelf Valley and off New Jersey in May, June, and September, and at the hurricane Belle XBT and bottom-oxygen stations, suggest an onshore movement of water beneath the pycnocline. This agrees with the sluggish onshore set found in the current meter records by Mayer et al. (1979). The thickness of the subpycnocline layer is about 4 m less in May and June off New Jersey than off Long Island, though the bottom of the pycnocline is at a shallower depth off New

Jersey. This difference in thickness was less in September. In both localities bottom water was effectively isolated from the surface by a relatively strong pycnocline. This pycnocline did not appear to be significantly stronger than normal, particularly early in the year.

The occurrence of bottom water that was 2.5^{o} C warmer than normal in April and 3° to 4° C colder than normal in the latter half of the season indicates advection of bottom water into the region. If the continental shelf were the source of this water, then it must have been subjected to anomalous conditions upstream when it was at the surface. Hurricane Belle had some effect on the water column down to at least 25 m.

1393 - Stebbing, A. R. D. (chairman), B. Akesson, A. Calabrese, J. H. Gentile, A. Jensen, and R. Lloyd. 1980. Bioassay report to monitoring the biological effects of pollution in the sea. In: International Council for the Exploration of the Sea. Rapports et Process-Verbaux des Reunions. Vol. 179. Charlottenlund Slot, Denmark.

1394 - Steele, J. H., A. D. McIntyre, R. Johnston, I. G. Baxter, G. Topping and H. D. Dooley. 1973. Pollution studies in the Clyde sea area. Mar. Pollut. Bull. 4(10):153-157.

1395 - Steimle, F. W. Jr., and R. B. Stone. 1973. Abundance and distribution of inshore benthic fauna off southwestern Long Island. NOAA Tech Report. Highlands, NJ. NMFS SSRF-673. 50p.

1396 - Steimle, F. W., Jr. 1978a. Hydrographic data from New York Bight: July-November 1976. NOAA DR ERL MESA-35. Boulder, CO. 202p. NTIS No. PB 284447/AS.

Abstract: During late June-early July 1976, sport divers and commercial fisherman reported extensive mortalities of demersal finfish and shellfish off the north-central New Jersey coast. The NMFS laboratory at Highlands, N.J. investigted the reports immediately and determined that an oxygen depletion problem existed in the sub-thermocline water layer off N.J. To further investigate and monitor the phenomenon and assess impacts to resources, a series of cruises collected data in the New York Bight from July through November 1976. This report is a compilation of hydrographic data collected during these cruises.

1397 - Steimle, F. W., Jr. 1978b. Dissolved oxygen levels in New York bight waters during 1977. Division of Environmental Assessment. Sandy Hook Lab. Highlands, NJ. 30p.

Abstract: Anoxia in bottom waters of the New York Bight, and associated mass mortalities of marine organisms in 1976, caused concern that anoxia may become a chronic problem in the Bight. The Northeast Fisheries Center (NEFC), National Marine Fisheries Service (NMFS), established a series of periodic surveys early in 1977 to further understand the hydrologic "climate" of the Bight and to monitor dissolved oxygen (DO) levels. The Bight did not become anoxic in 1977, although a band of low DO levels was found along the New Jersey coast during the summer.

1398 - Steimle F. W., Jr., and Carl J. Sindermann. 1978. Review of oxygen depletion and associated mass mortalities of shellfish in the Middle Atlantic Bight in 1976. Marine Fisheries Review. 40(12):17-26. MFR paper 1356.

Abstract: In summer and autumn of 1976, mass mortalities of shellfish occurred in a 165-km long corridor of severe oxygen depletion paralleling the New Jersey coast from 5 to 85 km from shore. Mortalities of surf clams, Spisula solidissima, the most severely affected species, were estimated, in excess of 140,000 t. Alteration of normal migration patterns of lobsters and several species of finfish was also noted. A series of anomalous meteorological and hydrogical events (particularly early warming of surface waters resulting in early thermocline development and massive shelf-wide phytoplankton bloom) superimposed on an already stressed coastal area, was considered to be responsible. The occurrence is particularly significant because the continental shelf of the Middle Atlantic Bight, from Cape Cod to Cape Hatteras on the east coast of the United States, contains the largest known stocks of ocean shellfish of any comparable coastal area of North America.

1399 - Steimle, F. W., Jr., and D. J. Radosh. 1979. Effects on the benthic invertebrate community. In: R. L. Swanson and C. J. Sinderman (editors), Oxygen Depletion and Associated Benthic Mortalities in the New York Bight, 1976. NOAA Professional Paper No. 11. U.S. Department of Commerce. Rockville, MD. pp. 281-293.

Summary: The oxygen depletion event of 1976 killed many benthic invertebrates, especially surf clams off central New Jersey. Some organisms, mostly polychaetes, showed tolerance. Recolonization and stabilization of the benthic invertebrate population appeared to be incomplete 1 year after disturbance. Several years may be required for some species-e.g., those with nonplanktonic larval dispersal-to return to preanoxic levels and for the benthic invertebrate community to stabilize to a preanoxic state.

1400 - Steimle, F. W., Jr., Janice Caracciolo, and John B. Pearce. (In Press). Impacts of dumping on New York Bight Apex benthos. In: G. F. Mayer, A. Calabrese, F. A. Cross, D. C. Malins, D. W. Menzel, J. S. O'Connor and F. J. Vernberg (editors), Ecological Stress in the New York Bight: Science and Management. Estuarine Research Federation. Columbia, SC.

Abstract: The discharge of materials in the Hudson-Raritan estuary and New York Bight Apex has altered some Apex benthic habitats physically and chemically, introducing high concentrations of a variety of pollutants (e.g., heavy metals, petroleum hydrocarbons, and organic carbon) that are known to be lethal or stressful to several groups of benthic invertebrates. Benthic invertebrate data from five survey cruises conducted during 1973-1974 at 66 stations in the Apex are reviewed, with evidence of dumping impacts discussed. The benthic community, in the vicinity of two dumpsites receiving New York harbor dredge spoils and metropolitan area sewage sludge, exhibited a reduction in the number of species present, especially small amphipod crustaceans that are highly sensitive to pollution and are important food items of many fish species.

1401 - Stern, A. M., and C. R. Walker. 1978. Hazard assessment of toxic substances: Environmental fate testing of organic chemicals and ecological effects testing. ASTM Spec. Techn. Publ. 657. pp. 81-131.

1402 - Stern, D. A., and C. L. Grant. 1981. A laboratory investigation of heavy metal absorption on marine dredge spoils. Bull. Envir. Contam. Toxicol. 26:213-218.

1403 - Sternberg, R. W., J. S. Creager, J. Johnson, and W. Glassley. 1979. Stability of dredged material deposited seaward of the Columbia River mouth. In: H. D. Palmer and M. G. Gross (editors), Ocean Dumping and Marine Pollution. Dowden, Hutchinson and Ross, Inc. Stroudsburg, PA. pp. 17-49.

Abstract: A two-part study was conducted in a region seaward of the Columbia River where disposal of large quantities of dredged material has occurred over the last several decades. The first part included repeated bathymetric surveys and sampling for distribution and seasonal variations of sediment texture and mineral composition throughout the study area, and especially at designated disposal sites. Near bottom hydraulic conditions (waves tides, currents, turbidity) were also measured at several sites and during all seasons. The second part was related to an experiment in which 459,000 m^3 of material dredged from the Columbia River estuary were discharged at a designated site, which was monitored before, during, and after disposal. Sedimentological aspects of the study were to identify and map all deposits of dredged material and to recognize seasonal and long-term changes. The objectives of the hydraulic aspects were to document ambient near-bottom conditions, and their effect on the deposit at the experimental site.

Deposits of dredge material can be identified relative to the surrounding sediments, because they tend to maintain their identity for many years. Sedimentological evidence suggests that the deposits disperse northward at approximately 0.5 km per year. At the experimental disposal site the volume of the bottom deposit was 61% of the total material dumped. Calculations of bedload transport rates, based on seasonal measurements of bottom currents, suggest that 635 m^3 of material (0.2% of the total deposit) spread northward from the site at about 0.4 km per year. This is similar to the rates determined by analysis of sediment mineralogy.

1404 - Stewart, H. B., Jr. 1958. Upstream currents in New York Harbor. Science. 127(3306).

1405 - Stewart, K. R., and L. K. Koditschek. 1980. Drug-resistance transfer in *Escherichia coli* in New York Bight sediment. Mar. Pollut. Bull. 11(5):130-133.

1406 - Stirling, E. A. 1975. Some effects of pollutants on the behaviour of the bivalve *tellina tennuis*. Mar. Pollut. Bull. pp. 122-124.

1407 - Stirn, J. 1975. Criteria for marine waste disposal in Yugoslavia. In: E. A. Pearson, and E. de Fraja Frangipane (editors), Marine Pollution and Marine Waste Disposal. (Supplement to Progress in Water Technology.) Proceedings of the 2nd International Congress. San Remo. 17-21 December 1973. Pergamon Press. New York, NY. pp. 57-66.

1408 - Stockton, William D., and Richard H. Backus. 1951. The distribution of salinity in New York Harbor and its approaches. In: John C. Ayers et al., The Hydrography of the New York Area. Status Report No. 12. (Contract N6 DNR 264, task 15). Cornell Univ. Ithaca, NY.

1409 - Stockton, William D., and John C. Ayers. 1952a. The distribution of salinity in the waters of New York Bight, Block Island Sound and Newport Bight; Cruise July-September 1951. In: John C. Ayers et al., The Hydrography of the New York Area. Status Report No. 16. (Contract N6 DNR 264, task 15). Cornell Univ. Ithaca, NY.

1410 - Stockton, William D., and John C. Ayers. 1952b. The distribution of temperature in New York Bight, Block Island Sound and Newport Bight; Cruise Stirni I, July - September 1951. In: John C. Ayers et al., The Hydrography of the New York area. Status Report No. 20. (Contract N6DNR 264, task 15). Cornell Univ. Ithaca, NY.

1411 - Straub, Conrad P. 1972. Radioactive wastes. Jour. Water Poll. Control Fed. 44(6):1115-1123.

1412 - Strickland, J. D. H. 1969. Biological effects on receiving waters: the plankton. U.S. National Academy of Sciences Background papers on coastal wastes management. Washington, DC. p. XVI-1 to XVI-23.

1413 - Stubberud, G. 1977. Procedures and arrangements for the discharge of noxious liquid substances. In: IMCO. Report of the Symposium on Prevention of Marine Pollution from ships, Acapulco, Mexico, 22-31 March 1976. U.S. Coast Guard. Office of Marine Environment and Systems. (CG-WEP-1-77). Washington, DC. 17p.

1414 - Stubblefield, W. L., M. Dicken, and D. J. P. Swift. 1974. Reconnaissance of bottom sediments of the inner and central New Jersey Shelf. MESA Data Report No. 1. Boulder, CO. 39p. NTIS No. COM 75-10285/AS.

1415 - Stubblefield, W. L. 1975. Temporal and spatial substrate variation in the New York Bight Apex. Geological Society of America. 7(7):1285-1286.

Abstract: Quarterly monitoring of selected areas in the New York Bight Apex, over a one and one-half year span, indicates a pronounced temporal and spatial variation within the upper few millimeters of the substrate. By means of sidescan sonar, bottom grab sampling, and bottom photography, bottoms ranging from coarse, clean sand to muddy, very-fine sand were observed. A temporal variation became apparent when the sampling stations were reoccupied with the aid of precision navigation. The monitoring, which included samples taken shortly before and after a decadal storm, suggests that the substrate mobility is most prevalent in the vicinity of Long Island and New Jersey shorelines and in Christiaensen Basin which marks the head of the Hudson Shelf Valley. Sidescan records from nearshore Long Island indicate development of sand wave-like forms during the winter and subsequent degradation during the summer months. The sand wave development is probably associated with the peak-flow events of the winter storms. The features are of negligible relief, and may be degraded by the action of bottom wave surge and benthic organisms. Christiaensen Basin, which is characterized by muddy, very-fine sand in its center, is the settling site for much of the sewage sludge material presently being dumped by New York City. Sidescan records show that the Basin is characterized by a patchy bottom pattern. The patches are irregular in shape, frequently elongated to the northwest and vary in shape and position between observations which support the suggestion of temporal mobility of the bottom sediments.

1416 - Stubblefield, W. L., R. W. Permenter, and D. J. P. Swift. 1977. Time and space variation in the surficial sediments of the New York Bight Apex. Estuar. and Coast. Mar. Sci. 5:597-607.

Abstract: Sidesan sonar records, grab samples, and bottom photographs were collected along two transects in the New York Bight Apex, on a quarterly basis for six quarters. Sampling was designed to determine the natural variability of bottom deposits in time and space, and the effect of ocean dumping on these patterns.

The Hudson Shelf Valley and the Christiaensen Basin at its head are floored by fine muddy sand and mud, while the high areas on either side are covered by medium-grained sand. Bottom sands on the seaward side are relatively uniform. On the New Jersey side, and near the Long Island coast, sand ribbon-like patterns with spacings of 10-200 m appear. Separate dumpsites for sewage sludge, dredge spoil, cellar dirt, and acid waste occur within the area.

The distribution of grain sizes and bedform patterns are stable over time, indicating that the bottom is in a state of textural near-equilibrium with the hydraulic climate. The effects of ocean dumping are most obvious in the vicinity of the dredge spoil dumpsite, which has shoaled 15 m within the last years. Here an aureole of anomalously fine sediment is spreading over a botton locally composed of artificial rubble. Cholera Bank, the area where the highly mobile sewage sludge is dumped, is floored by sand; no permanent sludge deposits form. The bottom muds of the adjacent Christiaensen Basin may, however, be contaminated with this material.

1417 - Stubblefield, W. L., J. W. Lavelle, T. F. McKinney, and D. J. P. Swift. (In Press). Sediment response to the hydraulic regime on the central New Jersey shelf. Jour. Sed. Petrology.

1418 - Stultz, B. E. 1974. Environmental effects of past deep water dumps. In: G. A. Young (compiler), Proc. 1st conf. Environmental Effects of Explosives and Explosions. May 30-31 1973. White Oak, MD. U.S. Naval Ordinance Laboratory. NOLTR 73-223. pp. 33-38.

1419 - Subcommittee on Oceanography, and Subcommitte on Fisheries and Wildlife Conservation and the Enviroment. 1976. Ocean dumping hearings. Gov. Printing Office. Washington, DC. 86p.

Summary: Joint Congressional Hearings on the Ocean Dumping Act and Assessment of the EPA's implementation of the act in the New York Bight. Includes testimony form Gerald Hansler, Regional Aministrator, EPA in New York City; Scott B. Lilly, Power Authority of the State of New York, Col. Thomas C. Hunter, Jr, U.S. Army Corps of Engineers, New York; John P. Mugler, Jr, Environmental Quality Program Office, NASA, Charles Samowitz, Commissioner, Dept. of Water Resources, N.Y.C., and CDR Lawrence Swanson, MESA New York Bight Project, NOAA.

1420 - Swanson, R. L. 1975. Status on NOAA's Ocean Dumping Research in the New York Bight. NOAA/MESA, New York Bight Project. State Univ. of New York. Stony Brook, NY. 31p.

Summarizing results from several investigations in the New York Bight. Sources of contamination are dumped wastes: dredge spoils, sewage sludges, acid wastes and toxic chemicals. Evidence for consequential degradation of Bight Apex - dissolved oxygen, nutrients and phytoplankton productivity. Organic carbon and metals found in sediment. The topographic low at Hudson Shelf Valley is preferential settling basin for sediments and contaminants.

1421 - Swanson, R. L. 1976. Tides. MESA New York Bight Atlas Monograph 4. MESA New York Bight Project and New York Sea Grant Institute. Albany, NY. 34p.

Abstract: ATLAS MONOGRAPH 4 surveys recent information about tides in the Bight and looks at some implications for coastal planners. As the two daily tides sweep into the region, the continental shelf exerts a drag, pulling tides parallel to the New Jersey coast. The tide range approaching New York City increases from both directions, northward along the New Jersey shore and westward along Long Island. Tide tables predicting tide levels are highly reliable, with seasonal modifications-lower levels in winter, higher levels in summer. One cannot predict when storms will coincide with high tide to cause possible flooding. Accurate information about gradually rising sea levels and about tides is especially crucial for prospective offshore activities and for questions about estuarine pollution and coastal boundaries.

1422 - Swanson, R. Lawrence. 1977a. Status of ocean dumping research in New York Bight. ASCE Jour. of the Waterway. Port, Coastal and Ocean Division Vol. 103. No. WWI. Proc Paper 12722. pp. 9-24.

Abstract: The New York Bight extends seaward from Long Island and New Jersey to the edge of the continental shelf. Wastes form approx 20,000,000 people are discharged to the Bight Apex. These wastes arrive by a variety of routes: Ocean dumping, sewer outfalls, atmospheric fallout, river discharge, land runoff, thermal discharges, vessel wastes, and occasional spills. Biotic aberrations in the Apex are more prevalent than in comparable coastal environments. Locally, ocean disposal of sewage sludge and dredge material can be associated with these impacts. Comparable studies of existing ocean dumping practices, and proposed alternative ocean disposal sites, nearly 65 nautical miles (120 km) out on the continental shelf were completed.

1423 - Swanson, R. Lawrence. 1977b. Long Island beach pollution: June 1976. Special Report. MESA New York Bight Project. State Univ. of New York. Stony Brook, NY.

Abstract: An assessment of the nature and possible sources of floating trash and pollution that were washed up in large quantities on most of Long Islands beaches during June 1976.

1424 - Swanson, R. L. 1978. NOAA's New York Bight Marine Ecosystem Analysis Project. ICES. C.M. 78/C:18.

1425 - Swanson, R. L., H. M. Stanford, J. S. O'Connor, S. Chanesman, C. A. Parker, P. A. Eisen, and G. F. Mayer. 1978. June 1976 pollution of Long Island ocean beaches. American Soc. of Civil Eng. Jour. of the Envir. Eng. Div. No. EE6. Proc. Paper 14238. 104:1067-1085.

Abstract: In June 1976, Long Island's ocean beaches were inundated with floating wastes including sewage-related material, trash, and garbage. The material exited the Hudson-Raritan estuary as a result of high spring runoff and intensive May rains. Once introduced into New York Bight waters, it was transported to the beaches by persistent southerly winds. The major sources of the floatable wastes were urban runoff and sewage treatment plant bypassing feeding through the combined sewer system. Assessment was accomplished by comparing waste materials found on the beaches with the types and volumes of materials associated with suspected sources. Continual washups of floatable material can be expected, particularly during summer months, until better source and screening controls are implemented.

1426 - Swanson, R. L. 1979. NOAA testimony at EPA Regional II public hearings on issuing of sewage sludge dumping permits. Oct. 2. Newarks, N.J. and Oct. 4, N.Y., NY. 25p.

1427 - Swanson, R. L., and C. J. Sindermann (editors). 1979. Oxygen depletion and associted benthic mortalities in New York Bight, 1976. NOAA Professional Paper No. 11. U.S. Dept. of Commerce. Rockville, MD. 345p.

Summary: The investigations of the 1976 environmental event in the Middle Atlantic Bight make this event one of the best documented examples of mass mortality in the sea, and of the impacts of such events on resource and food-chain species. It is a textbook-type study that focuses on solving many interrelated problems. Scientific studies on oxygen depletion in the Bight and adjacent coastal waters are continuing, especially since the possibility of repetition of the event at some level of intensity exists for future years.

This report brings together the results of many studies on the 1976 oxygen-depletion event. It examines the geographical extent of oxygen depletion and the environmental conditions preceding and during the event, and compares the 1976 conditions with historical information. Particularly useful in this regard were NMFS resource assessment data and the 1974-76 MESA current meter and water column chemistry data. The latter data sets for 1976 were used in the form of a model to diagnose the fluxes of dissolved oxygen into the affected region. An assessment is made of the causes including the effects human activities might have had. Impacts on fishery resources and associated socioeconomic aspects are examined. Finally, monitoring and prediction of future events are discussed.

1428 - Swanson, R. Lawrence, Carl J. Sindermann, and Gregory Han. 1979c. Oxygen depletion and the future: an evaluation. In: R. L. Swanson and C. J. Sindermann (editors), Oxygen Depletion and Associated Benthic Mortalities in New York Bight, 1976. NOAA Professional Paper No. 11. U.S. Dept. of Commerce. Rockville, MD. pp. 355-345.

Summary: Severe oxygen depletion in bottom waters of the New York Bight in 1976, especially off the New Jersey coast, was determined to have been caused from anomalous natural events, and probably will occur again. The phenomenon apparently cannot be controlled, but some reduction in its severity might be achieved by reducing the input of wastes and other substances from human activities. The intensive field investigations and study of the 1976 event revealed shortcomings in the data collected and research conducted before this event. It is clear that our understanding of dynamic processes in coastal waters must be improved, particularly the following: 1) development of the pycnocline-its timing, characteristics, and influence on ecosystem functions; 2) occurrence of large phytoplankton blooms-their relation to oceanic and climatic conditions; and 3) functions of the ecosystem-as they relate to C/O/N cycles and inputs to coastal waters from human activities.

In light of these needs, as they relate to D.O. depletion and possible mass mortalities, and to improve our understanding about and management of coastal waters, certain recommendations are made and discussed.

1429 - Swanson, R. L., H. M. Stanford, and D. M. Goodrich. 1980. A monitoring plan for the New York Bight. ICES. C.M. 80/E:48.

1430 - Swanson, R. L., M. A. Champ, T. P. O'Connor, P. K. Park, G. F. Mayer, J. S. O'Connor, H. M. Stanford, R. Erdheim, and J. L. Verber. (In Press). Sewage sludge dumping in the New York Bight Apex: A comparison with other proposed ocean dumpsites. In: Duedall, Ketchum, Park, and Kester (editors), near shore waste disposal. Vol. 6. Wiley-Interscience Inc., New York, NY. 42p.

1431 - Swartz, S. M., and B.H. Brinkhnis. 1978. The impact of dredged holes on oxygen demand in the lower bay, New York Harbor. Marine Sciences Research Center. State Univ. of New York. Stony Brook, NY. Special Report 17. Ref. 78-5. 80p.

Abstract: The purpose of this investigation was to determine the effects, if any, of dredged holes in the seabed of the Lower Bay of New York Harbor on oxygen dynamics in the water column and surficial sediments. The data for this study were collected during five cruises in the Lower Bay between February, 1978 and August, 1978. Samples of surficial sediments were analyzed for organic carbon, acid-soluble sulfide, and oxygen consumption rate. In addition, the water column at many of the stations was sampled at various depths to profile temperature, salinity, and dissolved oxygen.

The results show that the presence of dredged holes affects oxygen demand of the sediments and oxygen concentrations of the overlying waters. These effects are variable, and dependent on the location of the dredges holes. Lower oxygen concentrations were generally found in dredged holes on the West Bank. There was no effect of dredged holes on oxygen dynamics on the East Bank. This difference is ascribed to hydrodynamic dissimilarities between the East and West Banks, and the accumulation of fine organic matter only on the West Bank.

1432 - Swift, D. J. P., D. B. Duane, O. H. Pilkey (editors). 1972a. Shelf sediment transport, process and pattern. Dowden, Hutchinson and Ross. Stroudsburg, PA. 656p.

1433 - Swift, Donald J. P., David B. Duane and Thomas F. McKinney. 1973. Ridge and swale topography of the Middle Atlantic Bight, North America: Secular response to the Holocene hydraulic regime. Marine Geology. 15:227-247.

Abstract: The ridge and swale topography of the Middle Atlantic Bight was orginally interpreted as a relict strand plan whose ridges reflect stillstands of the returning Holocene sea. However, close examination indicates that the ridges appear to be instead longitudinal bed forms, resposes to a regime of intermittent, south-trending storm currents. Ridges may be initiated on the shore face and detached as the coast retreats to form fields of isolated ridges, or they may be molded into the shelf-transverse sand massifs that mark the retreat paths of littoral-drift depositional centers at estuary mouths and off cuspate forelands. The ridge and swale topography is thus a stable end configuration toward which a variety of near-shore constructional topographies have converged during the Holocene transgression.

Morphologic evidence for readjustment of ridge topography to the deepening shelf flow field during the Holocene trangression is discernable. However, the extent to which the offshore topography continues to respond to hydraulic regime is unclear. The role of helical flow structure in the storm flow field remains to be documented. Resoultion of these problems will require more detailed information of hydaulic process and substrate reponse on storm-dominated shelves.

1434 - Swift, D. J. P. 1975. Sediment response to the hydraulic regime on the central New Jersey shelf. Jour. Sedimentary Petrology. 45:337-358.

1435 - Swift, D. J. P., G. L. Freeland, P. E. Gadd, G. Han, J. W. Lavelle, and W. L. Stubblefield. 1976. Morphologic evolution and coastal sand transport, New York-New Jersey Shelf. In: M. G. Gross (editor), Amer. Soc. Limnol. Oceanogr. Spec. Symp. 2:69-89.

1436 - Swiss, J. J., D. W. McLeese, J. F. Payne and R. F. Addison. 1980. Regulated levels of schedule I substances in the ocean dumping control Act - a review, (Ocean Dumping Report 3). Department of Fisheries and Oceans, Ottawa. 73p.

1437 - Sylvester, A. J., and G. C. Ware. 1977. Laboratory studies on the effect of metals on oxygen uptake by sewage sludge in brackish water. Mar. Pollut. Bull. 8(2):45-47.

1438 - Sylvester, R. O. 1962. Sludge disposal by dilution in Puget Sound. Jour. Water Poll. Control Fed. 34(9):891-900.

1439 - Takenouti, A. Y. , H. Natsume, and T. Miyata. 1979. On the sinking and diffusion of bauxite residue discharged into the sea. Mar. Sci. Commun. 5(3):195-215.

1440 - Talbot, Gerald B. 1954. Factors associated with fluctuations in abundance of Hudson River shad. U.S. Fish and Wildlife Ser. Bull. 56(101):371-413.

1441 - Talbot, J. W. 1972. Transport and dispersion of soluble material (sludge disposal). U.K. Department of the Environment. London. Out of sight - out of mind. Vol. 2. pp. 209-271.

1442 - Talbot, J. W., B. Harvey, R. A. Eagle, and M. S. Rolfe. (In Press). Hydrographic and benthic studies in the Thames Estuary 1972. U.K. MAFF Tech. Report.

1443 - Tarzwell, C. M. 1969. Waste management in the marine environment. Proc. Civil Engng. in the Oceans - II. Miami Beach, FL. pp. 477-485.

1444 - Taylor, D. 1979. The distribution of heavy metals in the United Kingdom coastal waters of the North sea. Int. Conf. Man and Control of Heavy Metals in the Envir. September. London.

1445 - Templeton, W. L. 1980. Radiological aspects of seabed dumping in the deep oceans. Conference Paper. Battelle Memorial Inst. Pacific Northwest Labs. Seattle, WA. pp. 23-33.

1446 - Tenny, M. W., and T. G. Cole. 1968. The use of fly ash in conditioning biological sludges for vacuum filtration. Jour. Water Poll. Control Fed. 40(8), Part 2.

1447 - TerEco Corporation. 1975. Sea-level monitoring of the incineration of organic chloride waste by M/T Vulcanus in the northern Gulf of Mexico. Shell waste burn No. 2. Tereco Corporation. College Station, TX. 53p.

1448 - Theis, T. L. 1975. The potential trace metal contamination of water resources through the disposal of fly ash. The Second National Conference on Wateruse. Sponsored by Amer. Inst. Chem. Eng. and EPA. Palmer House-Chicago. 4-8 May 1975.

1449 - Thistlethwayte, D. K. B. 1971. Disposal of Sewage and Other Water-Borne Wastes. (Second edition). Ann Arbor Science Publishers, Inc. Ann Arbor, MI. 412p.

1450 - Thomas, J. P., and J. B. Pearce. 1975. NOAA ship Delaware II cruise 74-2 report of May 13-21, 1974 survey of deepwater dumpsite 106. In: U.S. NOAA May 1974 Baseline investigation of deepwater dumpsite 106. p. 13-19. U.S. National Oceanic and Atmospheric Administration. NOAA Dumpsite Evaluation Report 75-1. Rockville, MD.

1451 - Thomas, J. P., W. Phoel, and F. W. Steimle, Jr. 1976a. New York Bight Apex data on total oxygen consumption by the seabed, March 1974 - February 1975. NOAA DR ERL MESA-6. Boulder, CO. 92p. NTIS No. PB 271342/AS.

Abstract: During this period, seabed oxygen consumption and related bottom water hydrographic data (temperature, salinity, dissolved oxygen concentrations and percent oxygen saturation) were measured on four cruises in the Apex of the Bight to determine baseline values and distribution. Data from the approximately sixty stations per cruise are summarized in this report.

1452 - Thomas, J. P., W. Phoel, J. E. O'Reilly, and C. A. Evans. 1976b. Seabed oxygen consumption in the Lower Hudson estuary. In: H. McKeon and G. J. Lauer (editors). Fourth Symposium on Hudson River Ecology. Hudson River Environmental Society, Inc. New York. Paper No. 12.

1453 - Thomas, J. P., W. C. Phoel, F. W. Steimle, Jr., J. E. O'Reilly, and C. A. Evans. 1976. Seabed oxygen consumption - New York Bight. In: M. G. Gross (editor), Amer. Soc. Limnol. and Oceanogr. Spec. Symp. 2:354-369.

Abstract: Seabed oxygen consumption rates, temperature, salinity, and dissolved oxygen were measured during five cruises in the New York Bight Apex between March 1974 and August 1975. The area sampled included the waste disposal site for sewage sludge, dredge spoils, and industrial acid wastes. Samples were collected and incubated on shipboard at in situ temperature during oxygen uptake measurements.

Total oxygen consumption by the seabed ranged from 1-68 ml O_2 $m^{-2}h^{-1}$. An average of $444x10^6$ liters O_2 (183 tonnes C) are estimated to be consumed by the Apex (1,577 km^2) seabed each day ranging from $288x10^6$ in February to $689x10^6$ in August. In winter the highest rates of uptake were measured in the Christiaensen Basin adjacent to the sewage sludge disposal site, in the topographically high dredge spoil disposal area west of the Christiaensen Basin, and in the Hudson Shelf Valley. In summer the highest rates were measured in the dredged spoil area. Rates in the Christiaensen Basin, however, were low compared with the surrounding areas and were more like winter rates. This difference may have been caused by differential sedimentation rates of oxidizable organic carbon to the seabed, mediated by the presence or absence of a thermocline. The highest rates were measured near a municipal sewage outfall off Asbury Park, N.J. No discernible effects on seabed oxygen consumption were observed near the acid waste disposal area.

Rates of oxygen uptake by the bottom water and by the entire water column were measured and compared with oxygen uptake rates by the sediment. Most (93%-98%) oxygen uptake in the Apex occurs in the water column and not on the seabed.

1454 - Thomas, James P. et. al. 1979. Biological processes productivity and respiration. In: R. L. Swanson and C. J. Sinderman (editors), Oxygen Depletion and Associated Benthic Mortalities in New York Bight, 1976. NOAA Professional Paper No. 11. U.S. Department of Commerce. Rockville, MD. pp. 231-260.

Summary: Between August 24 and September 9, 1976, about 2 months after the onset of oxygen depletion, data concerning primary production, water-column and seabed oxygen consumption, nutrients, organic carbon, phytoplankton identification and abundance, chlorophyll a, and bacteria were collected to document conditions.

1455 - Thompson, G. B., and J. Ho. 1981. Some effects of sewage disharge upon phytoplankton in Hong Kong. Mar. Pollut. Bull. 12:168-173.

1456 - Thorslund, A. E. 1975. Disposal of contaminated spoil in a stratified Fjord - An example from the Swedish West Coast. Vatten. 31(2):133-138.

1457 - Thurberg, Frederick P. and, Randolph O. Goodlett. 1979. Impact on clams and scallops. Part 2. Low dissolved oxygen concentrations and surf clams - A Laboratory study. In: R. L. Swanson and C. J. Sinderman (editors), Oxygen Depletion and Associated Benthic Mortalities in New York Bight, 1976. NOAA Professional Paper 11. U.S. Dept. of Commerce. Rockville, MD. pp. 277-280.

Summary: Under laboratory conditions, the surf clam survived low levels of D.O. for extended periods of time. Levels below 1.4 ml/l were nearly always fatal. No deaths were recorded after 8 weeks at 2.1 ml/l D.O., and clams placed in water at 0.7 ml/l after previous exposure to 2.1 ml/l survived for 8 weeks, indicating that a gradual shift to anaerobic pathways is possibly advantageous. Flowing water exposures permitted better survival than did static water systems. Metabolic studies indicated that animals held under low oxygen conditions consumed oxygen at higher rates than normal.

1458 - Tibby, R. B., J. E. Foxworthy, M. Oguri and R. C. Fay. 1965. The diffusion of wastes in open coastal waters and their effects on primary biological productivity. Symp. Comm Int. Explor. Scient. Mer Medit. Monaco. pp. 95-113.

1459 - Tietjen, John H. 1980. Population structure and species composition of the free-living nematodes inhabiting Sands of the New York Bight Apex. Estuar. and Coastal Mar. Sci. 10:61-73.

1460 - Tietjen, John H. (In Press). Potential roles of nematodes in polluted ecosystems and the impact of pollution on meiofauna. In: G. F. Mayer, A. Calabrese, F. A. Cross, D. C. Malins, D. W. Menzel, J. S. O'Connor and F. J. Vernberg (editors), Ecological Stress in the New York Bight: Science and Management. Estuarine Research Federation. Columbia, SC.

Abstract: Meiofauna are small benthic animals (usually less than 2 mm) whose densities on the shelf generally range from $10^5/m^2$ to $10^7/m^2$. Through their grazing of bacterial populations, they maintain the bacteria in a state of active growth and thus contribute significantly (along with the protozoa and macrofauna) to the organic decomposition process. Through direct excretion of phosphorus and nitrogen, meiofauna and other benthic animals also may contribute to the regeneration of nutrients, most of which are rapidly utilized by bacteria and microalgae.

In sediments with high organic enrichments, such as in the New York Bight Apex and western Long Island Sound, meiofaunal densities and biomass are not depressed (compared to those of sediments with no organic enrichment). Where macrobenthic densities have been reduced by pollution, such as in the New York Bight Apex, meiofauna and protozoa may become the dominant benthic animal influence in detritus decomposition and remineralizaton. Pollutants that reduce meiofaunal densities or alter the manner in which the animals function thus have an impact on the organic decomposition process.

In the New York Bight Apex, increased levels of heavy metals in the dredge spoil and sewage sludge dumpsites are correlated with changes in the species composition and decreased species diversity of nematodes, the dominant taxon present. The replacement fauna contains species that have been shown in laboratory studies to have slower metabolic rates than the original fauna. Such metabolically less active animals could contribute to long-term accumulation of organic matter in the deeper basins of the New York Bight and Long Island Sound.

1461 - Tijssen, S. B., and A. Eijgenraam. 1980. Diurnal oxygen rhythm in the southern Bight of the North Sea: Net and gross primary production in April 1980 in a _phaeocystis_ bloom. ICES. C.M. 80/C:17.

1462 - Timoney, J. F., Jennifer Port, Janis Giles, and J. Spanier. 1978. Heavy-metal and antibiotic resistance in the bacterial flora of sediments of the New York Bight. Appl. and Envir. Microbiology. 36(3):465-472.

Abstract: The New York Bight extends seaward some 80 to 100 miles (ca. 129 to 161 km) from the Long Island and New Jersey shorelines to the edge of the continental shelf. Over 14×10^6 m^3 of sewage sludge, dredge spoils, acid wastes, and cellar dirt are discharged into this area each year. Large populations of _Bacillus_ sp. resistant to 20 μg of mercury per ml were observed in Bight sediments contaminated by these wastes. Resistant _Bacillus_ populations were much greater in sediments containing high concentrations of Hg and other heavy metals than in sediments from areas further offshore where dumping has never been practiced and where heavy-metal concentrations were found to be low. Ampicillin resistance due mainly to B-lactamase production was significantly ($P<0.001$) more frequent in _Bacillus_ strains from sediments near the sewage sludge dump site than in similar _Bacillus_ populations from control sediments. _Bacillus_ strains with combined ampicillin and Hg resistances were almost six times as frequent at the sludge dump site as in control sediments. This observation suggests that genes for Hg resistance and B-lactamase production are simultaneously selected for in _Bacillus_ and that heavy-metal contamination of an ecosystem pressure for antibiotic resistance in bacteria in that system. Also, Hg resistance was frequently linked with other heavy-metal resistances and, in a substantial proportion of _Bacillus_ strains, involved reduction to volatile metallic Hg (Hg^0).

1463 - Timoney, John F., and Jennifer Graves Port. (In Press). Heavy metal and antibiotic resistance in _Bacillus_ and _Vibrio_ from sediments of New York Bight. In: G. F. Mayer, A. Calabrese, F. A. Cross, D. C. Malins, D. W. Menzel, J. S. O'Connor and F. J. Vernberg (editors), Ecological Stress in the New York Bight: Science and Management. Estuarine Research Federation. Columbia, SC.

Abstract: A large proportion of the sedimentary _Bacillus_ from the waste disposal area of the New York Bight was resistant to 20 μg of mercury (Hg) per ml and to high concentrations of cadmium (Cd) and zinc (Zn). Concentrations of heavy metals were much higher in these than in control sediments further offshore where dumping of wastes has never been practiced and where heavy metal concentrations were low. Ampicillin resistance, mainly because of B-lactamase production, was significantly ($P < 0.001$) more prevalent in _Bacillus_, from the dump area than from control sediments. Also, _Bacillus_ with combined mercury and ampicillin resistance were six times more frequent in dump sediments than elsewhere.

Mercury resistance also was prevalent in Vibrio from dump sediments. The resistance pattern mercury-kanamycin-streptomycin was found in 10.2% (60 out of 586) of Vibrio or Vibrio-Aeromonas from dumpsite sediments, but was found in only 1.6% (8 out of 510) of similar strains from control sediments. This resistance was transferable by conjugation and was probably plasmid-mediated. An important mechanism of mercury resistance in both Bacillus and Vibrio was reduction to volatile metallic mercury (Hg).

The results of this study suggest that heavy metal contamination of an ecosystem can exert a selection pressure for antibiotic resistance in bacteria in that environment.

1464 - Tokos, Joseph J. 1978. The distribution of chlorophyll-a in the New York Bight as determined by continuous in-vivo fluorometry. August, 1977. Masters Thesis. State Univ. of New York. Stony Brook, NY. 89p.

Abstract: An extensive, high resolution survey of phytoplankton chlrorphyll-a and temperature in both the horizontal and vertical dimensions was conducted during August, 1977 in the New York Bight. Most phytoplankton biomass is located within the 30-40 m isobath. Mapping activities, two offshore transects and three time-series indicate the nearshore Bight was divided into three distinct regions: a chlorophyll-a band (5-10 km x 50-60 km) in frontal system along the New Jersey coast, a low chlorophyll mixed environment near Long Island and a high concentration surface population in the Hudson River discharge. The New Jersey frontal system (the major conformational feature) has not previously been reported. The three areas appear related to the effects of prevailing wind stress on local variations in coastline orientation and bottom topography. Results encourage repeated mapping surveys in order to more accurately assess phytoplankton response to physical forcing functions within this highly diverse environment.

1465 - Tomczak, G. 1968. Versuch einer klassifikation industrieller abfallprodukte ein bezug auf die Moglichkeit einer versenkung auf see. (Attempt of a classification of industrial wastes with regard to a possible dumping at sea.) Helgolander wiss. Meeresunters. 17:6-24.

1466 - Topping, G., and A. D. McIntyre. 1972. Benthic observations on a sewage sludge dumping ground. ICES. C.M. 72/E:30. 4p.

1467 - Topping, G. 1974. The atmospheric input of some heavy metals to the Firth of Clyde and its relation to other inputs. ICES. C.M. 74/E:32.

1468 - Topping, G. 1976. Sewage and the sea. In: R. Johnson (editor), Marine Pollution. Academic Press. New York, NY. pp. 303-351.

1469 - Torpey, W. N. 1967. Response to pollution of New York harbor and Thames Estuary. Jour. Water Poll. Control Fed. 39(11):1797.

1470 - Trefry, J. H., and B. J. Presley. 1976. Heavy metal transport from the Mississippi River to the Gulf of Mexico. Health Books. Lexington, MA. pp. 39-76.

1471 - Tromp, D. 1976. Aspects of the concentration of discharged noxious substances in the wake of the ship. In: IMCO Report of the symposium on prevention of marine pollution from ships. Acapulco, Mexico. 22-31 March 1976. U.S. Coast Guard, Office of Marine Environment and Systems. Washington, DC. (C-WEP-1-77). 38p.

1472 - Tromp, D. 1980. Coastal pollution in northern Europe. Paper No. 1. Published in the Proc. of Coastal Discharges: Engineering aspects and experience. Institution of Civil Engineers. London. 12p.

1473 - Turekian, Karl K. 1974. Heavy metals in estuarine systems. Oceanus. 18(1):32-33.

Abstract: Man's activities have clearly increased the flux of metals to coastal waters delivered by streams, sewer outfall and the atmosphere. The metals deposited in dredge spoils has little chance of leaching out of the sediment. Problems involve mobilized toxic organic compounds and changes in the physical nature of the substrate.

1474 - Udell, Harold P. 1952. Bacterial pollution of Raritan and Lower Bays and its relation to shelfish, 1951. Fed. Sec. Agency-Public Health Service. 86p.

1475 - Udell, Harold P., Thomas Doheny, John Zarudsky, and Joan Keene. 1974. Ocean dumping of sewage sludge; its effect on the South Shore of Long Island. A status Report. Town of Hempstead. Dept. of Conservation and Waterways. Point Lookout, NY. February-June 1974. 21p.

Abstract: Ocean bottom sediments were sampled from an area extending from East Rockaway Inlet south to the ocean sewage sludge dumpsite, easterly to Jones Inlet and southerly eight miles. Physical, chemical, and microbiological characteristics of the sediments were determined. Upon evaluation of these characteristics, giving consideration to metal content as Cd, Cr, Cu, Ni, Pb, and Zn; organic matter, and bacterial concentrations--it was found that sewage sludge as deposited at the designated dump site, loses its identity seven and a half (7.5) miles south of Rockaway Inlet. Sediments beyond a three mile radius north and east of the dump site contain background levels of heavy metals and organic matter; however, in the metals, organic matter and bacteria are slightly in excess of background. This results from material being flushed from the west region of Hempstead Bay by diurnal tidal transport. Material carried from the bay is carried east by the reversing current, and then westerly by long-shore currents. Thus, deposition of material from the bay takes place in a confined area offshore of Atlantic Beach.

1476 - U.K. Department of the Environment. 1972-1976. Out of sight out of mind. Report of a Working Party on Sludge Disposal in Liverpool Bay. 4 volumes. Vol. 1 (1972): Main Report. 36p. Vol. 2 (1972): Appendices. 485p. Vol. 3 (1973): Report for 1972-73. 104p. Vol. 4 (1976): Report for 1973-75. 188p. M.M.S.O. London.

1477 - U.K. Department of the Environment. 1977. Sewage sludge disposal data and technical notes. Water Engineering. 1. R&D Div. Tech. Note 6. H.M.S.O. London. 44p.

1478 - U.K. Department of the Environment. 1978. Sewage sludge disposal data and reviews of disposal to sea. National Water Council Standing Technical Committee Report No. 8. H.M.S.O. London. 49p.

1479 - U.K. Department of the Environment. 1979a. Report of the sub-committee on the disposal of sewage sludge by incineration. National Water Council Standing Technical Committee Report No. 14. H.M.S.O. London. 83p.

1480 - U.K. Department of the Environment. 1979b. Sewage sludge disposal in Liverpool Bay. Research into effects 1975 to 1977. Part 1 - General. National Water Council Standing Technical Committee Report No. 16. H.M.S.O. London. 44p.

1481 - U.K. Department of the Environment. 1979c. Report of the sub-committee on the disposal of sewage sludge to sea 1975-78. National Water Council Standing Technical Committee Report No. 18. H.M.S.O. London. 65p.

1482 - U.K. Department of the Environment. 1981a. Report of the sub committee on the economics of sewage sludge disposal. National Water Council Standing Technical Committee Report No. 19. H.M.S.O. London. 107p.

1483 - U.K. Department of the Environment. 1981b. Report of the sub-committee on the disposal of sewage sludge to land. National Water Council Standing Technical Committee Report No. 20. H.M.S.O. London. 83p.

1484 - U.K. Department of the Environment. 1981c. Report of the standing committee on the disposal of sewage sludge. National Water Council Standing Technical Committee Report No. 21. H.M.S.O. London. 30p.

1485 - U.K. Department of the Environment. 1979. Monitoring the marine environment - the way ahead. The Second Report of the Marine Pollution Monitoring Management Group 1977-1978. Central Directorate on Environmental Pollution Report No. b. 23p.

1486 - U.K. Ministry of Defense. 1970. Report of a working party on the transport of nerve agents from Nancekuke to Porton and the disposal of effluent from nerve agent production. H.M.S.O. London. 5p.

1487 - Ulbricht, K. A., D. Schmidt, and U. Horstmann. Mass appearance of blue green algae in the Baltic sea: Evaluation of multispectral LANDSAT scenes by image processing. Earth Observation from Space and Management of Planetary Resources. pp. 77-89.

Abstract: Accumulation of blue green algae, *Nodularia spumigena* and *Alphanisomenon flosaquae*, was detected from digital image processing of LANDSAT scenes of the Baltic sea. White traces appear in some spectral regions (MSS 4 and MSS 5), which are shown to be due to algae. Blue green algea participate in the fixing of nitrogen from the air and its abnormal growth is probably related to an increase in phosphate concentrations due to sewage discharge having a high phosphorus content in the regions near the sea coast. It is also concluded that space detected traces are to be complimented by in situ measurements in order to get reliable quantitative information.

1488 - Underwood, Peter C. 1976. The spatial distribution of phytoplankton biomas in the surface waters of the New York Bight from April 1975 to September 1975. Masters Thesis. State Univ. of New York. Stony Brook, NY. 84p.

Abstract: The monthly distribution of chlorophyll *a* in the surface waters of the New York Bight from April 1975 to September 1975 is described, and a method for ulitization of a continuous record of surface in vivo chlorophyll fluorescence to increase sampling resolution and decrease between station interpolation is presented. The spring bloom in the New York Bight is characterized by shelfwide increase in chlorophyll *a* concentration. As the bloom progresses and the nutrient supply is utilized, the distribution becomes characterized by patches of high concentation imbedded in a background "matrix" of lower concentration. Evidence is presented to show that these highs reflect input of nutrients from outside sources.

1489 - UNESCO/FAO/IMCO. 1975. Scientific criteria for the selection of sites for dumping of wastes into the sea. GESAMP. Reports and studies. No. 3. London. 22p.

1490 - UNESCO/FAO/IMCO. 1980. Monitoring biological variables related to marine pollution. GESAMP. Reports and studies No. 12. London. 22p.

1491 - University of Rhode Island. 1968. Progress report-systems analysis for shipboard municipal incineration. Contract No. U.R.I. 00557-03.

1492 - University of Southern California, Los Angeles. 1979a. Theory of spectral radiance of pollutants at sea. Volume 1. Final Report. Inst. for Marine and Coastal Studies. NASA-CR-158791. 42p.

Abstract: Remote measurement of soluble pollutants dumped in the sea, that change the color of the water, is addressed. The sensor is a spectral radiometer that flies over the polluted area and compares its spectral radiance (color) to that of surrounding clean seawater. The goal is to infer the concentration of pollutants using the measured radiance of the sea compared to laboratory measurements of reflection and transmission spectra of the pollutants. The subject is treated in three steps: (1) the quantities involved are defined and means for measuring them are described; (2) the equations for remote sensing with a low-flying aircraft are derived, in which the absorption and radiance of intervening air is negligible; and (3) high-flying aircraft and satellites are applied, in which case the radiance of intervening air is the major problem.

1493 - University of Southern California. 1979b. Satellite passive remote sensing of off-shore pollutants. Volume 2. Final Report. Inst. for Marine and Coastal Studies. NASA-CR-158792. 46p.

Abstract: Satellite detection and monitoring of off-shore pollutants are discussed. Summaries of satellite sensor performance in three spectral bands (visible, infrared, and microwave) are presented. The bulk of the report gives all the calculations, trade-offs and limitations of the three sensor systems. It is asserted that the problem of pollution monitoring is not a sensor problem but a problem of mathematical modelling and data processing.

1494 - U.S. Army Corps of Engineers. 1978. 1977 Report to Congress on administration of ocean dumping activities, Public Law 92-532 (MPRSA of 1972). Department of the Army. Washington, DC. 47p.

1495 - U.S. Army Corps of Engineers. 1980. Final summary U.S. Army Corps of Engineers-Dredged Material Research Program. Shore and Beach 48(1):35-41.

1496 - U.S. Coast and Geodetic Survey. Undated. Sea disposal of low level packaged radioactive waste. C&GS Project No. 10.000-827.

1497 - U.S. Coast Guard. 1977. U.S. Coast Guard utilization of remote sensing techniques for ocean surveillance. Oceans 1977. Annual Combined Conference. MTS/IEEE. Los Angeles, CA (USA).

1498 - U.S. Code of Federal Regulations. 40 CFR 220-229. Title 40-Protection of the Environment. Chapter I - Environmental Protection Agency. Subchapter H - Ocean Dumping.

1499 - U.S. Commission on Natural Resources (Ocean Disposal Study Steering Committee). 1976. Disposal in the marine environment; an oceanographical assessment. U.S. National Academy of Sciences. Washington, DC. 76p.

1500 - U.S. Congress. 1972. Public Law 92-532 (H.R. 9727). The Marine Protection, Research, and Sanctuaries Act of 1972. October 23.

1501 - U.S. Congress. 1973. Ocean dumping. House of Representatives Committee on Merchant Marine and Fisheries. Subcommittee on Oceanography and Subcommittee on Fisheries and Wildlife Conservation and the Environment. 93rd. Cong. 1st Sess. June 5, 1973. Serial No. 93-14. Government Printing Office. Washington, DC. 1973. 97p.

1502 - U.S. Congress. 1974. Public Law 93-254 (H.R. 5450). To Amend the Marine Protection, Research, and Sanctuaries Act of 1972. March 22.

1503 - U.S. Congress. 1975. Public Law 94-62 (H.R. 4710). To Amend the Marine Protection, Research, and Sanctuaries Act of 1972. July 25.

1504 - U.S. Congress. 1976. Public Law 94-326 (S. 3147). To Amend the Marine Protection, Research, and Sanctuaries Act of 1972. June 30.

1505 - U.S. Council on Environmental Quality. 1970. Ocean dumping: A national policy. U.S. Government Printing Office. Washington, DC. 45p.

1506 - U.S. Department of Commerce. 1971. Deep ocean dumping of baled refuse. Sub-Council Report. 15p.

1507 - U.S. Department of Commerce. 1972. The Efects of Waste Disposal in the New York Bight. Final Report in nine sections. National Marine Fisheries Service (NMFS), National Oceanic and Atmospheric Administration (NOAA). U.S. Department of Commerce. Highlands, NJ.

1508 - U.S. Department of Commerce. 1974a. Report to the Congress on ocean dumping and other man-induced changes to ocean ecosystems, October 1972 through December 1973. First Annual Report for Public Law 92-532, Title II. U.S. GPO. Washington, DC. 96p.

1509 - U.S. Department of Commerce. 1974b. The chemical, physical, and sedimentary oceanographic aspects of the New York Bight Apex as they relate to the offshore dumping of wastes - an interim report. NOAA. Atlantic Oceanographic and Meteorological Laboratories. Miami, FL. 3 Parts. 124p.

Abstract; Chemical - Sewage contributes a significant amount of carbohydrate to sediment organic matter. Data indicate that Long Island Mud is possibly derived in part from sewage. Physical - Seward flow of brackish water in surface layers and return flow of external water in bottom layers. There exists a clockwise eddy outside the river influenced region. Sedimentary - 450 metric tons sewage sludge is dumped per day. Northwest of dumpsite, Christiaensen basin is a natural zone of mud deposition and is significantly contaminated with sewage sludge.

1510 - U.S. Department of Commerce. 1974c. Bibliography of the New York Bight. Two volumes, Part I - List of Citations. Part 2 - Indexes. NOAA Environmental Data Service (Special Report). Rockville, MD. 677p.

1511 - U.S. Department of Commerce. 1974d. Middle Atlantic coastal fisheries center cruise report: NOAA Ship Delaware II (May 13-21, 1979). NOAA. NMFS Northeast Region. Highlands, NJ. 16p.

Abstract: This cruise was part of a joint effort with the NOAA Ship *Albatross* IV to characterize Deep Water Dump Site 106. The *Albatross* IV investigated the water column while the *Delaware* II investigated the benthos (see separate cruise report for the *Albatross* IV).

1512 - Department of Commerce. 1975a. May 1974 baseline investigation of Deepwater Dumpsite 106. NOAA/ODP Dumpsite Evaluation Report 75-1. Rockville, MD. 388p.

1513 - U.S. Department of Commerce. 1975b. Ocean dumping in the New York Bight. NOAA TR ERL 321-MESA 2. Boulder, CO. 78p.

Summary: The New York Bight extends seaward over 15,000 square miles (39,000 km^2) from Long Island and New Jersey to the edge of the continental shelf, some 80-100 nautical miles (150-180 km) offshore. Wastes from 20 million people are discharged to the Bight. These wastes arrive by a variety of routes: ocean dumping, outfall sewers, air pollution, river discharge, land runoff, thermal discharges, vessel wastes, and occasional spills. Although impacts of these wastes on the marine environment are not clearly understood, there is evidence that the waters, bottom sediments, and living resources are under stress.

In 1973 the amount of raw and digested sewage sludge was 150 million ft^3 (4.3×10^6 m^3). An average of 260 million ft^3 (7.4×10^6 m^3) per year of dredge spoils were dumped each year between 1965 and 1970. During the same period an average of 72 million ft^3 (2.0×10^6 m^3) per year of waste acid and an average of 16 million ft^3 (0.5×10^6 m^3) per year of construction and demolition debris were dumped into the New York Bight. The hazards of this dumping are not known, however above normal incidence of fin-rot disease in fish in the area and the closing of the area to shellfishing are indicatings that something is wrong. The amount of sludge that moves northward to the vicinity of Long Island Beaches is unknown; there is no evidence of massive shoreward movement of the sludge, or of imminent bacteriological hazard to the beaches. Meanwhile it is recommended that interim use of alternative dump sites be avoided and that land-based disposal alternatives be developed.

Before final decisions can be made to solve the problems identified, further studies of various alternative solutions are required.

1514 - U.S. Department of Commerce. 1975c. Mutagenicity of marine pollutants could be affecting inshore and offshore marine fisheries. NOAA. NMFS. Middle Atlantic Coastal Fisheries Center. Highlands, NJ. 77p.

Summary: Chemicals introduced in the last 30 or so years and prevalent now throughout the terrestrial and aquatic environment may have either or both, individually or collectively powerful mutagenic, that is, hereditary altering effects, just as does radiation. The diversity of mutagenic chemicals is so great that no group term would comprise them all. A few combinations of mutations under certain conditions of exposure will have a particular selective value and hence be conserved by natural selection. Almost all mutations, however, are at least harmful.

Salts of heavy metals are one of three groups of substances now recognized as having a significant influence on chromosome breakage. Chromosome damage may occur at sub-toxic doses. That certain pesticides - herbicides, insecticides, fungicides - may be classified as environmental mutagens, has been amply demonstrated. Pesticides are a non-homogeneous group of chemicals whose effects on chromosomes range from no observable effect to one greater than that for well-known powerful mutagens experimentally used to alter genes. Such recognized environmental mutagens as heavy metals and pesticides, as well as many other chemicals, are also major marine pollutants.

Consideration of specific consequences of contaminant-induced genetic damage summarized in this report cannot but lead one to suppose that insidious genetic changes could, in combination with other factors as overfishing, significantly reduce important commercial fish populations before the nature of the damage came to light. The biological potential for ultimate genetic recovery of these populations would mean little to the immediate economics of the situation. Continued changes could, in combination with other factors as overfishing, significantly reduce important commercial fish populations before the nature of the damage came to light. The biological potential for ultimate genetic recovery of these populations would mean little to these immediate economics of the situation. Consideration of the effect of mutagenic marine pollutants on spawn alone raises cause for concern. Fish gametes are spawned out into polluted waters where uptake of pollutants by the gametes is added to their already existing cell load of mutagens from the parent fish. Fertilization and completion of meiosis of fish eggs and early cleavage division of the embryo, which takes place in polluted waters after spawning, are very genetically sensitive stages. Most mutations are probably lethal, and further, the expression of normal, unaltered genes and chromosomes is intimately interwoven into the very development, health, vigor, and reproductive performance of any organism.

1515 - U.S. Department of Commerce. 1976a. Evaluation of proposed sewage sludge dumpsite areas in the New York Bight. NOAA TM ERL MESA 11. Boulder, CO. 212p.

Abstract: Study of significant features of the New York Bight environment and the two proposed alternate sewage dumpsites. Dumpsites composed of medium size clean sands (no mud deposits). Sediment transport to the southwest. Shoreward transport of surface waters occurs during summer. Potential biological impacts are an increase of fin rot, changes in migratory patterns of bottom fishes, changes in food web from added detritus and bacteria.

1516 - U.S. Department of Commerce. 1976b. Report to the Congress on ocean dumping research January through December 1975. NOAA Ocean Dumping Program. Rockville, MD. 51p.

Abstract: This is the third annual report to the Congress on ocean dumping research required by P.L. 92-532, Title II, Section 201. The report describes progress made by Federal agencies in ocean dumping research in 1975, elements of interagency coordination, and future program direction. It is intended to provide a framework for assessing strengths and weaknesses in the Federal effort. Although not specifically required by the legislation, the report also includes a summary of 1975 activities directed to fulfilling the requirements of Title II Section 203, research into disposal alternatives to ocean dumping.

The principal research efforts of involved Federal agencies during calendar year 1975 described in this report are organized by category of ocean dumped material rather than agencies. This organization of information is intended to convey an impression of the total research effort being applied to each major type of material presently dumped in the ocean. The three categories of ocean dumped materials discussed in this report are, dredged material, municipal wastes, and industral wastes.

1517 - U.S. Department of Commerce. 1976c. Passage of anticyclonic Gulf Stream eddies through deepwater dumpsite 106 during 1974 and 1975. NOAA/ODP Dumpsite Evaluation Report 76-1. Rockville, MD. 39p.

1518 - U.S. Department of Commerce. 1976d. The New York Bight Project 1975. NOAA, MESA Special Report. MESA New York Bight Project. State Univ. of New York. Stony Brook, NY.

1519 - U.S. Department of Commerce, Maritime Administration. 1976e. Final environmental impact statement. Maritime administration chemical waste incinerator ship project. Vol. 1. Washington, DC. 383p.

1520 - U.S. Department of Commerce. 1977a. Baseline report of environmental conditions in deepwater dumpsite 106. NOAA/ODP Dumpsite Evaluation Report 77-1. Rockville, MD. Vol. 1. Physical Characteristics. 218p.

1521 - U.S. Department of Commerce. 1977b. Baseline report of environmental condition in deepwater dumpsite 106. NOAA/ODP Dumpsite Evaluation Report 77-1. Rockville, MD. Vol. 2. Biological characteristics. 269p.

1522 - U.S. Department of Commerce. 1977c. Baseline report of environmental conditions in deepwater dumpsite 106. NOAA/ODP Dumpsite Evaluation Report 77-1. Rockville, MD. Vol. 3. Contaminant Inputs and Chemical Characteristics and Appendices. 314p.

1523 - U.S. Deparment of Commerce. 1977d. New York Bight project annual report for FY 1976-76T. NOAA TM ERL MESA-25. Boulder, CO. 91p. NTIS No. PB 283196/AS.

Abstract: The Annual Report for Fiscal Year 1976-76T describes MESA New York Bight Project activities between July 1, 1975, and September 30, 1976. Specifically, it summarizes research efforts sponsored by the Project and reviews significant technical, operational, and administrative achievements during the period. It is used as a management tool by the MESA Program Office and by administrators of the Environmental Research Laboratories (ERL) and National Oceanographic and Atmospheric Administration (NOAA). It provides information to the U.S. Congress and to the Office of Management and Budget on the use of funds allocated for research into ocean dumping and marine ecosystems analysis. Finally, the Report describes Project activities to groups and individuals with interests in the Bight, including the MESA New York Bight Advisory Committee, Project investigators, users of MESA-generated information, and members of the concerned public.

1524 - U.S. Department of Commerce. 1977e. Report to congress on ocean dumping research January - December 1976. NOAA Ocean Dumping Program. Rockville, MD. 21p.

Abstract: Fourth annual report to Congress, Research on sludge dumping in New York Bight and industrial waste disposal at Deepwater Dumpsite 106. Reports on assessing inputs, dredged material dumping, Long Island beach pollution, the New Jersey anoxia event, and remote surveillance and monitoring.

1525 - U.S. Department of Commerce. 1978a. Delaware II Clam Survey 78-01, Information Report (covering the Mid-Atlantic Shelf Long Island to Cape Hatteras) 4 Jan.-11 Feb. 1978. NOAA/NMFS. Highlands, NJ. 20p.

1526 - U.S. Department of Commerce. 1978b. Report to congress on NOAA ocean dumping research January-December 1977. NOAA Ocean Dumping Program. Rockville, MD. 38p.

Abstract: Report of NOAA research on sewage sludge, industrial wastes and dredged materials. Progress report on development of remote sensing techniques to monitor dispersion of dumped wastes. Toxic metals (chromium, iron, copper, silver) are present in high concentrations at dumpsites. Anoxic dredged materials contain highly toxic sulphur compounds.

1527 - U.S. Department of Commerce, NOAA National Environmental Satellite Service. 1979a. A summary of remote sensing investigations in the New York. Bight Environmental Research Institute of Michigan. Ann Arbor, MI. 157p.

Abstract: A number of experiments were conducted using specific remote sensing techniques and directed toward gathering and analyzing information on significant ocean parameters observed in the New York Bight. These remote sensing experiments were conducted by or for the National Environmental Satellite Service of NOAA in April 1973 and April 1975. The data were collected from multi-altitude remote sensing operation (intermediate and high altitude aircraft, and the Landsat-1 satellite) and from surface operations to obtain concurrent in situ data. The resulting data were analyzed to obtain information on chlorophyll-a distribution, suspended particulate concentration, Secchi disk transparency, sea surface temperature, water mass classification, and circulation, both at the surface and at 10 meters depth. The results provide significant information on various water constituents, including those resulting from sewage and acid dumping, tidal processes, pollutant dispersal, and other dynamic processes occurring in the study area.

With proper planning and execution, remote sensing operations can replace or supplement information from other sources on these phenomena. Current remote sensing technology can provide synoptic and repetitive coverage of large ocean areas and permit quantitative estimation of important parameters. However, future applications should continue to incorporate operational refinements and improved algorithms to increase the accuracy and reliability of remote sensing results. Operational techniques should stress the use of adequate atmospheric corrections, careful planning of the number, location, and timing of water samples collected, and sample coordination of the efforts of the remote sensing technologists and the personnel of the investigative team.

1528 - U.S. Department of Commerce. 1979b. Chemical pollutants of the New York Bight; priorities for research. J. S. O'Connor and H. M. Standord (editors). NOAA Special Report. MESA New York Bight Project. Boulder, CO. 217p.

Abstract: An assessment of research priorities of long-term, low concentration hazards. Does not consider acute problems of spills. Major threats (Category A) are chlorinated pesticides, lead, mercury, polynuclear, aromatic hydrocarbons, polychlorinated biphenyls, and plutonium.

1529 - U.S. Department of Commerce. 1980a. Report to the congress on ocean pollution and offshore development - October 1977 through September 1978. NOAA Ocean Dumping Program. Rockville, MD. 90p.

Abstract: Report of NOAA's studies of long-range effects. Chlorinated hydrocarbons (CHC's) cause changes in cell division and photosynthesis. PCB's reduce primary productivity and change food web structure.

In FY 78 compilation of "Perspectives on the NY Bight," which deals with seven care issues. Five expended water column characteristics cruises were made. Work on two diagnostic models of water circulation. Summaries of other NY Bight Project accomplishments.

1530 - U.S. Department of Commerce. 1980b. Biological/chemical survey of Texoma and Capline sector salt dome brine disposal site off Louisiana, 1978-1979. Volume 1. (Living and dead benthic (Maro-meio-) Communities. NOAA National Marine Fisheries Service. Galveston, TX. 132p.

1531 - U.S. Department of Commerce. 1980c. Report of the Interagency Ad Hoc Work Group for the chemical waste incinerator ship program. Maritime Administration. Washington, DC. 275p.

1532 - U.S. Department of Commerce. 1981. Assessment report on the effects of waste dumping in 106-mile ocean waste disposal site. NOAA Dumpsite Evaluation Report 81-1. Boulder, CO. 319p.

Summary: This report contains the results and analyses of field and laboratory studies on the effects of ocean dumping at 106-Mile Ocean Waste Disposal Site (106-Mile Site), previously called Deepwater Dumpsite 106 (DWD-106). This deep ocean dumpsite is approximately 90 nautical miles east of Cape Henlopen, Delaware. These studies occurred between mid-1976 and early 1978 by the National Oceanic and Atmospheric Administration (NOAA), which has basic legislative authority to study the environmental effects of ocean waste disposal. The study program was designed to assess the impacts of dumping at 106-Mile Site by combining direct field evaluations of dumping episodes with controlled laboratory studies of the biological effects of dumped materials and of their physical and chemical behavior in seawater. Major new wastes, including acid-iron waste, have been disposed of at 106-Mile Site since March 1977, and NOAA is continuing to address both the effect of this new disposal and the cumulative impact of dumping at the site.

Part I of this report--is a summary, analysis, and interpretation of individual investigations carried out under NOAA funding by various federal and univeristy scientists.

Part II--Waste Characteristics and Inorganic Chemistry--emphasizes field and laboratory studies of the heavy and transition metal in the major wastes dumped at 106-Mile Site.

Part III--Waste Dispersion and Physical Oceanography--contains several reports on dump tracking, followed by two studies of the physical oceanography of the 106-Mile Site region.

Part IV--Biological Effects--considers field and laboratory studies of the effects of the wastes on bacteria, plankton, fish eggs, and fish.

1533 - U.S. Department of the Interior. 1969a. Report of the Department of the Interior Working Group on possible effects of disposal of chemical warfare munitions in the sea. Washington, DC. 23p. (mimeo).

1534 - U.S. Department of the Interior. 1970a. The national estuarine pollution study. Report of the Secretary of the Interior to the 91st Congress. Senate Document #91-58. p. 274, 275-276, 296-297, 301-302.

1535 - U.S. Department of the Interior. 1970b. Evaluation of influence of dumping in the New York Bight with a brief review of general ocean pollution problems. Ad Hoc Committee Interagency Report, Roland F. Smith (Chairman). Washington, DC. 65p. and Appendices.

1536 - U.S. Department of State. 1972. Proposed ocean dumping convention. (Legislative action). U.S. National Technical Information Service. Government Reports Announcements. 72(22):122. November 25.

1537 - U.S. Environmental Protection Agency. 1972. Ocean Disposal Practices and Effects-A Report of the President's Water Pollution Control Advisory Board. September 26-29. Washington, DC.

1538 - U.S. Environmental Protection Agency. 1973a. Environmental impact statement on wastewater treatment facilities. Construction grants for the Lower Raritan River Basin and for the South Shore of the Raritan Bay. EPA Region. II. New York, NY. 188p.

1539 - U.S. Environmental Protection Agency. 1973b. Combined sewer overflow study for the Hudson River conference. Office of Research and Monitoring. Washington, DC. 287p.

Abstract: A detailed examination was conducted of ten (10) combined sewer overflow systems within that portion of the Hudson River Basin lying within the Interstate Sanitarian District. The work included the identification and study of these combined sewer systems in order to determine their location, physical characteristics, and service areas.

The procedure employed included the physical examination of each system's regulators to determine their location, type, dimensions, and condition. A study of available records was made to determine, where possible, trunk line flow, intercepter line design capacity, and characterization of the drainage area served by each regulator which included population and land use. Ten (10) summary tables and forty (40) regulator location figures are included in the report. Dry weather and wet weather sampling was also conducted. Bypass loadings for several pollution parameters have been calculated during storm flow conditions based upon this sampling. Recommendations for minimizing combined sewer overflows are included.

This report was submitted in fulfillment of Contract Number 68-01-0055, under the partial sponsorship of the Office of Enforcement and General Counsel, Environmental Protection Agency and is published by the Office of Research and Monitoring, Environmental Protection Agency.

1540 - U.S. Environmental Protection Agency. 1973c. Ocean dumping in the New York Bight-facts and figures. EPA Region II. Surveillance and Analysis Division. New York, NY. 15p.

Abstract: What goes into the sea is everybody's business. At present, solid and liquid wastes are being shipped to sea for disposal in such quantities that marine disposal has become of extreme concern, particularly in the New York Bight area where more than 70% of all sludge dumping in the nation occurs just 12 miles off the coasts of New York and New Jersey. In addition, almost 60% of all industrial wastes disposed of in this manner in the U.S.A., are dumped in approved sites in waters contiguous to both States. It is important to note that waste acid makes up over 90% of the total industrial wastes going to sea in this area. Tables I through IV summarize pertinent and "eye opening" facts associated with this practice.

1541 - U.S. Environmental Protection Agency. 1973d. Bibliography on ocean waste disposal. Washington, DC.

1542 - U.S. Environmental Protection Agency. 1973e. Annual report by the U.S. EPA on administration of the ocean dumping permit program under Public Law 92-532, the Marine Protection, Research, and Sanctuaries Act of 1972. U.S. Government Printing Office, Washington, DC.

1543 - U.S. Environmental Protection Agency. 1973f. Recommended procedures for diposal of storage of pesticides and pesticide containers. Office of Solid Waste Management Programs. Washington, DC.

1544 - U.S. Environmental Protection Agency. 1974a. Briefing report - Ocean dumping in the New York Bight since 1973. EPA Region II. Surveillance and Analysis Division. New York, NY. 47p.

Abstract: This briefing document broadly summarizes the position and activities of Region II in carrying out its responsibilities for resolving the complex environmental problems associated with the pratice of ocean disposal of municipal sludges and industrial wastes in the New York Bight. Our responsibilities began on April 5, 1973, the effective date of the Marine Protection, Research, and Sanctuaries Act of 1972.

1545 - U.S. Environmental Protection Agency. 1974b. Progress report on bacterial water quality at western Long Island beaches and northern New Jersey beaches. May 15, 1974. EPA Surveillance and Analysis Division. Edison, NJ. (mimeo). 3p.

1546 - U.S. Environmental Protection Agency. 1974c. Progress report on bacterial water quality at western Long Island beaches and northern New Jersey beaches, July 1, 1974. EPA Region II. Surveillance and Analysis Division (mimeo). Edison, NJ. 2p.

1547 - U.S. Environmental Protection Agency. 1974d. Ocean disposal in the New York Bight. Technical Briefing Report No. 1. EPA Region II. New York, NY. 54p.

1548 - U.S. Environmental Protection Agency. 1974e. Second annual report by U.S. EPA on administration of the ocean dumping permit program under the Marine Protection, Research, and Sanctuaries Act as Amended. EPA. Washington, DC.

1549 - U.S. Environmental Protection Agency. 1974f. Ocean dumping in the New York Bight, facts and figures; surveillance and analysis division. EPA. Region II. New York, NY.

1550 - U.S. Environmental Protection Agency. 1975a. Ocean disposal in the New York Bight technical briefing report No. 2. EPA Region II. New York, NY. 86p.

Abstract: The EPA, Region II, has expressed its desire to prevent or strictly regulate ocean dumping. Over the past two years, the region has implemented a series of programs aimed at phasing out ocean dumping. The initial phase of this comprehensive program was to establish a permit program to regulate dumping. Another step was to initiate a program to identify environmental acceptable, technically feasible, and viable alternatives to ocean disposal.

A comprehensive sampling program indicates that water along Long Island and N.J. beaches are safe for swimming and the leading edge of the sludge mass is still 5 1/2 - 6 miles from the L.I. shoreline.

1551 - U.S. Environmental Protection Agency. 1975b. Ocean dumping in the United States third annual report of the Environmental Protection Agency on administration of Title I, Marine Protection, Research, and Sanctuaries Act of 1972 as amended. Washington, DC. 58p.

1552 - U.S. Environmental Protection Agency. 1976a. Designation of a site in the Gulf of Mexico for incineration of chemical wastes, draft environmental impact statement. Office of Water and Hazardous Materials. Washington, DC.

1553 - U.S. Environmental Protection Agency. 1976b. Designation of a site in the Gulf of Mexico for incineration of chemical wastes; final environmental impact statement. Office of Water and Hazardous Materials. Washington, DC.

1554 - U.S. Environmenal Protection Agency. 1976c. Ocean dumping in the United States - 1976. Fourth annual report of EPA on title I of Marine Protection, Research, and Sanctuaries Act of 1972. Office of Water and Hazardous Materials. Washington, DC.

1555 - U.S. Environmental Protection Agency. 1976d. Ocean dumping of sewage sludge in the New York Bight. Draft Environmental Impact Statement. EPA Region II. New York, NY.

1556 - U.S. Environmental Protection Agency. 1977a. Ocean dumping, final revision of regulations and criteria. U.S. Federal Register, Vol. 42, No. 7, p. 2485. (January 11).

1557 - U.S. Environmental Protection Agency. 1977b. Ocean Dumping in the United States - Fifth Annual Report of the Environmental Protection Agency on Administration of Title I, Marine Protection, Research, and Sanctuaries Act of 1972, as amended. Washington, DC. 65p.

Abstract: This is the fifth annual report of the Environmental Protection Agency (EPA) to the Congress on the implementation of Title I of the Marine Protection, Research, and Sanctuaries Act of 1972, as amended (referred to in this report as "the Act"). See Appendix A. The Act became effective April 23, 1973, and since that time all ocean dumping of waste materials transported for the purpose of dumping has been regulated under permits issued by EPA except for dredged material, which is regulated by the U.S. Army Corps of Engineers (COE).

This report covers EPA Headquarters and Regional permit operations, research projects by EPA's Office of Research and Development (ORD), and other program activities during the calendar year 1976. A calendar of Headquarters and Regional activities during 1976 is provided as Table 1.

Previous annual reports by EPA included information on COE activities related to the issuance of permits for the ocean dumping of dredged material and on surveillance an monitoring activities of the U.S. Coast Guard (USCG) on ocean dumping operations. Under amendments to the Act passed in 1976, both the U.S. Army Corps of Engineers and the U.S. Coast Guard will submit separate reports of their activities in implementing Title I of the Act. This EPA report, therefore, does not contain a discussion of the activities of these two agencies under the Act, except as these activities impact the resonsibilities of EPA.

Program responsibilities under the Act are divided among EPA Headquarters and the seven EPA coastal Regions and supported by related ORD research activities. The Regions are responsible for all activities relating to the issuance of special and interim permits for dumping in the respective Regions. The Regions are also delegated some responsibility for the management of ocean dumping sites. EPA Headquarters is responsible for all other program activities, including the designation of ocean dumping sites, issuance of emergency, research, and general permits, and coordination of Regional activities.

1558 - U.S. Environmental Protection Agency. 1977c. Ocean dumping in the United States - Sixth annual report of the Environmental Protection Agency on Administration of Title I. Washington, DC. 53p.

Abstract: This is the Environmental Protection Agency's (EPA) sixth annual report to the Congress on the implementation of Title I of the Marine Protection, Research, and Sactuaries Act of 1972, as ameded (MPRSA). The report covers the Agency's authorities and responsibilities in carrying out the ocean dumping program and reviews those program activities conducted within EPA Headquarters, the Regions, and the Office of Research and Development during the calendar year 1977.

During 1977, the amount of ocean dumping declined by 12 percent from the level of dumping in 1976. Region II (New York) continues to be the most active area, with approximately 7,300,000 tons of material dumped in and adjacent to the New York Bight and off the coast of Puerto Rico under permits issued for municipal and industrial wastes, construction debris, and incineration of wood debris. In the Permit Program Section, Table V summarizes the total amount of dumping during 1977 (by geographic area) and presents an annual comparison of amounts dumped under EPA permit during preceding years.

A summary of information on ocean dumping permits issued by Contracting Parties to the International Ocean Dumping Convention is included in the section discussing the deliberations of the Convention. Accomplishments are being made in establishing agreements on major items of high priority of work in clarifying the provisions and requirements of the Convention.

A joint EPA-U.S. Air Force project on incineration at sea of Herbicide Orange was successfully completed during August and September 1977. A brief description of the project is included in the section on Incineration at Sea, and a detailed report is being prepared for publication in the near future. The November 1977 amendment to the MPRSA mandates the termination of ocean dumping of sewage sludge by December 31, 1981, placing even greater emphasis on alternative to ocean dumping and compliance with implementation schedules in transitioning to the chosen alternative. The section on Implementing Alternatives to Ocean Dumping gives the descriptions of the available alternatives, the associated technologies and guidelines that have been developed through related environmental legislation, and discusses the continuing studies directed toward futher development.

1559 - U.S. Environmental Protection Agency. 1977d. Multi-Federal agency monitoring strategy in the New York Bight - Calendar year 1977.

Abstract: A review of monitoring plans by Federal agencies (EPA, NOAA, USCG, NASA) in the New York Bight during Calendar year 1977.

1560 - U.S. Environmental Protection Agency. Draft. Dispersion of sewage sludge discharged into New York Bight. Physical oceanographic data - December 1974. EPA Environmental Research Lab. Corvallis, OR.

Abstract: This volume contains physical oceanographic data collected at the sewage sludge disposal site near the Appex of New York Bight December 18 through 21, 1974. An optical tracer method was used to measure the water column distribution of waste material with time after discharge. Profiles with depth were taken for 2 to 4 hours after waste discharge. Ambient temperature-salinity-density profiles and current measurements were also taken.

This report covers a period from June 1974 to May 1975 and work was completed as of May 1976.

1561 - U.S. Environmental Protection Agency. 1978b. Dispersion of sewage sludge discharged into New York Bight. Physical Oceanographic Data-December 1974. EPA Environmental Research Lab. Corvallis, OR. 53p.

Abstract: This volume contains physical oceanographic data collected at the sewage sludge disposal site near the Appex of New York Bight December 18 through 21, 1974. An optical tracer method was used to measure the water column distribution of waste material with time after discharge. Profiles with depth were taken for 2 to 4 hours after waste discharge. Ambient temperature-salinity-density profiles and current measurements were also taken.

This report covers a period from June 1974 to May 1975 and work was completed as of May 1976.

1562 - U.S. Environmental Protection Agency. 1978c. Dispersion of sewage sludge discharged into New York Bight. Vol. II. Physical Oceanographic Data and Laboratory Analyses - 1975. EPA Environmenal Research Laboratory, Corvallis, OR. 202p.

Abstract: This volume contains data on the dispersion of sewage sludge subsequent to its disposal at a site near the Appex of New York Bight. Cruises were made in May, July, and October, 1975. An optical tracer method was used to measure the water column distribution of waste material for 2 to 4 hours after discharge. Direct measurements of the concentration of suspended material were made. Ambient temperature-salinity-density profiles were taken. Currents were measured by moored and profiled instruments and by drogue tracking. Laboratory analyses on the settling characteristics, densities, and optical properties of sewage sludges from the New York City area are presented.

1563 - U.S. Environmental Protection Agency. 1978d. Decision on proposals to relocate sewage sludge in the Mid-Atlantic Bight. Washington, DC. 36p.

1564 - U.S. Environmental Protection Agency. 1978e. The effects of sewering on Long Island's shellfishing industry. A Supplement to the 1972 Final Environmental Impact Statement on Waste Water Treatment Facilities Construction Grants for Nassau and Suffolk Counties. Region II. New York, NY. 91p.

Abstract: Specific effects of present sewering programs on the shellfishing industry are not assessable with the information available. Measures such as groundwater recharge and streamflow augmentation are being developed to satisfy wastewater treatment needs, and maintain fresh to salt water balance in the bay.

1565 - U.S. Environmental Protection Agency. 1978f. Final environmental impact statement on the ocean dumping of sewage sludge in the New York Bight. Region II. New York, NY. 226p. + appendices.

Abstract: The results of this reexamination are set forth in detail in the final EIS. Basically, they confirm EPA's finding in the draft EIS that the existing site is not a threat to public health or water quality, and can reasonably continue in use provided that it is closely monitored and that an alternate site is available should monitoring ever indicate the need for relocation. Since there is no present need for relocation, immediate use of an alternate site would result in the unnecessary contamination of still another area of the New York Bight. In addition to reexamining the present situation, the final EIS explains the steps that EPA is taking to insure that sludge dumping is phased out by December 31, 1981, a deadline that has progressed from matter of EPA policy to a legal requirement under the Marine Protection, Research, and Sanctuaries Act.

Description of environment, proposed areas for sites, impacts of proposed action, relative regulations and reports.

1566 - U.S. Environmental Protection Agency. 1980a. (Draft). Environmental impact statement (EIS) for proposed North Atlantic incineration site designation. Washington, DC. 20460. 6 Chapters & 4 Appendixes.

Abstract: This Environmental Impact Statement (EIS) provides the public with information required in the decision making process with respect to designation of a site in the North Atlantic Ocean for at-sea incineration of certain wastes. It evaluates the types of industrial materials (principally organochlorine and other organohalogen wastes) which may be disposed at the site in an environmentally sound manner, and provides guidance for the EPA to manage the site by means of the ocean dumping permit program.

1567 - U.S. Environmental Protection Agency. 1980b. Annual report to congress Jan-Dec. 1979. On Administration of the Marine Protection, Research, and Sanctuaries Act of 1972, as Amended (P.L. 92-532) and Implementing the International London Dumping Convention. EPA Office of Water Programs. Washington, DC. 41p.

Abstract: List of ocean dumping permits granted by EPA in 1979. Summaries of baseline and monitoring surveys of ocean sites for site designation. Winter and summer surveys at each site. Enforcement is by Coast Guard. Violations have decreased considerably in the last three years.

1568 - U.S. Environmental Protection Agency. 1980c. Final environmental impact statement for 106-Mile ocean waste disposal site designation. EPA Marine Protection Branch. Washington, DC. 480p.

1569 - U.S. Federal Register. 1973a. Ocean dumping interim regulations governing transportation for dumping and dumping of material into ocean waters. Title 40. Subchapter H - Ocean Dumping (Rules and Regulations). April 5. 38FR 8726-7830.

1570 - U.S. Federal Register. 1973b. Ocean dumping criteria. Title 40. Subchapter H - Ocean Dumping (Rules and Regulations). May 16. 38FR 12872-12877.

1571 - U.S. Federal Register. 1973c. Ocean dumping final regulations and criteria. Title 40. Subchapter H - Ocean Dumping (Rules and Regulations). October 15. 38FR 28610-28621.

1572 - U.S. Federal Register. 1974a. Environmental impact statements, procedures for the voluntary preparation (notices). October 21. 39FR 37419-37422.

1573 - U.S. Federal Register. 1974b. Environmental impact statement, statement of policy (notices). May 7. 39FR 16186-16187.

1574 - U.S. Federal Register. 1976. Ocean dumping proposed revision of regulations and criteria. Title 40, Subchapter H - Ocean Dumping (Proposed Rules). June 28. 41FR 26644-26667.

1575 - U.S. Federal Register. 1977. Ocean dumping final revision of regulations and criteria. Title 40. Subchapter H - Ocean Dumping (Rules and Regulations). January 11. 42FR 2462-2490.

1576 - U.S. General Accounting Office. 1977. Report to the congress: Problems and progress in regulating ocean dumping of sewage sludge and industrial wastes. U.S. Controller General. Washington, DC. CED-77-18. 61p.

1577 - U.S. General Accounting Office. 1978. Report to the congress: Sewage sludge -- how do we cope with it? U.S. Controller General. Washington, DC. 37p.

Abstract: This report addresses problems municipalities face in selecting and implementing sludge management systems that dispose of sludge in a safe, beneficial, and cost-effective manner and actions the Federal Government should take to help them. Sewage sludge disposal is a growing problem because the Nation's sludge volume is increasing dramatically; it is expected to double in size by 1987. At the same time, certain sludge diposal methods are being phased out, and use of others is being restricted. Also, development and implementation of new disposal methods is being hampered for a number of reasons.

1578 - U.S. House Committee on Merchant Marine and Fisheries. 1971a. Report No. 92-361. Marine Protection, Research, and Sanctuaries Act of 1971 (to accompany HR, 9727). July 17.

1579 - U.S. House Committee on Merchant Marine and Fisheries. 1971b. Ocean Dumping of Waste Materials. Hearings before the Subcommittee on Fisheries and Wildlife Conservation and Subcommittee on Oceanography on HR 4273 et al. April 5, 6, 7. Serial No. 92-2. Washington, D.C.

1580 - U.S. House Committee on Merchant Marine and Fisheries. 1972. Report No. 92-1546. Marine Protection, Research, and Sanctuaries Act of 1972 (to accompany HR 9727). Conference Report. Committee of Conference. October 9.

1581 - U.S. House Committee on Merchant Marine and Fisheries. 1973. Report No. 93-568. Ocean Dumping Convention Implementation (to accompany HR 5650). October 10.

1582 - U.S. House Committee on Merchant Marine and Fisheries. 1975a. Marine Protection, Research, and Sanctuaries Act Authorization (to accompany HR 5710). May 15.

1583 - U.S. House Committee on Merchant Marine and Fisheries. 1975b. Ocean dumping. Joint Hearings before the Subcommitte on Fisheries and Wildlife Conservation and the Environment and the Subcommittee on Oceanography on HR 5710 and HR 6282. April 24 and 25. Serial No. 94-10. Washington, DC.

1584 - U.S. House Committee on Merchant Marine and Fisheries. 1975c. Ocean dumping oversight. Hearings before the Subcommittee on Fisheries and Wildlife Conservation and Environment and the Subcommittee on Oceanography on Oversight of the Marine Protection, Research, and Sanctuaries Act of 1972. May 22 and 31. Serial No. 93-38. Washington, DC.

1585 - U.S. House Committee on Merchant Marine and Fisheries. 1976a. Marine Protection, Research, and Sanctuaries Act Authorization - Fiscal Year 1977 (to accompany HR 11505). April 27.

1586 - U.S. House Committee on Merchant Marine and Fisheries. 1976b. Ocean dumping - part 3. Joint Hearings before the Subcommittee on Fisheries and Wildlife Conservation and the Environment, and the Subcommittee on Oceanography on Oversight of the Marine Protection, Research, and Sanctuaries Act of 1972 and Investigating the Waste Disposal in the New York Area. July 24 - Hempstead, N.Y. September 30 - Washington, DC. Serial No. 94-41. Washington, D.C.

1587 - U.S. House Committee on Merchant Marine and Fisheries. 1976c. The environmental effects of dumping in the oceans and great lakes. Hearings before the Subcommittee on the Environment and the Atmosphere. September 17, 18, 19, 24, 26. No. 55. Washington, DC.

1588 - U.S. House Committee on Merchant Marine and Fisheries. 1976d. Ocean dumping - part 2. Joint Hearings before the Subcommittee on Fisheries and Wildlife Conservation and the Environment, and the Subcommittee on Oceanography on Oversight of the Marine Protection, Research, and Sanctuaries Act of 1972. January 23, 29. February 27 - Washington, D.C. March 5 - New York, N.Y. Serial 94-25. Washington, DC.

1589 - U.S. House of Representatives. 1928. Examination of rivers and harbors: Examination and survey of Hudson River Channel, N.Y. House Document No. 450, 33, 70th Congress, 2nd session.

1590 - U.S. House of Representatives. 1935. Examination of rivers and harbors: New York and New Jersey channels, preliminary examination and survey. House Document No. 133, 20. 74th Congress. 1st session.

1591 - U.S. House of Representative. 1970. Waste disposal in the coastal waters of New York Harbor. Hearing before the subcommittee on Rivers and Harbors of the committee on Public Works, on H.R. 15915. February 23.

1592 - U.S. Maritime Administration 1980. Report of the Interagency Ad Hoc Work Group for the Chemical Waste Incinerator Ship Program. Washington, DC. NTIS No. PB81 112849. 275p.

1593 - U.S. Academy of Sciences - National Academy of Engineering, Committees on Oceanography and Ocean Engineering. 1969. Background papers on coastal wastes management. Vol. 1. 391p. Washington, DC.

1594 - U.S. Academy of Sciences. 1976. Waste management for the coastal zone; concepts for the assessment of ocean outfalls. Marine Board. Washington, DC.

1595 - U.S. National Industrial Pollution Control Council. 1971. Deep ocean dumping of bailed refuse. Sub-council report. February, 1971. Washington, DC. 16p.

1596 - U.S. National Science Foundation. 1976. Anoxia on the Middle Atlantic shelf during the summer of 1976. Report of a Workshop, October 15 and 16, 1976. International Decade of Ocean Exploration. National Science Foundation. Washington, DC.

1597 - U.S. Navy Department. Office of Oceanographer of the Navy. 1972. Environmental condition report for numbered deep water munitions dumps sites. 360p.

1598 - U.S. Public Health Service. 1917. Investigation of the pollution of certain tidal waters of New Jersey, New York and Delaware.

1599 - U.S. President's Water Pollution Control Advisory Board. 1972. Ocean disposal practices and effects. Report of meeting held in New York on September 26-29, 1972. U.S. Environmental Protection Agency. Washington, DC. 76p.

1600 - U.S. Senate Committee on Commerce. 1971. Ocean waste disposal; Hearings before the Subcommittee on Oceans and Atmosphere, on S. 307, S. 1082, S. 1238, and S. 1286. March 2, 3, April 12, 21, 22, and 28, 1971. Serial No. 92-11. Washington, DC.

1601 - U.S. Senate Committee on Commerce. 1974. Ocean pollution; hearings before the Subcommittee on oceans and atmosphere on S. 1067, S. 1070, and S. 1351. June 12, 13, and 28, 1973. Serial no. 93-46. Washington, DC.

1602 - U.S. Senate Committee on Commerce. 1975. Marine Protection, Research, and Sanctuaries Act of 1972. Hearings before the Subcommittee on Oceans and Atmosphere. May 20, 1975. Serial No. 94-32. Washington, DC.

1603 - U.S. Senate Committee on Commerce. 1976a. To extend the Marine Protection, Research, and Sanctuaries Act. Hearings before the Subcommittee on Oceans and Atmosphere on S. 3147, April 12, 1976. Serial No. 94-83 (incorrectly printed 91-83). Washington, DC.

1604 - U.S. Senate Committee on Commerce. 1976b. Ocean dumping regulations; an appraisal of implementation. Committee Print for Use by the Committee on Commerce and the National Ocean Study pursuant to S. Res. 222. April 1976. Washington, DC.

1605 - U.S. Senate. 1971. Marine Protection and Research Act of 1971 (to accompany H.R. 9727). Senate Committee on Commerce. November 12. Executive Report No. 92-451.

1606 - U.S. Senate. 1973a. Convention on the prevention of marine pollution. Message from the President of the United States, Senate Foreign Relations Committee. February 28.

1607 - U.S. Senate Executive Report No. 93-13. 1973b. Convention on the prevention of marine pollution (to accompany Ex. C. 93-1). Senate Foreign Relations Committee. August 1.

1608 - U.S. Senate Report No. 93-726. 1974. Ocean dumping convention implementation (to accompany H.R. 5450). Senate Committee on Commerce. March 6.

1609 - U.S. Senate Report No. 94-271. 1975. Marine Protection, Research, and Sanctuaries Act Authorization (to accompany H.R. 5710). Senate Committee on Commerce. July 7.

1610 - U.S. Senate Report No. 94-860. 1976. Marine Protection, Research, and Sanctuaries Act Authorization - Fiscal Year 1977 (to accompany S. 3147). Senate Committee on Commerce. May 13.

1611 - Usry, J. W., and J. B. Hall, Jr. 1975a. NASA operations - remote sensing experiments in the New York Bight, April 7-17. NASA Tech. Memo. 72802.

1612 - Uthe, J. F., H. C. Freeman, S. Mounib, and W. L. Lockhart. 1980. Selection of biochemical techniques for detection of environmentally induced sublethal effects in organisms. International Council for the Exploration of the Sea (ICES). Rapports, et Proces-Verbaux des Reunions. Vol. 179. Charlottenlund Slot, Denmark.

1613 - Vaccaro, R. F., G. D. Grice, G. T. Rowe, and P. H. Wiebe. 1972. Acid-iron waste disposal and the summer distribution of standing crops in the New York Bight. Water Res. 6:231-256.

Abstract: Ecological consequences arising form the disposal of 50 million tons of acid-iron industrial waste in the coastal waters off New York over the past 22 yr. were assessed. Most of the data were obtained at two identical grids of stations which enabled comparisons of hydrographic, chemical, and biological conditions within the acid-iron disposal area with similar parameters in a nearby control area. Supplementary information on benthos and sediment was obtained at other locations peripheral to the two station grids and in Hudson Gorge, and these were used to construct a synoptic picture of the physio-chemical conditions and standing crops in the New York Bight.

At each grid station the hydrographic measurements made were temperature, salinity, and light penetration; chemical observations consisted of dissolved oxygen, dissolved and suspended iron, total inorganic nitrogen, and phosphate while chlorophyll a, zooplankton and benthos biomass provided a measure of the abundance of standing crops. Trace metal spectra (Fe, Zn, Co, Cu, Pb, Cr, Ni, and Cd) were determined on selected zooplankton, benthos, and sediment samples. Laboratory toxicity studies were conducted on phytoplankton and zooplankton species at several concentrations of acid-iron waste in seawater.

The maximum concentration of iron in the water column (832 g 1^{-1}) occurred as suspended material within a restricted area of the acid grid. In terms of raw acid-iron effluent this suggests a maximum in situ concentration of 1 part waste in 39,000 parts of seawater, thereby providing a useful guide for the design of laboratory toxicity studies. Despite the abundance of suspended iron in the overlying water of the acid-grid the average concentrations of iron in the sediments of both the acid and control grids were remarkably similar, while sediments from the nearby Hudson Gorge were notably richer in iron. However, a comparison of previous measurements in the study area dating back to 1948 indicates that there has been no accumulation of iron within the sediments below the disposal area or Hudson gorge over the past 22 yr.

The phytoplankton toxicity experiment conducted with an acid-iron waste concentration four times greater than that observed in the field shows no adverse effect on phytoplankton growth or diversity. Similar experiments with copepods caused either failure of these organisms to reproduce or a delay in the time required to transform eggs into adults.

Although the average zooplankton abundance within the control grid exceeded that of the acid grid by about 30 percent, the range of values describing zooplankton abundance in the two areas was similar. This difference was attributed to a transitory large scale patchiness in the area and not to toxicity of acid-iron waste. A positive correlation was found between Fe:C in zooplankton and the amount of particular iron present in the seawater.

The average number of benthic animals on the bottom of the acid area was significantly less than in the sediment of the control grid but there was no difference in biomass or species diveristy between the two areas. As was the case with zooplankton the higher Fe:C in the benthos corresponded to the higher iron in the sediment of Hudson Gorge and acid-grid.

The heavy metal content of zooplankton, benthos, and sediment showed that samples from the acid grid were significatnly richer in these elements than the comparable control area samples. However, a broader comparison showed that samples from Hudson Gorge contained the maximum amounts of lead and chromium in benthos as well as the maximum concentrations of all eight metals in the sediment. These data are consistent with the possibility that entrapment in the gorge sediments may be the ultimate fate of the heavy metal enrichment in the New York Bight area and that sources of heavy metals other than acid-iron waste may be substantial.

The remaining data reviewed in this study did not demonstrate any adverse in situ effects of acid-iron waste on the distribution of such pattern as dissolved oxygen, chlorophyll a and plant nutrients. Present indications are that the disposal of acid-iron waste in the New York Bight appears to influence standing crops in minor ways considering the magnitude and nature of the waste material involved.

1614 - Vaccaro, R. F., J. M. Capuzzo, and N. H. Marcus. 1980. The oceans and U.S. sewage sludge disposal strategy. ICES. C.M. 80/E:24.

1615 - Vaccaro, Ralph F., and Mark R. Dennett. 1981a. Environmental response of marine bacteria to waste disposal activities at 106-Mile site. In: Assessment Report on the Effects of Waste Dumping in 106-Mile Ocean Waste Disposal Site. NOAA/ODP Dumpsite Evaluation Report 81-1. Rockville, MD. pp. 295-308.

1616 - Vaccaro, Ralph F., Judith M. Capuzzo, and Nancy H. Marcus. 1981b. The oceans and U.S. sewage sluge disposal strategy. Oceanus. 24(1):55-59.

Abstract: Review of present and future concerns, and site characteristics affecting sludge disposal with recommendation for the U.S. to take a broader look at deep-water sewage sludge ocean dumping as a future waste management alternative.

1617 - Vaccaro, Ralph F., and Mark R. Dennett. 1981c. The bacterial bioassay and laboratory assessments of waste disposal activities at DWD-106. In: B. H. Ketchum, D. R. Kester, and P. K. Park (Editors), Ocean Dumping of Industrial Wastes. Plenum Press. New York, NY. pp. 381-397.

Abstract: Changes in the bacterial uptake of ^{14}C glucose in seawater are used to quantify some sublethal consequences of Edgemoor and Grasselli waste disposal at Deep Water Dumpsite 106.

The fractional amounts of waste in seawater which led to a 50 percent reduction in ^{14}C uptake ranged from 0.01-0.02 percent for Edgemoor waste and from 0.10-0.20 percent for Grasselli waste.

Both Edgemoor and Grasselli wastes cause an inhibitory bacterial response which exceeds that associated with their respective acid and caustic chemical compositions. Heavy metals appear to be the principle toxic components of Edgemoor waste whereas organic species appear to produce the toxicity of Grasselli waste.

Resistance to chemical alteration as demonstrated by ultraviolet radiation and persulfate oxidation may imply an environmental persistence for Grasselli waste.

Edgemoor waste reacts with seawater with a precipitation of its heavy metal content. Such behavior is certain to influence its distribution kinetics and its impact on life processes of the ocean.

Mixtures of Edgemoor and Grasselli wastes impart an inhibitory response which is measurably less than that anticipated from the sum of their individual effects. This suggests the possibility of positive benefits from their coordinated release within the Dumpsite area.

An unambiguous approch to improve bioassay interpretations in potentially reactive, multi-waste situations was developed. The method relies upon a graphical solution to differentiate between the net response induced by mixed wastes and a hypothetical response corresponding to the sum of the individual response patterns observed independently.

1618 - Vaissiers, R., J. Semeria, L. Schonmers, and M. Boisson. 1973. A propos d'une experience de localisation par traceurs et photographie aerienne, de rejets en mer dans la region de Monaco. [Concerning an experiment on localization, by tracers an aerial photography, of discharges to the sea on the region of Monaco]. CIESMH. Journees d'etudes sur les pollutions marines. Commission Internationale pour l'Exploration scientifique de la Mer Mediterranee. Monaco.

1619 - Valencia, M. J. 1979. South China Sea: Present and potential coastal area resource use conflicts. Ocean Management. 5(1):1-38.

1620 - Valerio, C., and Y. Vuillaume. 1978. Teledetection and water resources. Operation Pilote Interministerielle de Teledetection, Paris (France). Earth Observation from Space and Management of Planetary Resources. pp. 97-102.

Abstract: The role of interministerial collaboration in the Teledetection Interministerial Pilot Operation is pointed out and the main phases of the program are described. Two examples involving water pollution are given, covering the subjects of aircraft monitoring with teledetection of hydrocarbon pollution on the sea and simulation of sea water pollution by Rhodamine diffusion in the Propiano bay.

1621 - Valiela, Ivan, Susan Vince, and John M. Teal. 1976. Assimilation of Sewage by Wetlands. In: M. Wiley (editor), Estuarine Processes. Vol. I: Uses, Stresses, and Adaptation to the Estuary. Academic Press. New York, NY. pp. 234-253.

Abstract: Wetlands have attracted attention as potential components of waste treatment systems because of typically large plant production, high decomposer activity, anaerobic condition, and large adsorptive areas in the sediments. These properties seem to provide wetlands with the ability to degrade and climinate contaminants in waste waters.

The effluent-properties of a variety of wetland habitats in very diverse parts of the world (salt and brackish marshes, holders, cypress, papyrus and mangrove swamps, freshwater marshes and bogs) have been studied. Although each habitat type differs, current results from marshes show that nitrogenous and phosphorus nutrients are removed from contaminated or waste waters. Heavy metals from sludges are retained by marsh muds, but the effectiveness varies considerably for different metals. Petroleum hydrocarbons accumulate in marsh muds and may be actively decomposed, particulary at lower levels of contamination. Chlorinated hydrocarbons have high affinities for marsh sediments and can be altered by microorganismal activity. There is preliminary evidence that counts of coliform bacteria may be reduced in tidal water that floods marshes.

The waste processing ability of coastal wetlands must have an upper threshold. Little is known about these upper limits but studies in chronically contaminated sites show that marshlands function well even in severely polluted areas.

1622 - Van Dam, G. C. 1975. The Hague outfall. Progr. Water Technol. Suppl. Pap. No. 39:393-401.

1623 - Van de Velde, O. 1978. Occurrence and origin of non-bio-degradable contaminants in surface waters of the Netherlands. Pergamon Series. Environm. Sci. No. 1. pp. 135-140.

1624 - Venrick, E. L., T. W. Backman, W. C. Bartram, C. J. Platt, M. S. Thornhill, and R. E. Yates. 1973. Man-made objects on the surface of the central North Pacific Ocean. Nature. London. 241:271.

1625 - Verber, J. L., R. W. Buelow, and B. H. Pringle. 1970. Ocean disposal of waste material - off Delaware Bay and New York Bight. Proc. 2nd Mid-Atlantic Industrial Waste Conf. 18-20 Nov. 1968. Drexel Institute of Technology. Philadelphia, PA. pp. 127-153.

1626 - Verber, James L. 1976. Safe shellfish from the sea. In: M. G. Gross (editor), Amer. Soc. Limnol. and Oceanogr. Spec. Symp. 2:433-441.

Abstract: The National Shellfish Sanitation Progran was initiated in 1925 after a widespread typhoid fever outbreak caused by raw oysters. Specific recommendations made in 1925 established guidelines for classifying offshore waters for clam harvesting and stated that shellfish being harvested must not be exposed to fecal contamination. In 1974, sea clams accounted for 59% of all shellfish (oysters, clams, and mussels) harvested from United States waters. The Food and Drug Administration is responsible for classifying the offshore water beyond 5.5 km for harvesting shellfish. In recent years, the sea has been used increasingly for sewage sludge and industrial waste disposal. Depletion of the resources in the original major sea clam harvest area off New Jersey and increased demand (43 million kg were harvested in 1974) have caused the industry to expand its area of operation to the southeast in search of more productive shellfish beds. There too, heavy population and industrial growth have increased the amounts of chemical wastes, sewage sludge, and other waste material being disposed of at sea. Warning notices to harvesters, closing areas of the New York Bight to shellfishing, have been posted since 1970. High bacterial levels are found in both the sewage sludge and dredge spoil sites.

1627 - Vind, H. P. 1975. The survival of sewage bacteria at various ocean depths. U.S. Naval Construction Battalion Center. Civil Engineering Laboratory. Port Hueneme, CA. Technical Note N-1396. 24p.

1628 - Vollbrecht, K. 1980. Schwebstoff-und Schwermertallführung der Unterelbe. Ein Beitrag zur Frage der Selbstreinigungskraft der Flüsse im Astuarbereich. Dt. Hydrogr. Z. Vol. 33(3):91-109 (English Summary).

1629 - Von Westernhagen, H., V. Dethlefsen, and H. Rosenthal. 1979. Combined effects of Cd, Cu, and Pb on developing herring eggs and larvae. ICES. C.M. 79/E:12. 35p.

Abstract: Eggs of Baltic herring were incubated (10^0C; 16 ppt) in seawater containing mixtures of Cd, Cu and Pb at concentrations of 0.56-5.0, 0.0167-0.15, 0.56-5.0 mg metal/1, and embroyonic survival until hatching, viable hatch, and uptake of metals by embryos and early larvae was measured. Negative effects of metals on embryonic survival and viable hatch were additive in the case of Cu and Cd. Pb did not exert detrimental effects. Uptake of metals with exposure time was nonlinear in eggs and linear in larvae. Total uptake of Cu and Cd by eggs was subjected to antagonistic or synergistic action of the other two metals present. Accumulation of Pb by eggs was enhanced by the presence of Cu as well as Cd.

1630 - Waddington, J. I., G. A. Best, J. P. Dawson, and T. Lithgow. 1973. PCB's in the Firth of Clyde . Mar. Pollut. Bull. 4(2):26-29.

1631 - Waldhauer, R., A. Matte, and R. E. Tucker. 1978. Lead and copper in the waters of Raritan and lower New York Bays. Mar. Pollut. Bull. 9:38-42. GC 1080 M 35.

1632 - Waldichuck, M. 1960. Containment of radioactive wastes for sea disposal and fisheries off Canadian Pacific Coast. Proceedings of 1959 Monaco Conference. AEC. Vol III.

1633 - Waldichuck, M. 1965. Estimation of flushing rates from tide height and current data in an inshore marine channel of the Canadian Pacific coast. Advances in Water Pollution Research. Proc. 2nd int. Conf. Tokyo, August 1964. Pergamon Press. New York, NY. Vol.3:133-166.

1634 - Waldichuck, M. 1968. Waste disposal in relation to the physical environment - oceanographic aspects. Syesis. 1:4-27.

1635 - Waldichuck, M. 1977. Global marine pollution: An overview. Techn. Ser. Intergovernm. Oceanogr. Comm. No. 18. 96p.

1636 - Waldichuck, M. 1978. Disposal of mine wastes into the sea. Mar. Pollut. Bull. 9(6):141-143.

1637 - Walker, Henry A., Saul B. Saila, and Eric L. Anderson. 1979. Exploring data structure of New York Bight benthic data using post-collection stratification of samples, and linear discriminant analysis for species composition comparisons. Estuar. and Coast. Mar. Sci. 9:101-120.

Abstract: Techniques for exploring data structure from four cruises on a rectilinear sampling grid system on the New York Bight are present. The problem of microenvironmental variability is handled by post-collection stratification of samples based on the physical characteristics of each grab, rather than classical spatial strata definitions. The three physical variables used to define strata are: mean particle size, percent organic matter, and total concentration of selected heavy metals. Variations in species densities within strata over time are investigated. Multivariate discriminant techniques are used to search for major statistical differences in a species abundance between strata. A flexible sampling program based on post-collection stratification is suggested to detect shifts in mean species abundance within strata over time.

1638 - Waller, Robert. 1968. Deep sea disposal of drummed wastes. E. I. du Pont de Nemouis & Company, Inc. La Porte, TX.

1639 - Walsh, James P. 1981. U.S. policy on marine pollution: Changes ahead. Oceanus. 24(1):18-24.

Abstract: Overview of U.S. Legislation for marine pollution with discussion of driving forces, changing social views concerning uncertainty and risk with shifts through governmental regulations from laissez-faire to environmental laws with policy trends shifted away from society at large to producers and consumers suggested directions for future changes.

1640 - Walsh, John J., Terry E. Whitledge, Steven O. Howe, Creighton D. Wirick, Louis J. Castiglione, and L. A. Codispoti. 1976. Transient forcings of the lower trophic levels during the spring bloom within the New York Bight. In: M. G. Gross (editor), Amer. Soc. Limnol. and Oceanogr. Spec. Symp. 2:273-274.

1641 - Walsh J. John, Terry E. Whitledge, Frank W. Bravenik, Creighton D. Wirick, Steven O. Howe, Wayne E. Esaias, and Jon T. Scott. 1978. Wind events and food chain dynamics within the New York Bight. Limnol. and Oceanogr. 23(4):659-683.

Abstract: Time series of wind, current, nutrients, chlorophyll, and zooplankton are used to examine the effect of storm events on the food chain dynamics of the New York Bight. Storms cause dilution of phytoplankton concentration in the vertical plane, but lead to aggregation of chlorophyll in the horizontal field. Nutrients are made available with onshore flow in response to wind events favorable for upwelling. A series of nutrient budgets suggest that storm-induced mixing and upwelling of nitrate may satisfy at least 33% of the productivity demand of this system. Examples of the biological response to storms are drawn from 30 cruises during January, March, April-May, and August-September 1974, 1975, 1976, and 1977 under mixed and stratified conditions of the water column. The interaction of storms and seasonal stratification suggests predictable structure and frequency of chlorophyll distribution across the shelf which may influence both the survival strategies of herbivores and the loci of energy transfer to the rest of the food chain.

1642 - Walter, L. 1961. Composition of sewage and sewage effluents. Water, and Sewage Works. Part 2. pp. 478.

1643 - Ward, P. S. 1976. Say good-bye to ocean dumping. Jour. Water Pollut. Control Fed. 48(3):426-528.

1644 - Warsh, C. E. 1975a. Physical oceanography historical data for deepwater dumpsite 106. U.S. NOAA May 1974 Baseline investigation of deepwater dumpsite 106. NOAA Dumpsite Evaluation Report 75-1. U.S. Department of Commerce. NOAA Ocean Dumping Program. Rockville, MD. pp. 105-140.

1645 - Warsh, C. E. 1975b. Physical oceanographic observations at deepwater dumpsite 106-May 1974. U.S. NOAA May 1974 Baseline investigation of deepwater dumpsite 106. NOAA Dumpsite Evaluation Report 75-1. U.S. Department of Commerce. NOAA Ocean Dumping Program. Rockville, MD. pp. 141-187.

1646 - Wasp, E. J., et al. 1969. Bulk transport of waste slurries to inland and ocean diposal sites. Summary. Bechtel Corporation, San Francisco, CA. 32p.

1647 - Wastler, T. A., C. K. Offutt, C. K. Fitzsimmons, and P. E. DesRosiers. 1975a. Disposal of organochlorine wastes by incineration at sea. Division of Oil and Special Materials control. Office of Water and Hazardous Materials. EPA. Washington, DC. EPA-430/9-75-014.

1648 - Wastler, T. A., and C. K. Offutt. 1975b. Chemical need for the regulation of ocean disposal. In: T. M. Church (editor), Marine Chemistry in the Coastal Environment. America Chemical Society. Washington, DC. ACS Symposium Series 18. pp. 416-423.

1649 - Waterloopkundig Laboratorium, Delft, in co-operation with Rijswater-Staat-Math. FYS. AFD. and Farbenfavriken Bayer AG. 1970. Untersuchungen über die Vermischung con dunnsaure mit meerwasser im schraubenstrahl des kustenmotorschiffes "Kathe H". M939. 20p.

1650 - Watermann, B., and V. Dethlefsen. 1980. Preliminary report on histology and incidence of pseudobranch tumors in Atlantic cod (<u>Gadus morhua</u>) in the German Bight. ICES. C.M. 80/E:25.

1651 - Water Pollution Control Federation (Research Committee). 1971. A review of the 1970 literature on wastewater and water pollution control. Jour. Water Pollut. Control Fed. 43:933-1426.

1652 - Water Resources Engineers, Inc. 1975. A conceptual representation of the New York Bight ecosystem. NOAA TM ERL MESA-4. Boulder, CO. 251p. NTIS No: PB 252343/AS.

1653 - Watling, L., W. Leathem, P. Kinner, C. Wethe, and D. Maurer. 1974. Evaluation of sludge dumping off Delaware Bay. Mar. Pollut. Bull. 5(3):39-42.

1654 - Watson, J. D., and D. M. Watson. 1972a. Marine disposal of sewage and sewage sludge. J. D. and D. M. Watson, Ltd., London. 52p.

1655 - Watson, J. D., and D. M. Watson. 1972b. Marine disposal of sewage and sewage disposal. J. D. and D. M. Watson, Ltd. London. 25p.

1656 - Weichart, G. 1968a. Veranderungern des Meeres durch Abfälle. (Chemical alternation of seawater due to waste discharge.) Foderation Europaischer Gewasserschütz (FEG). Schütz der Meereskusten gegen Verunreinigung. Symposium. 5-7 Oktober 1967. Hamburg, FEB, Zurich. 37p.

1657 - Weichart, G. 1968b. Veranderunger des Meeres durch Abfälle und die darauf basierende Einführung von fünf Gefahrenklassen für das Einbrigen von Abfällen ins Meer. (Changes of the sea by wastes, and the introduction of five danger classes for the disposal of wastes in the sea, based on such changes.) Helgolander wiss. Meeresunters. 17:398-410.

1658 - Weichart, G. 1972a. Chemical and physical investigations on marine pollution by wastes of a titanium dioxide factory. In: M. Ruivo (editor), Marine Pollution and Sea Life. London, Fishing News (Books) Ltd. London. pp. 186-188.

1659 - Weichart, G. 1972b. Neuere entwicklungen in der meereschemie. Naturwissenschaften. 59:16-19.

1660 - Weichart, G. 1973a. Pollution of the North Sea. Ambio. Vol. 11(4):99-106.

1661 - Weichart, G. 1973b. Verschmutzung der Nordsee. Naturwissenschaften. 60: 469-472.

Abstract: The North Sea is one of the earth's most polluted sea areas. Most of the pollutants are deposited into it from the mainland by rivers and direct introduction, by dumping from ships, and from the atmosphere. The southern North Sea exhibits an especially high degree of pollution, because the burden from the adjacent countries and from international shipping is greatest in this area.

1662 - Weichart, G. 1974. The North Sea. Environment. 15(1):29-33.

1663 - Weichart, G. 1975. Investigations on the Fe concentration in the waters of the German Bight in connection with the dumping of wastes from titanium dioxide production. ICES. C.M. 75/C:23.

1664 - Weichart, G. 1977. The pH in the waters of the German Bight as influenced by primary production and respiration and acid TiO_2 wastes. ICES. C.M. 77/C:3.

1665 - Weidemann, H., and H. Sendner. 1972. Dilution and dispersion of pollutants by physical processes. In: M. Ruivo (editor), Marine Pollution and Sea Life. Fishing News (Books) Ltd. London. pp. 115-121.

Abstract: The behaviour of pollutants in the marine environment is strongly dependent on their nature and their density.

Insoluble substances with lower density will float on the surface. They may be solids such as plastics or wood, or liquids such as mineral oil. If the substance is neutrally buoyant, it will float between surface and sea-bed. Many organic particles and colloidal suspensions belong to this group. If the substance is heavier than sea water, it will sink with a velocity depending on its density, shape, and size. Most inorganic material and some insoluble liquids are of that kind.

Soluble subtances will at first show the same behaviour, but sooner or later, depending on their solubility rate, will be completely dissolved. This also applies to all classes of substances mentioned above. Gases may dissolve completely or partly in sea water; their origin will usually be in the atmosphere, but can also be in chemical or biochemical processes in the sea.

The main difference between insoluble and soluble substances becomes evident by their duration of stay in the sea and location of their deposition. Floating insoluble substances will usually sooner or later be deposited on shore from which they may again be removed by waves, surf, or currents.

Neutrally buoyant and soluble substances, however, may stay in the sea for an almost unliminted period. The best examples are the natural sea salts dissolved over millions of years from the earth. A few natural physical processes may remove small fractions of dissolved substances. Freezing processes in arctic or antarctic regions separate salt from fresh water, but the salt usually remains in the sea water. Separation of sea salts occur in lagoons when they are cut off from the ocean and if their location is in an arid climate, but this process obviously had a greater importance during the early stages of the earth's history.

Dissolved gases, however, are different from other soluble substances because the vapour pressure always balances that of the atmosphere.

1666 - Wenner, C. A. 1975. Reproduction in fishes of DWD 106. U.S. NOAA May 1974 baseline investigation of deepwater dumpsite 106. NOAA Dumpsite Evaluation Report 75-1. U.S. Department of Commerce. NOAA Ocean Dumping Program. Rockville, MD. pp. 253-256.

1667 - West, Richard H., Patrick G. Hatcher, Donald K. Atwood. 1976. Polychlorinated biphenyls and DDTs in sediments and sewage sludge of the New York Bight. U.S. Department of Commerce. NOAA. ERL-AOML. Miami, FL. 42p.

Abstract: Sediment samples taken on a cruise in September, 1973, in the New York Bight Apex were analyzed for polychlorinated biphenlys and DDT's. In addition, several samples of sewage sludge were obtained from treatment plants in the New York Metropolitan area which dispose of their sewage via ocean dumping. Results of the analyses and data reduction are presented in tabular form and some preliminary conclusions stated. It is noted that although sediments close to the Hudson-Raritan outflow contain organics which are richer in PCB's than sediments in the rest of the Bight, the high organic inputs at the sewage sludge dump site result in this area having the major input of PCB's and DDT's to the Bight Apex.

1668 - West, R. H., and P. G. Hatcher. 1980. Polychlorinated biphenyls in sewage sludge and sediments of the New York Bight. Mar. Pollut. Bull. 11(5):126-129.

1669 - Westman, J. R., and M. H. Bidwell. 1948. Waste disposal and the fisheries of the salt water adjacent to the greater New York metropolitan district. N. Y. State Conservation Dept. Bureau of Marine Fisheries.

1670 - Wetherall, Neal Thomas. 1973. Recovery and Enumeration of Lancefield's Group D Streptococci from Estuarine and Marine Waters. Masters Thesis. Adelphi University. Garden City, NY. 115p.

1671 - Wezernak, C. T. 1973a. Monitoring of dumping by means of satellite remote sensing. Ambio. 2(3):84-86.

1772 - Wezernak, C. T., and F. J. Thomson. 1973b. Barge dumping of wastes in the New York Bight. In: W. A. Finch, Jr. (editors), Proc. Symp. Earth Resources Technology Satellite-1 held at the Goddard Space Flight Center. Greenbelt, Maryland, 29 September 1972. National Aeronautics and Space Administration. Greenbelt, MD. pp. 142-145.

1773 - Wezernak, C. T., D. R. Lyzenga, and F. C. Polcyn. 1975. Remote sensing studies in the New York Bight. Report (COM-75-11358).

1674 - White, I. C., M. S. Rolfe, and P. A. Hardiman. 1974. Disposal of wastes at sea. III. The field assessment of effects. ICES. C.M. 74/E:25. 22p.

1675 - White, I. C. 1976. The environmental effects of drilling fluids. ICES. C.M. 76/E:21.

1676 - White, H. H., T. P. O'Connor, and P. Kilho Park. 1980. Biological effects of the marine dumping of industrial wastes at U.S. deep ocean dumpsites. ICES. C.M. 80/E:2.

1677 - Whitledge, Terry E. 1980. Water column monitoring cruise 1 - New York Bight 21-25 April 1980 data report. Division of Oceanographic Sciences. Brookhaven National Laboratory. Upton, NY.

1678 - Whitte, W. G., J. W. Usry, C. H. Whitlock, and E. A. Gurgnus. 1979. Spectral measurements of ocean-dumped wastes tested in the marine upwelled spectral signature laboratory. Tech. Pap. U.S. Natl. Aero. Space Adm. NASA-Hampton, VA. 34p.

1679 - Wiebe, P. H., G. D. Grice, E. Hoagland. 1973. Acid-iron waste as a factor affecting the distribution and abundance of zooplankton in the New York Bight. II. Spatial variations in the field and implications for monitoring studies. Estuar. and Coast. Mar. Sci. 1:51-64.

Abstract: A study was undertaken in the New York Bight in an effort to understand small scale variations of single species populations and coastal zooplankton communities as they relate to the disposal of acid wastes. Two grids of eight locations each, one day and one night station per location, were placed so that one covered the acid grounds and the other a similar area functioning as a control 9 km to the northeast. Thirty-nine taxonomic categories of zooplankton were counted from oblique net tow samples collected at the 32 stations. Biomass was determined from length measurements of individuals of 24 taxa.

Species composition of the samples was typical of neritic waters of the northeast Atlantic coast. The spatial distribution of the majority of the species was markedly aggregated, but no trend was observed which would suggest that the acid wastes were an important factor in shaping the distributions. Species did not show collective agreement as to the area in which a higher average abundance for each occurred; and no significant trends in percent similarity or diversity (Simpson's D and the information theory H') were evident. Although Vaccaro et al. (1972) found zooplankton biomass to be approximately 30% higher from the control area than from the acid grounds, comparison of the biomass difference between the two areas on a species basis showed that 95% of the overall difference was accounted for by only three species, *Pseudocalanus sp.* and its copepodids, *Calanus finmarchicus* copepodids and *Temora longicornis*. The acid-iron wastes appeared to be a minor factor affecting the distribution and abundance of zooplankton species during the time of this investigation.

Empirical measures of the variability of single species populations and community indices presented in the text may be useful guides for future surveys or monitoring studies.

1680 - Wiebe, P. 1976. The biology of cold-core rings. Oceanus. 19(3):69-76.

1681 - Wiemeyer, S. N., D. M. Swineford, and Paul R. Spitzer. 1978. Organochlorine residues in New Jersey osprey eggs. Bull. of Environ. Contam. and Toxicol. 19:56-63. 56-63.

Abstract: Presents data on levels of organochlorine pesticides and polychlorinated biphenyls (PCB's) in the eggs of New Jersey ospreys, the changes in eggshell thickness, and relates this information to similar data for declining and stable osprey populations in other areas.

1682 - Wilk, S. J, and M. J. Silverman. 1976. Fish and hydrographic collections made by the research vessels *Dolphin* and *Delaware II* during 1968-1972 from New York to Florida. NOAA Tech. Report NMFS SSRF-697. Seattle, WA. 159p.

1683 - Wilk, Stuart J., Wallace W. Morse, Daniel E. Ralph, and Thomas R. Azarovitz. 1977. Fishes and associated environmental data collected in New York Bight, June 1974-June 1975.

Abstract: Tabulations of fishes and associated environmental observations are given for 700 trawl stations made during 30 collecting intervals in the New York Bight from June 1974 to June 1975. Summary tables included give the following information: collecting-interval data (vessel, dates, stations sampled, gear, and area); station data (date, location, time of day, total catch, and environmental observations); and catch data for 127 species, representing 67 families (location, number, and weight).

The 10 most frequently collected species were: *Merluccius bilinearis* (456 stations), *Scophthalmus aquosus* (419), *Raja erinacea* (411), *Urophycis Chuss* (409), *Pseudopleuronectes americanus* (363), *Hippoglossina oblonga* (325), *Lophius americanus* (305), *Perprilus triacanthus* (284), *Paralichthys dentatus* (272), and *Squalus acanthias* (224).

1684 - Wilkniss, P. E. 1973. Environmental condition report for Deep Water Dump Area A. U.S. Naval Research Laboratory. Washington, DC. (Ocean Dumping of Munitions DWD-106). NRL Report 7553. 118p.

1685 - Williams, B. R. H. 1974. Disposal of liquid wastes to estuaries and the sea by pipelines. In: B. Laverick (editor), Industrial Pollution Control Year Book. Fuel and Metallurgical Jour. Ltd. London. pp. 178-182.

1686 - Williams, S. J. 1973. The geologic framework of inner New York Bight - Its influence on positioning offshore engineering structures. Geol. Soc. Amer. Abstr. 5(2):239.

Abstract: Much interest has been expressed recently in evaluating the feasibility of placing man-made engineering works (e.g., airports, deep draft terminals, nuclear power plants) on the inner continental shelf proximal to New York Megalopolis. Major impetus for such proposals is the lack of available space in the highly urbanized area, coupled with economic factors and increasing environmental concerns. Prior to making final site selection and commencement of engineering design, a knowledge of the shallow subbottom structure and stratigraphy, surficial sediment character, and geologic shelf history is necessary.

Results of an ICONS program reveal this region straddles two distinct physiographic provinces which are underlain by gently SE dipping Coastal Plain strata which have been differentially eroded and covered with variable thickness of Pleistocene-Holocene stratified sand and gravel. Shrewsbury Rocks extend offshore from Long Branch, N.J., in a NE direction and form a sea floor cuesta marking the physiographic boundary between the deeply eroded and subsequently filled subbottom to the north and the nearly outcropping truncated edges of Coastal Plain strata to the south. The buried submarine Hudson Channel has been traced on geophysical records from The Narrows to its shelf head (a natural deep channel) south of Sandy Hook, N.J. Other buried channels which drained the terminal moraine to the north are evident south of Rockaway Beach. Holocene transgression has served to rework exiciting sea floor sediments to yield the present distribution and to supply littoral currents with material for the northward growth of Sandy Hook Spit and westward growth of Rockaway Beach.

1687 - Williams, S. Jeffress, and David B. Duane. 1974. Geomorphology and sediments of the inner New York Bight continental shelf. U.S. Army Corps of Engineers. Tech. Memo No. 45.

Abstract: Approximately 445 miles of continuous seismic reflection profiles and 61 virbrating cores were obtained from the Inner New York Bight which encompasses about 250 square miles offshore from northern New Jersey and western Long Island. The major physiographic features include Sandy Hook and Rockaway Beach, both prograding barrier islands, Shrewsbury Rocks and the Hudson (submarine) Channel. Shrewsbury Rocks mark the demarcation between two distinct geomorphic provinces. The area north of Shrewsbury Rocks is underlain by Coastal Plain strata which have been deeply eroded by Pleistocene glacial processes and covered by sand and gravel outwash. South of Shrewsbury Rocks, Coastal Plain strata have been evenly truncated and covered by a veneer of residual material. Three primary types of bedding have been observed on the seismic records. Coastal Plain strata exhibit a monoclinal regional southeast dip; steeply inclined crossbeds are restricted to an elongate basin east of Sandy Hook, considered to be of fluvial origins. The third type is Pleistocen-Holocene stratified fluvial sands and gravels which are regionally discontinuous and exhibit gentle seaward dip. Cores reveal that fine-to-medium sand is the predominant sediment type on the inner shelf. Isolated patches of coarse sand rounded sea gravels are present off Long Island where fluviant materials are exposed. Course sediment off New Jersey is judged to be residual from sea floor outcrops of Coastal Plain strata. Very fine sand, silt, and muds comprise the sea floor at the head of the Hudson Channel and along the body.

Sand suitable for beach nourishment projects is found in abundance throughout the shallow shelf parts of the Inner New York Bight. Sea floor topography is fairly flat and sand occurs as blanket deposits. It is estiamted that over 2 billion cubic yards of clean sand is available for retreival by present dredging techniques.

1688 - Williams, S. J. 1975. Construction in the coastal zone: A potential use of waste materials. Marine Geology. 18(1):1-15.

1689 - Williams, S. Jeffress. 1976. Geomorphology, shallow sub-bottom structure, and sediments of the Atlantic Inner Continental Shelf Off Long Island. CERC-Coastal Engineering Research Center. Technical Paper No. 76-2. Fort Belvoir, VA.

Abstract: About 800 square miles of the Atlantic inner continental shelf off Long Island, New York, were studied by CERC to obtain information on the sea floor morphology, sediment distribution, and shallow sub-bottom stratigraphy and stucture. This information is used for delineating sand and gravel resources and deciphering shelf geologic history. Basic survey data by CERC consist of 735 miles of high-resolution continuous seismic profiles and 70 vibratory cores; additional data were available from 82 sediment cores and 225 miles of seismic records. Data coverage extends from Atlantic Beach east to Montauk and in Gardiners Bay, and from the shoreface seaward about 10 miles to water depths of 105 feet.

Three primary acoustic horizons are evident on the seismic profiles and have been identified by correlation with cores, land borings, and surface exposures of the reflectors. Granitic bedrock is the oldest and deepest horizon underlying Long Island, but its recognition on the seismic records, due to limited sub-bottom penetration, is confined to northern Gardiners Bay. The bedrock surface slopes southeast and exhibits considerable relief where glacial ice has enlarged pre-Pleistocene drainage channels. Upper Cretaceous and Tertiary semiconsolidted clastic sediments overlie the bedrock and dip and thicken to the southeast. The surfaces of these strata, which are present throughout the study area and project north under Long Island, and the second major horizon.

The third seismic horizon is Pleistocene erosion surface cut by fluvial and glacial agents into the older rock units. Depth of this surface varies from -50 to -300 feet MSL off the western and eastern Long Island shelf to sea floor outcropping in parts of the central Long Island inner shelf. Pleistocene detritis consists primarily of blanketlike deposit of outwash sand and gravel; however, radiocarbon dates show that Holocene-age barrier-lagoonal sequences and estuarine sediments cover parts of the Long Island shelf.

Much of the surficial sand on the inner shelf is suitable as fill for beach restoration, except for that of the shoreface region (0 to -30 feet MSL) which contains fine sand and that of major parts of Gardiners Bay which contain organic-rich silt and clay. Topographic highs on the sea floor in the form of linear shoals, and broad deltalike platforms in eastern Long Island appear most suitable for sand recovery. The sea floor in most potential borrow areas is flat and sand occurs as blanket deposits. Potential sand reserves within about 12 feet of the sea floor in the region are estimated to be more than 8 billion cubic yards.

1690 - William S. Jeffress. 1977. Geologic effects of ocean dumping on New York Bight inner shelf. Amer. Assoc. of Petrol. Geol. Bull. 61(5):841.

Abstract: A detailed study intended to inventory marine sand and gravel resources was conducted using 824 km of high-resolution seismic-reflection subbottom profiles and 61 vibratory sediment cores. Analyses of these data together with bathymetric maps from 1845 to 1973 and ancillary deep-boring logs have revealed that much of the natural bathymetry and geologic conditions on the inner New York Bight shelf have been modified significantly by ocean disposal of waste solids since about 1888. Innermost parts of the Hudson Shelf channel have aggraded by approximately 15 m, and broad hills mask the former natural-channel morphology. Composition of the waste solids varies considerably; much of the early material was natural soil and excavation debris from construction of buildings, subway tunnels, and navigation channels. More recent wastes consist primarily of fine-grained dredge spoil, treated sewage sludge, and industrial acids. The natural geologic stratigraphy shows considerable variation also, making the distinction between natural and anthropogenic sediment difficult. Upper Cretaceous clastic strata crop out at Shrewsbury Rocks and variable thickness of stratified Quaternary sand and gravel predominate over the shelf. Cores from the Hudson Channel depression contain fine-grained sediments which appear to be natural relict-estuarine deposits. Deposition rates of solid wastes have been far greater than for natural sediment; most of the waste solids have remained in the original dump sites and are modified only slightly by present shelf processes.

1691 - Williams, S. Jeffress. 1979. Geologic effects of ocean dumping on the New York Bight inner shelf. In: H. D. Palmer and M. G. Gross (editors), Ocean Dumping and Marine Pollution. Douden, Hutchinson and Ross, Inc. Stroudsburg, PA. pp. 51-72.

Abstract: High resolution seismic reflection records, sediment cores and deep borings, and comparison of bathymetric charts from 1945 to 1973 provide evidence that ocean dumping of assorted solid materials has significantly filled parts of the Hudson shelf channel, and is an important geologic process. Ocean disposal of natural and man-made wastes was officially initiated seaward of New York Harbor in 1888 to relieve health problems, congestion, and accelerated shoaling of navigation channels long associated with uncontrolled disposal within the city and adjacent waterways. Records show that about 850 million m^3 of liquid and solid wastes have been dumped in the past 85 years. This has resulted in creation of several mounds with relief of about 15 m covering an area of about 9 thousand hectares. The calculated volume of anthropogenic solids filling the Hudson channel is 318 million m^3. Much of the material is similar in character to indigenous sediment. The results indicate most materials except sewage sludge are fairly stable and remain in the original dump sites. In spite of large volumes of sludge dumped at the same site since 1924, no evidence of significant accumulation on the seafloor has been found.

1692 - Willis, S. E. 1961. Large scale refuse disposal. Paper read at the R.S.H. Sessional meeting. Jan. 25. Chelmsford.

1693 - Wilson, J. S. 1979. Maintenance dredging in New York Harbor-a developing dilemma. American Seaport. 41(6):21-22.

1694 - Wilson, K. W. 1973a. Effects of red mud on marine animals. Mar. Pollut. Bull. 4(11):169-171.

1695 - Wilson, K. W., and R. A. A. Blackman. 1973b. Observations on the effects of red mud on marine animals. ICES. C.M. 73/E:15.

1696 - Wilson, K. W., and R. G. J. Shelton. 1973c. On the occurrence of lymphocystis with notes on other pathological conditions in the flatfish stocks of the northeast Irish sea. Aquaculture. (2):395-410.

1697 - Wilson, K. W., I. C. White, in Coll, with N. Cartwright. 1974a. A review of the biological effects of acid-iron wates from titanium dioxide production in the U.K. ICES. C.M. 74/E:40.

1698 - Wilson, K. W., and R. A. A. Blackman. 1974b. Observations on the effects of red mud on marine animals (Re-issue of. Doc. C.M. 1973/E.15). ICES. C.M. 74/E:44.

1699 - Wilson, K. W., and R. A. A. Blackman. 1974c. Red mud. Mar. Pollut. Bull. 5(8):127-128.

1700 - Wilson, K. W. 1975. The role of toxicity tests for the prevention of marine pollution by dumping from ships. In: F.A.O. Lectures presented at the fourth FAO/SIDA Training Course in aquatic pollution in relation to protection of living resources. Bioassays and toxicity testing. Lysekil, Sweden. 13 Oct-29 Nov. 1975. Rep. FF-INT.173. Suppl. 1. pp. 199-210.

1701 - Windom, H., F. Taylor, and R. Stickney. 1973. Mecury in North American plankton. Jour. du Conseil. 35(1):18-21.

Abstract: Concentrations of mercury in North Atlantic plankton (on a dry weight basis) vary from less than 0.2 to about 0.4 ppm in unpolluted areas to as high at 5.3 ppm in polluted areas near shore. No correlation in mercury content with species composition of the plankton samples was observed. Variations in concentration appear to be a function of distance from coastal pollution sources.

1702 - Windom, H. L., and R. A. Duce. 1976. Marine pollutant transfer. Heath Books. Lexington, MA. 391p.

1703 - Winter, J. 1972. Long-term laboratory experiments on the influence of ferric hydroxide flakes on the filter-feeding behaviour, growth, iron content and mortality in Mytilus edulis L. In: M. Ruivo (editor) Marine Pollution and Sea Life. FAO. Fishing News (Books) Ltd. London. 392-396.

1704 - Wohlschlag, Donald E., and Faust R. Parker, Jr. 1981. Metabolic sensitivity of fish to ocean dumping of industrial wastes. In: B. H. Ketchum, D. R. Kester, and P. K. Park (editors), Ocean Dumping of Industrial Wastes. Plenum. New York, NY. pp. 455-471.

Abstract: Respiratory metabolic stress studies of an ocean-dumped waste on Lutjanus campechanus, red snapper, indicated that sublethal concentrations of 0.2% v/v dilution of the ponded "biotreated" waste would have a depressing effect on swimming activity on respiratory metabolism at maximum sustained swimming rates and on the metabolic scope for activity, which is the difference between the active and the standard (maintenance) metabolic rates. At 20°C the effects were lessened, presumably because the fish were already under some thermal stress and possibly because of the thermal liability and volatility effects on the toxicants.

An evaluation of the general technique of utilizing metabolic scope attenuation with stress reveals that weight and length effects on both respiratory metabolism and swimming propensities need careful future consideration. Several suggestions are made for increasing the sensitivity and precision of future applications of this technique.

The possibility of modifying the technique for a continuously monitoring system is advanced for cases in which the chemical nature, concentration, and/or uptake rates of the toxic materials by fish are not known. The value of acquiring data from stress-metabolism experiments that can also be energy-related to other population and ecosystem characteristics is also emphasized.

1705 - Wohlschlag, Donald E., and Faust R. Parker, Jr. (In Press). Fish: Response to ocean dumped pharmaceutical wastes. In: I. W. Duedall, B. H. Ketchum, P. K. Park, and D. R. Kester (editors), Wastes in the Ocean, Vol. I: Industrial and Sewage Wastes in the Ocean. John Wiley and Sons, Inc. New York, NY.

Abstract: Metabolic levels and swimming performances of Lutjanus campechanus (red snapper) at 20°C and a salinity of 35 ppt in dilutions of a composite pharmaceutical waste indicated fatal toxicity for fish required to swim in 24 h at a concentration of 0.25% (vol/vol) in seawater. For fish held two days at 0.125% and 0.0625%, both swimming performance and active metabolism dropped as waste concentration increased. Standard (maintenance) metabolism was fairly constant. Metabolic scope--the difference between active and standard metabolic rates--declined in proportion to the sublethal waste concentration for two-day exposures.

Exposure-recovery experiments at 28°C and a salinity of 35 ppt with Cynoscion nebulosus (spotted seatrout) were conducted with composite and six individual industry wastes. The first visible signs of stress appeared in 2 hr or less at concentrations between 0.5% and 0.00625%, depending upon the initial toxicity of each waste source. When the fish were transferred to clean seawater the metabolic levels were highly variable. The standard metabolism always increased compared to controls, and swimming performance declined. The active metabolic levels did not decrease uniformly because of the stimulatory effects of the Upjohn/Waste and possibly of the Merck, Bristol, and Pfizer wastes. The metabolic scope, except in the Upjohn experiments, remained depressed at the end of the recovery period. The unusually great depression of scope in the composite experiment may have been caused by negative interactions among waste components. Signs of morbidity after a 2- to 4-day recovery indicated that initial exposure concentrations were too high and too long for complete recovery. The extent of cumulative or delayed lethality in ocean dumping needs to be investigated in terms of population suppression effects on fishes that may be exposed in short-term pulses of subacute toxic levels of wastes every few days.

1706 - Wolfe, Douglas A., et al. (In Press). Summary report of the panel on the effects of ecosystems. In: G. F. Mayer, A. Calabrese, F. A. Cross, D. C. Malins, D. W. Menzel, J. S. O'Connor, and F. J. Vernberg (editors), Ecological Stress in the New York Bight: Science and Management. Estuarine Research Federation. Columbia, SC.

Abstract: The New York Bight receives a variety of wastes dumped into the Hudson River and the Hudson-Raritan estuaries, as well as direct barged inputs of dredge spoils, sewage sludge, and chemical wastes. These pollutant additions have caused widespread degradation of the marine and estuarine environments with effects being most severe in the western portions of Raritan Bay and the New York Bight Apex. Beyond the inner Apex, there is an uneven, seasonably variable gradient of decreasing impacts.

The scientific evidence that the changes are caused by pollutant additions is overwhelming and conclusive; it is supported by extensive field observations of ecological degradation and measurements of occurrences of pollutants, laboratory experiments on organism responses to toxicant exposures, and comparisons with observations on disturbed and undisturbed systems in other coastal areas. The evidence shows a progressive deterioration of the marine environment which is caused by the discharge of pollutants into the area. The trend of these changes is away from the traditional aesthetic and resource values associated with the marine environment.

Unfortunately, the New York Bight has been subjected to such a wide variety of pollutants for such a long time that it is not possible at this time to ascribe specific effects to specific pollutants. The observed changes represent cumulative effects of combinations of toxic materials; nutrient and organic loading, and habitat modifications. Thus, the relative contributions and importances of the various toxicant additions cannot be readily separated from each other, and selection of optional approaches for stopping and reversing the environmental degradation is therefore exceedingly difficult.

Despite the current shortcomings of ecological science, environmental managers should develop management alternatives for alleviating the impacts of pollution in the New York Bight. For more effective management-directed planning of future environmental-research in the New York area, environmental managers and environmental scientists should approach environmental problem-solving jointly and cooperatively. Managers and scientists should develop mutually agreeable action plans which consider the full range of potential options available for environmental management and the specific risks and costs associated with each option. Proposed courses of action should be field tested on a limited, experimental scale to assess their probable ecological consequences prior to full-scale implementation.

1707 - Wong, Kuo Chuin, and R. E. Wilson. 1979. An assessmnt of the effects of bathymetric changes associated with sand and gravel mining on tidal circulation in the lower bay of New York Harbor. Marine Sciences Research Center. State Univ. of New York. Stony Brook, NY. Special Report 18. Ref. 79-1. 24p.

Abstract: Present sand and gravel mining operations within the Lower Bay of New York Harbor are restricted to the east bank of Ambrose Channel and to the vicinity of Chapel Hill North Channel because of the concern that mining in other areas might adversely affect water quality and shore erosion. As part of an evaluation of environmental affects associated with expanded sand and gravel mining, we have simulated numerically tidal circulation patterns and tidal elevations in Lower Bay for a number of altered bathymetries corresponding to hypothetical mining operations. Results suggest that tidal currents will decelerate over the mined region and accelerate outside of them, and that the tidal stream will be deflected toward the region. It is also clear that the mining near the mouth of the Bay could increase tidal range along Staten Island substantially.

1708 - Wood, Peter C. 1970. The effects of pollution on marine organisms. Paper No. 5. Published in the Proceedings of the The Institute of Water Pollution Control on Water Pollution Control in Coastal Areas. Bournemouth. 19-21 May. pp. 3-8.

1709 - Wood, P. C. 1973a. The disposal of sludges to sea. Symposium on Disposal of Municipal Industrial Sludges and Solid Toxic Wastes. The Institute of Water Pollution Control. London. Paper No. 8. pp. 103-112.

1710 - Wood, P. C. 1973b. Disposal of sludge to sea. Paper presented at the National Symposium on the Disposal of Municipal and Industrial Sludges and Solid Toxic Wastes, 26th and 27th November. London. Paper No. 8. 11p.

1711 - Wood, P. C. 1975. The discharge of sewage from sea outfalls into the North Sea. Progr. Water Technol. Suppl. 1975. Pap. No. 1. pp. 1-7.

1712 - Woodhead, D. S. 1980. Marine disposal of radioactive wastes. Helgolaender Meeresuntersuchungen. 33(1-4):122-137.

1713 - Woods Hole Oceanographic Institution. 1950. Summary of findings on waste disposal in the New York Bight as of October 1949. Woods Hole Oceanogr. Inst. Ref. 49-49. Woods Hole, MA.

1714 - World Health Organization. 1979. Principles and guidelines for the discharge of wastes into the marine environment. World Health Organization, Regional Office for Europe. Copenhagen. 534p. [incl. metals; oil; pesticides; detergents; cyanides; fluorides; sewage; radioactivity, phosphorous; oxygen; industrial wastes; selenium; organesilicons; phenols].

1715 - World Water. 1979. Ocean dumping - is the U.S. right to ban it? World Wat. Sept. pp. 42-43.

1716 - Worthington, L. V. 1977. Intensification of the Gulf Stream after the winter of 1976-77. Nature. 270:415-417.

1717 - Wright, W. R., and C. E. Parker. 1976. A volumetric temperature/ salinity census for the Middle Atlantic Bight. Limnol. and Oceanog. 21(4):563-571.

1718 - Wuestefeld, Robert H. 1968. Waste disposal in the coastal waters of New York harbor. Paper presented at joint meeting. New York Water Pollution Control Assoc. Metropolitan Section. Amer. Soc. of Civil Engin. Metropolitan Section. December 5, 1968. (mimeo).

1719 - Wunderlich, Lewis Daniel. 1976. The hydrographic features of the New York Bight Apex. Masters Thesis - State Univ. of New York. Stony Brook, NY.

Abstract: An analysis is presented of the seasonal cycle of hydrographic properties in the New York Bight Apex. The study is based upon existing reports and data, principally those gathered by the National Marine Fisheries Service and the Marine Escosystems Analysis Program on monthly hydrographic sampling cruises during the years 1969, 1973-74. These data, including temperature, salinity, and density, are present graphically on a monthly basis through the use of isometric block diagrams and planimetric projections.

The profiles of temperature, salinity, and density illustrate the importance of the local exchange of heat and momentum between the atmosphere and surface waters of the Apex and the seasonal runoff pattern of the Hudson River in determining the hydrographic properties of the New York.

1720 - Yasso, W. E., and E. M. Hartman, Jr. 1975. Beach forms and coastal processes. MESA New York Bight Atlas Monograph 11. MESA New York Bight Project and New York Sea Grant Institute. Albany, NY. 51p.

1721 - Yde, par M., S. De Maeyer-Cleempoel, and Z. Dartevelle. 1980. La pollution fecale en Mer Du Nord au large d'ostende (Belgique). Rev. Int. Oceanogr. Med. Tome LIX. pp. 47-54.

1722 - Yentsch, C. S. 1977. Plankton production. MESA New York Bight Atlas Monograph 12. MESA New York Bight Project and New York Sea Grant Insitute. Albany, NY. 25p.

Abstract: The prinicipal mechanism for regulating primary production in the waters of the New York Bight is vertical mixing. The intensity of this mixing is caused by a unique blend between local meteorological conditions and the physiology of photosynthetic phytoplanton. In winter the distribution of phytoplankton is largely due to the intensity of vertical mixing and water depth; primary productivity in slope waters is limited by vertical mixing, which in effect reduces the amount of sunlight received by the phytoplankton. The spring bloom arises from the reduction in vertical mixing because of heat (via sunlight) being added to surface waters and the relaxation of winds. In summer, primary production is low through most of the Bight, apparently due to the limited quantities of N in the euphotic zone. Intense production occurs in autumn as a result of the breakdown of the summer themocline with the onset of winter temperatures and strong winds.

1723 - Young, D. L., and R. T. Barber. 1973b. Effects of waste dumping in the New York Bight on the growth of natural populations of phytoplankton. Environ. Pollut. 5: 237-252.

Abstract: In New York Bight (USA) waters from outside those areas highly affected by sewage sludge or dredge-spoil dumping, the growth of natural phytoplankton populations was initially exponential. Waters sampled from the disposal areas were temporally and spatially inconsistent in supporting initial rapid growth. Inhibition occurred only as a temporary lag. The dominant phytoplankter that grew exponentially was, with a single exception, a typical coastal bloom species. The temporary inhibition was more probably due to toxic organic materials than to heavy metals.

1724 - Young, D. R., D. J. MacDermott, and T. C. Heesen. 1975. Pollutant inputs and distribution off southern California. ACS Symp. Ser. 18. pp. 424-439.

1725 - Young, D. R., M. D. Moore, T. K. Jan, and R. P. Eganhouse. 1981. Metals in seafood organisms near a large California municipal outfall. Mar. Pollut. Bull. 12:134-138.

1726 - Young, David R. (In Press). A comparative study of trace metal contamination in the Southern California and New York Bights. In: G. F. Mayer, A. Calabrese, F. A. Cross, D. C. Malins, D. W. Menzel, J. S. O'Connor, and F. J. Vernberg (editors), Ecological Stress in the New York Bight: Science and Management. Estuarine Research Federation. Columbia, SC.

Abstract: Estimated annual inputs of trace metals to the New York Bight are four to thirty-four times higher than those for the Southern California Bight. Typical seawater concentrations of dissolved and particulate metals in major harbors adjacent to these regions are remarkably similar and usually are at least several times higher than those for noncontaminated control sites. However, reported concentrations of dissolved metals above the New York Bight dumpsites are sometimes one or more orders of magnitude above values measured in the vicinity of submarine municipal wastewater discharges off southern California. In contrast, sediment concentrations of metals around the two major southern California outfalls are two to five times higher than those at the New York dumpsites for dredged materials and sewage sludge. Metals levels in flocculent material inshore of one California outfall often are an order of magnitude above those at control sites; such particulates may be a significant vector for metals contamination of filter- and deposit-feeding invertebrates in this discharge zone. However, it is not known yet whether elevated tissue burdens are damaging to invertebrates from the vicinity of the outfalls. Fishes from the same area exhibit little or no metals contamination of their tissues; a similar finding is reported for the New York Bight. With the exception of mercury, concentrations of target trace metals do not appear to increase with trophic level in marine food webs.

1727 - Young, David R. Chlorinated hydrocarbon contaminants in the Southern California and New York Bights. In: G. F. Mayer, A. Calabrese, F. A. Cross, D. C. Malins, D. W. Menzel, J. S. O'Connor, and F. J. Vernberg (Editors), Ecological Stress in the New York Bight: Science and Management. Estuarine Research Federation. Columbia, SC.

Abstract: Submarine discharge of municipal wastewater has been the dominant source of several important chlorinated hydrocarbons to the Southern California Bight. Seawater concentrations of PCB's and DDT's near the urban coast seldom exceed a few parts per trillion; in contrast, concentrations in surface sediments in the vicinity of Los Angeles outfalls range to 10 parts per million and 100 parts per million, respectively. Contaminated particulates apparently are the dominant mechanism of input and dispersion, and accumulations of these materials on the sea bottom are believed to control levels in the benthos. PCB concentrations in muscle tissue of seafood organisms around major outfalls range to approximately 1 ppm, and values for liver tissues are an order of magnitude higher. Mussels suspended at several depths near one diffuser system revealed gradients of PCB and DDT in the water column, with a ten-fold increase from the surface to the bottom. These chlorinated hydrocarbons are concentrated by organisms at higher tropic levels. Also, relatively high PCB concentrations have been reported in the New York Bight. The severe contamination of Hudson River sediments by PCB wastes may lead to extensive and persistent contamination of the New York Bight and the Hudson River estuary.

1728 - Young, J. S. 1973. A marine kill in New Jersey coastal waters. Mar. Pollut. Bull. 4:70.

Abstract: A kill of lobsters around wrecks off the New Jersey coast seem not to be direct result of pollution but may have been due to an influx of water causing reduced oxygen levels combined with high temperatures. Mortalities of this kind have been observed before in the area, but it is not known if this is an annual event or due to abnormal circumstances. It could be worth keeping the situation under review in the future.

1729 - Young, James S., and John B. Pearce. 1975. Shell disease in crabs and lobsters from New York Bight. Mar. Pollut. Bull. 6(7):101-105.

Abstract: Dumping grounds in New York Bight receive very large quantities of sewage sludge. Lobsters and rock crabs collected in or near the dumping grounds sometimes show various pathological condition of the shell and gills. In this study the histopathology of 'shell disease', the caustive agents, and its effects on respiration are discussed in connection with a possible association with the disposal of solid wastes into the ocean.

1730 - Young, R. A., and J. B. Southard. 1975. Erodability of cohesive marine sediments: Fields and laboratory experiments. Geol. Soc. of Amer. Abstr. Vol. 7.

1731 - Young, R. A. 1977. Seaflume: A device for in situ studies of threshold erosion velocity and erosional behavior of undisturbed marine muds. Mar. Geol. 23:M11-M18.

1732 - Young, R. A., and J. B. Southard. 1978. Erosion of fine-grained marine sediments: Sea-floor and laboratory experiments. Bull. Geolog. Soc. of Amer. 89:663-672.

1733 - Young, R. A. 1978. Suspended-matter distribution in the New York Bight Apex related to Hurricane Belle. Geology. 6:301-304.

Abstract: Suspended matter and hydrographic properties of the New York Bight Apex were studied three-days after passage of Hurricane Belle (August 1976). Distributions of suspended matter were found to be similar to those present during previous periods of calm summer weather. By comparing the present poststorm observations with previous poststorm sampling, it is hypothesized that during summer a stratified water column can restrict vertical mixing of resuspended bottom material to the relatively thin near-bottom layer, whereas mixing througout the water column takes place during unstratified winter conditions. Clearing times for resuspended material, therefore, are shorter in summer than in winter. Both extent of vertical mixing and clearing times are important considerations for those concerned with problems of marine ecology.

1734 - Yost, E., R. Hollman, J. Alexander, and R. Nuzzi. 1974. An interdisciplinary study of the estuarine and coastal oceanography of Block Island Sound and Adjacent New York coastal waters. Scient. and Tech. Aerospace Rep. 12(20):2425 (Abs. only)

Abstract: The ERTS-1 photographic data products were analyzed by additive color viewing and electronic image analysis techniques. Satellite data were compared to water sample data collected simulataneously with the data of ERTS-1 coverage in New York Bight. Prediction of the absolute value of total suspended particles can be made by composities of positives of Multi Spectral Scanner (MSS) bands 5 and 6 which were precisely made using the step wedge supplied on the imagery. Predictions of the relative value of the extinction on coefficient can be made with MSS bands 4 and 5. Thematic charts of total suspended particles (particles/1) and extinction coefficient provide performance improvement data to scientists conducting state and federal water sampling programs in New York Bight.

1735 - Zapatka, T. F., and R. W. Hann, Jr. 1976. Evaluation of safety actors with respect to ocean disposal of waste materials. Texas A&M Univ. Sea Grant Report. College Station, TX. 102p.

1736 - Zapatka, M. C., and R. W. Hann, Jr. 1977. Technical and philosophical aspects of ocean disposal. Texas A&M Univ. Sea Grant Report. College Station, TX. TAMU-SG-78-203. 160p.

1737 - Zijlstra, K. C. 1968. Ablagerung von festen und flussigen abfassstoffen auf hoher see. (Discharge of solid and liquid waste at sea). Foderation Europaischer Gewasserschutz (FEG). Schuts der Mecreskusten gegen Verunreinigung. Symposium. 5-7 Oktober 1967. Hamburg, FEB, Zurich. pp. 66-70.

1738 - Zimmerman, Rae. 1973. Organizational evaluation of industrial water pollution control in the New York region. Water Resources Bull. 9(6):1210-1224.

Abstract: Nontechnological factors related to the behavior and characteristics of organizations involved in pollution control are explored here as to their importance as determinants or indicators of the level of pollution control effectiveness. Methods of evaluating the existing level of effectiveness are developed and tested using the response of a selected set of industrial establishments to state water pollution abatement action in the New York Region form 1966 to 1971. The compliance of 209 manufacturing establishments to state abatement orders in the New York Region is evaluated with respect to selected organizational and industrial characteristics and characteristics of the firms socioeconomic environment. The influence of state policies and programs is also discussed. It was found that the degree of compliance to state abatement action by industry in the region is positively related to organization size, the extent of waste generation, the wealth and size of the town in which the firm is located, and the availability of waste treatment facilities in the town. The age of the firm was not related to compliance. The structure of state abatement scheduling also influences the degree of compliance. The methods of analysis used here provide a basis for a comprehensive analysis of the effectiveness of pollution control as an alternative to the case-by-case approach that is currently being used.

1739 - Ziskowski, J. J., and R. A. Murchelano. 1975. Fin erosion in winter flounder. Mar. Pollut. Bull. 6(2):26-29.

Abstract: Diseased fish showing erosion of the fins have been recorded from polluted waters in several parts of the world. This article gives the results of a year's survey of fin erosion among fish in oceanic and estuarine waters of New York Bight. The pristine Great Bay was chosen as a control for the heavily contaminated Sandy Hook-Raritan Bay system. Almost all fin erosion was confined to demersal flat fish, particularly the winter flounder, a species which spends a long time in polluted estuaries near Sandy Hook.

1740 - Ziskowski, John J., V. T. Anderson, Jr., and R. A. Murchelano. 1980. A bent fin ray condition in winter flounder, _Pseudopleuronectes americanus_, from Sandy Hook and Raritan Bays, New Jersey, and Lower Bay, New York. Copeis. No. 4:895-899.

1741 - Zucchetti, P., M. Floccia, and G. Palmieri. 1974. Lo smaltimento dei liquami nel mare della provincia di Roma. [The disposal of wastes into the sea of the province of Rome.] Ann. Sclavo. 16:393-412.

1742 - Zurlo, N. 1972. Scarlino case wastes from titanium dioxide production. Ext. Ecologia (It). 2(4):21-25.

SUBJECT INDEX

MARINE POLLUTION/OCEAN DUMPING

General:

1551, 1554, 1558, 1567, 1576, 1677, 1599, 1614, 1616, 1619, 1624, 1625, 1628, 1634, 1635, 1649, 1659, 1660, 1661, 1685, 1688, 1701, 1702, 1708, 1715, 1724, 1735, 1736.

MUNICIPAL WASTES

Sewage Sludge:

27, 32, 53, 61, 62, 65, 93, 112, 132, 139, 141, 143, 197, 225, 238, 255, 272, 273, 274, 276, 280, 282, 284, 285, 287, 325, 333, 334, 343, 356, 378, 384, 386, 388, 391, 397, 398, 399, 400, 401, 402, 403, 404, 405, 406, 407, 409, 411, 415, 418, 420, 421, 422, 423, 424, 439, 444, 470, 471, 473, 476, 477, 486, 494, 509, 512, 515, 516, 518, 547, 554, 555, 561, 564, 573, 574, 575, 576, 577, 578, 579, 584, 617, 620, 661, 665, 675, 686, 691, 710, 736, 742, 756, 757, 780, 797, 807, 812, 856, 870, 871, 872, 873, 874, 875, 876, 877, 878, 890, 899, 904, 924, 925, 929, 947, 951, 962, 983, 1008, 1009, 1010, 1022, 1023, 1024, 1051, 1052, 1059, 1060, 1070, 1072, 1082, 1083, 1084, 1096, 1097, 1098, 1109, 1134, 1136, 1139, 1140, 1163, 1164, 1165, 1167, 1168, 1171, 1204, 1205, 1220, 1221, 1223, 1224, 1231, 1234, 1235, 1239, 1252, 1255, 1262, 1267, 1271, 1283, 1306, 1308, 1310, 1311, 1312, 1314, 1316, 1330, 1336, 1340, 1341, 1347, 1361, 1373, 1384, 1385, 1400, 1420, 1422, 1426, 1427, 1428, 1430, 1437, 1438, 1441, 1442, 1446, 1449, 1466, 1468, 1475, 1476, 1477, 1507, 1508, 1513, 1515, 1516, 1524, 1526, 1535, 1540, 1544, 1547, 1550, 1554, 1555, 1557, 1558, 1560, 1561, 1562, 1563, 1565, 1567, 1576, 1577, 1614, 1621, 1642, 1654, 1655, 1667, 1672, 1709, 1710, 1718, 1737.

Outfalls/Wastewater Discharges:

8, 24, 80, 83, 97, 101, 130, 131, 133, 159, 161, 187, 211, 223, 242, 249, 250, 340, 357, 360, 361, 393, 403, 404, 443, 447, 475, 500, 501, 511, 516, 524, 628, 629, 661, 665, 680, 681 683, 685, 693, 695, 707, 722, 765, 766, 832, 843, 846, 852, 863, 864, 910, 913, 917, 968, 971, 973, 974, 982, 988, 989, 1027, 1036, 1067, 1081, 1093, 1147, 1208, 1225, 1277, 1370, 1373, 1387, 1449, 1538, 1618, 1622, 1654, 1655, 1685, 1711.

Outfalls/Sewage Sludge Discharges:

24, 80, 223, 249, 661, 693, 722, 989, 990, 991, 992, 993, 1008, 1009, 1010, 1316, 1346, 1347, 1371, 1373, 1438, 1472, 1539, 1711, 1724, 1725, 1726.

Solid Wastes:

227, 228, 441, 464, 582, 653, 691, 848, 890, 923, 932, 1173.

Garbage/Baled Refuse:

186, 903, 1080, 1366, 1425, 1506, 1595, 1692.

INDUSTRIAL WASTES

General:

187, 256, 257, 271, 276, 280, 282, 284, 288, 304, 375, 378, 409, 411, 424, 451, 452, 453, 554, 555, 586, 600, 626, 627, 630, 682, 684, 726, 756, 757, 791, 792, 793, 794, 795, 796, 797, 812, 817, 857, 882, 890, 910, 916, 929, 947, 1051, 1052, 1053, 1056, 1078, 1123, 1125, 1129, 1140, 1150, 1167, 1168, 1195, 1230, 1238, 1256, 1264, 1287, 1322, 1323, 1352, 1359, 1361, 1362, 1363, 1369, 1374, 1387, 1413, 1436, 1439, 1465, 1471, 1494, 1507, 1508, 1513, 1516, 1524, 1526, 1531, 1535, 1544, 1547, 1550, 1554, 1557, 1558, 1567, 1576, 1656, 1657, 1658, 1663, 1672, 1679, 1685, 1797, 1703, 1735, 1736, 1737, 1741, 1742.

Acid Wastes:

20, 21, 22, 87, 88, 89, 123, 202, 217, 226, 256, 270, 280, 288, 310, 359, 428, 429, 430, 431, 432, 433, 434, 435, 452, 470, 471, 473, 486, 515, 516, 518, 547, 548, 554, 555, 600, 616, 626, 713, 736, 776, 777, 798, 799, 803, 805, 806, 810, 812, 816, 821, 822, 825, 826, 866, 885, 886, 897, 898, 913, 931, 947, 1035, 1077, 1078, 1102, 1129, 1134, 1167, 1213, 1257, 1263, 1264, 1352, 1362, 1363, 1531, 1535, 1544, 1547, 1554, 1557, 1567, 1656, 1657, 1658, 1663, 1664, 1679, 1697, 1703, 1742.

Alkaline/Organic Wastes:

257, 270, 358, 449.

Containerized Wastes:

160, 1193, 1227, 1228, 1229, 1294, 1368, 1638.

Coal Wastes:

129, 261, 357, 408, 410, 474, 554, 586, 618, 812, 829, 830, 1104, 1275, 1446, 1448.

Cryolite (Aluminum) Sludges/Red Muds:

1, 137, 368, 370, 387, 1071, 1694, 1695, 1698, 1699.

Dredged Materials:

27, 134, 182, 184, 189, 191, 213, 214, 215, 337, 344, 353, 378, 388, 442, 482, 488, 491, 493, 522, 523, 554, 555, 561, 564, 570, 686, 786, 812, 826, 831, 847, 879, 947, 1004, 1012, 1013, 1014, 1050, 1087, 1117, 1126, 1162, 1167, 1206, 1218, 1234, 1254, 1293, 1321, 1361, 1363, 1364, 1402, 1403, 1420, 1456, 1495, 1535, 1540, 1554, 1557, 1693.

Drilling Fluids/Muds and Cuttings:

969, 1079, 1287, 1675.

Fish (Seafood) Wastes:

286, 288, 363, 364, 365.

Munitions:

31, 36, 382, 657, 901, 996, 1143, 1144, 1280, 1298, 1377, 1418, 1533, 1597, 1684.

Ocean Incineration:

6, 59, 71, 93, 216, 266, 309, 320, 454, 457, 460, 461, 462, 463, 464, 465, 466, 490, 499, 590, 608, 773, 790, 812, 900, 960, 965, 1005, 1148, 1151, 1152, 1153, 1154, 1155, 1156, 1274, 1276, 1291, 1292, 1447, 1479, 1491, 1519, 1531, 1552, 1553, 1557, 1566, 1592, 1647.

Ocean Mining Wastes:

46, 178, 241, 290, 383, 689, 1160, 1636.

Petrochemical/Petroleum:

26, 27, 125, 126, 970, 1000, 1239, 1309, 1325, 1349.

Pharmaceutical:

95, 222, 287, 882, 883, 1150, 1207, 1246, 1322, 1323, 1359, 1516, 1524, 1526, 1676, 1704, 1705.

Radioactive Wastes:

33, 181, 268, 317, 352, 354, 385, 417, 445, 551, 656, 667, 668, 676, 727, 734, 833, 837, 964, 1007, 1011, 1088, 1131, 1209, 1226, 1227, 1228, 1229, 1247, 1249, 1290, 1302, 1303, 1332, 1343, 1360, 1368, 1411, 1445, 1496, 1632, 1712.

Contaminant Inputs/Load/Sources:

23, 27, 78, 91, 163, 173, 187, 283, 346, 448, 449, 553, 554, 555, 600, 627, 678, 704, 705, 706, 757, 820, 947, 963, 984, 1029, 1030, 1031, 1050, 1053, 1071, 1111, 1115, 1121, 1122, 1123, 1134, 1237, 1264, 1287, 1323, 1361, 1363, 1422, 1423, 1436, 1465, 1467, 1522, 1528, 1548, 1549, 1550, 1558, 1565, 1623, 1624, 1635, 1642, 1656, 1657, 1658, 1663, 1709, 1724.

LEGISLATION/REGULATIONS:

International Conventions/Commissions:

44, 51, 77, 78, 91, 322, 323, 671, 672, 674, 698, 699, 703, 704, 705, 706, 712, 812, 889, 1001, 1074, 1149, 1168, 1196, 1202, 1236, 1606, 1607, 1608.

Legislation/Regulations:

30, 31, 39, 41, 44, 45, 47, 48, 50, 51, 52, 56, 62, 64, 72, 131, 135, 377, 442, 478, 532, 654, 671, 672, 673, 674, 676, 770, 771, 774, 786, 921, 1003, 1094, 1236, 1364, 1498, 1500, 1501, 1502, 1503, 1504, 1536, 1542, 1548, 1551, 1556, 1557, 1558, 1569, 1570, 1571, 1572, 1573, 1574, 1575, 1576, 1577, 1578, 1579, 1580, 1581, 1582, 1583, 1584, 1585, 1586, 1587, 1588, 1589, 1590, 1591, 1600, 1601, 1602, 1603, 1604, 1605, 1606, 1607, 1608, 1609, 1610, 1648, 1714.

Ocean Dumping Criteria/Site Selection Studies

9, 12, 131, 344, 420, 442, 508, 654, 666, 720, 721, 723, 786, 850, 851, 853, 861, 879, 917, 922, 995, 1203, 1233, 1278, 1339, 1489.

Waste Management Strategies:

30, 38, 93, 159, 255, 372, 377, 378, 380, 428, 442, 478, 512, 514, 522, 630, 653, 654, 665, 720, 730, 735, 747, 771, 793, 863, 864, 916, 932, 1012, 1013, 1014, 1056, 1070, 1075, 1091, 1092, 1223, 1224, 1233, 1235, 1239, 1255, 1264, 1265, 1266, 1283, 1298, 1342, 1351, 1360, 1363, 1364, 1413, 1443, 1479, 1483, 1489, 1563, 1565, 1566, 1568, 1593, 1594, 1614, 1616, 1639, 1643, 1648, 1714, 1715, 1738.

CHEMICAL PROCESSES:

General:

14, 15, 16, 17, 18, 31, 95, 120, 154, 156, 161, 163, 210, 214, 226, 252, 283, 289 296, 297, 312, 324, 332, 356, 374, 391, 398, 399, 400, 401, 402, 403, 404, 405, 407, 493, 494, 500, 504, 505, 564, 573, 574, 575, 617, 620, 621, 622, 623, 624, 625, 627, 693, 694, 755, 812, 813, 868, 920, 935, 937, 939, 942, 944, 945, 982, 987, 993, 1008, 1028, 1035, 1050, 1073, 1125, 1166, 1214, 1217, 1218, 1219, 1246, 1273, 1327, 1328, 1330, 1338, 1461, 1509, 1642, 1656, 1657, 1658, 1659, 1660, 1661, 1662, 1663, 1664.

Heavy Metals:

15, 16, 17, 18, 31, 83, 92, 247, 264, 265, 277, 281, 302, 303, 312, 326, 347, 348, 349, 385, 386, 392, 412, 413, 414, 502, 509, 527, 528, 533, 534, 535, 536, 537, 538, 539, 540, 541, 542, 543, 544, 545, 564, 588, 592, 607,

613, 614, 615, 626, 669, 710, 736, 764, 779, 791, 792, 794, 795, 796, 812, 815, 876, 877, 878, 902, 971, 989, 1022, 1023, 1072, 1106, 1116, 1127, 1134, 1190, 1212, 1217, 1218, 1232, 1246, 1248, 1277, 1296, 1318, 1329, 1402, 1437, 1444, 1456, 1467, 1470, 1473, 1475, 1528, 1629, 1631, 1701, 1725, 1726.

Organochlorine Compounds:

161, 163, 168, 192, 193, 194, 306, 351, 414, 444, 589, 677, 678, 696, 812, 876, 1057, 1112, 1113, 1170, 1214, 1217, 1218, 1241, 1242, 1269, 1274, 1313, 1375, 1379, 1381, 1528, 1543, 1630, 1667, 1668, 1681, 1727.

PHYSICAL PROCESSES:

General:

148, 171, 172, 189, 205, 206, 207, 208, 209, 234, 236, 253, 263, 278, 279, 292, 294, 295, 298, 300, 305, 336, 338, 339, 398, 407, 433, 435, 475, 525, 529, 564, 604, 605, 606, 679, 681, 700, 701, 709, 738, 739, 745, 755, 800, 804, 812, 822, 854, 858, 880, 891, 892, 893, 894, 912, 949, 959, 967, 998, 1015, 1016, 1017, 1026, 1130, 1136, 1137, 1138, 1139, 1140, 1143, 1164, 1165, 1172, 1198, 1219, 1225, 1246, 1262, 1272, 1333, 1404, 1408, 1409, 1410, 1417, 1421, 1432, 1438, 1439, 1475, 1508, 1517, 1520, 1633, 1634, 1644, 1645, 1665, 1691, 1716, 1730, 1731, 1732, 1733, 1734.

Dispersion/Distribution/Transport Processes:

2, 8, 26, 27, 102, 103, 104, 105, 127, 144, 147, 149, 156, 166, 170, 171, 184, 189, 253, 278, 279, 334, 354, 429, 430, 432, 433, 435, 440, 445, 467, 468, 551, 564, 625, 700, 713, 743, 751, 752, 768, 800, 802, 812, 821, 840, 850, 854, 1130, 1140, 1252, 1253, 1272, 1273, 1282, 1384, 1403, 1438, 1439, 1441, 1458, 1471, 1517, 1560, 1561, 1562, 1665, 1733.

Currents:

147, 148, 149, 158, 171, 172, 174, 175, 176, 204, 235, 278, 279, 292, 295, 427, 529, 564, 601, 605, 660, 787, 812, 821, 822, 1026, 1633, 1716.

Sedimentation Processes:

3, 157, 166, 177, 178, 183, 188, 189, 190, 191, 192, 193, 194, 223, 240, 302, 303, 311, 316, 318, 331, 334, 345, 347, 352, 353, 385, 389, 390, 406, 444, 470, 471, 481, 482, 484, 487, 489, 496, 523, 552, 556, 557, 558, 559, 564, 589, 603, 614, 658, 659, 664, 746, 749, 763, 812, 831, 868, 877, 891, 985, 1072, 1076, 1087, 1103, 1106, 1113, 1126, 1144, 1162, 1164, 1165, 1218, 1248, 1315, 1317, 1380, 1403, 1414, 1415, 1416, 1417, 1432, 1434, 1435, 1687, 1689, 1730, 1731, 1732, 1733.

BIOLOGICAL/ECOLOGICAL PROCESSES

Biological/Ecological Effects (General):

Assimilative Capacity:

Bioavailability/Bioaccumulation:

26, 27, 256, 374.

Bioassays:

13, 25, 73, 98, 179, 231, 247, 439, 684, 812, 1100, 1150, 1152, 1153, 1154, 1210, 1276, 1297, 1322, 1393, 1401, 1700.

Field Studies/Monitoring/Data Reports:

420, 421, 422, 423, 431, 434, 436, 437, 458, 486, 514, 518, 521, 616, 632, 633, 634, 635, 636, 637, 638, 639, 640, 641, 642, 643, 644, 645, 646, 647, 648, 649, 650, 651, 652, 725, 736, 841, 842, 844, 869, 870, 872, 914, 1051, 1054, 1061, 1062, 1063, 1064, 1065, 1066, 1085, 1086, 1095, 1096, 1097, 1099, 1109, 1110, 1155, 1156, 1189, 1192, 1217, 1271, 1295, 1296, 1337, 1365, 1376, 1429, 1442, 1450, 1476, 1477, 1485, 1490, 1497, 1511, 1531, 1559, 1598, 1677.

Microorganisms:

25, 112, 119, 144, 198, 200, 201, 245, 246, 287, 319, 374, 475, 492, 519, 526, 527, 528, 564, 609, 629, 664, 778, 782, 813, 834, 835, 896, 899, 902, 1059, 1090, 1122, 1207, 1219, 1250, 1289, 1304, 1305, 1306, 1307, 1308, 1309, 1310, 1311, 1312, 1313, 1314, 1317, 1349, 1359, 1398, 1405, 1462, 1463, 1474, 1475, 1545, 1546, 1615, 1617, 1627, 1670, 1721.

Phytoplankton/Plants:

99, 155, 187, 212, 280, 290, 312, 315, 403, 404, 473, 511, 548, 690, 755, 775, 784, 785, 806, 812, 813, 934, 935, 937, 938, 940, 941, 942, 943, 944, 945, 946, 948, 987, 1046, 1047, 1048, 1049, 1059, 1069, 1093, 1112, 1135, 1142, 1170, 1240, 1241, 1242, 1268, 1288, 1297, 1322, 1323, 1412, 1455, 1458, 1461, 1464, 1487, 1488, 1613, 1701, 1722, 1723.

Zooplankton:

256, 257, 341, 473, 602, 689, 812, 881, 883, 1214, 1345, 1412, 1613, 1679.

Shellfish:

212, 349, 479, 480, 498, 545, 564, 592, 619, 776, 777, 779, 936, 1122, 1134, 1212, 1232, 1355, 1356, 1357, 1358, 1380, 1406, 1457, 1525, 1564, 1626, 1727.

Invertebrates:

185, 187, 197, 212, 259, 260, 280, 313, 358, 419, 439, 520, 561, 586, 588, 589, 742, 761, 762, 819, 828, 876, 877, 882, 885, 886, 910, 911, 926, 927, 954, 955, 956, 957, 958, 970, 972, 1000, 1002, 1022, 1023, 1024, 1175, 1176, 1177, 1178, 1179, 1180, 1181, 1182, 1183, 1184, 1185, 1186, 1187,

1188, 1190, 1191, 1217, 1256, 1258, 1259, 1260, 1261, 1267, 1279, 1281, 1284, 1285, 1286, 1293, 1295, 1304, 1305, 1307, 1311, 1340, 1395, 1400, 1442, 1459, 1460, 1466, 1613, 1637, 1729.

Finfish:

15, 113, 114, 115, 116, 117, 118, 140, 164, 165, 179, 212, 219, 220, 221, 239, 270, 275, 280, 312, 313, 326, 349, 358, 446, 497, 543, 564, 571, 572, 586, 587, 588, 589, 592, 677, 702, 788, 789, 816, 817, 849, 866, 883, 885, 886, 896, 905, 906, 907, 908, 909, 933, 939, 970, 971, 988, 990, 999, 1018, 1019, 1020, 1037, 1038, 1039, 1040, 1041, 1042, 1043, 1044, 1045, 1054, 1122, 1127, 1150, 1208, 1211, 1214, 1215, 1217, 1346, 1347, 1354, 1356, 1357, 1375, 1381, 1398, 1440, 1514, 1629, 1632, 1650, 1666, 1669, 1682, 1683, 1796, 1703, 1704, 1705, 1725, 1726, 1727, 1739, 1740.

ENGINEERING STUDIES

General:

2, 8, 24, 32, 35, 38, 40, 54, 68, 84, 127, 285, 458, 463, 512, 603, 662, 724, 745, 838, 857, 919, 1162, 1193, 1199, 1200, 1201, 1223, 1224, 1227, 1228, 1229, 1239, 1332, 1366, 1370, 1371, 1374, 1481, 1484, 1654, 1655, 1686, 1688.

Socioeconomic:

262, 459, 590, 836, 918, 1173, 1231, 1482.

Modeling/Predictive:

11, 97, 149, 180, 188, 203, 251, 258, 269, 289, 307, 325, 335, 356, 438, 467, 468, 550, 595, 695, 708, 768, 812, 836, 838, 839, 856, 888, 951, 972, 986, 992, 994, 1106, 1295, 1343, 1350, 1351, 1430, 1652.

Remote Sensing:

123, 169, 202, 217, 280, 291, 293, 340, 547, 591, 687, 692, 697, 700, 737, 748, 749, 750, 751, 752, 753, 754, 821, 822, 823, 824, 825, 856, 869, 897, 898, 928, 966, 1128, 1129, 1142, 1161, 1219, 1220, 1221, 1319, 1320, 1487, 1492, 1493, 1497, 1527, 1611, 1618, 1620, 1671, 1673, 1678, 1734.

Transportation/Ships/Barges:

35, 41, 54, 68, 127, 150, 151, 152, 153, 451, 768, 1441, 1478, 1646, 1702, 1730, 1731, 1732.

REGIONS

NORTH SEA

General:

45, 77, 78, 121, 122, 254, 285, 368, 369, 370, 371, 373, 448, 449, 510, 530, 593, 594, 664, 698, 699, 756, 764, 783, 1071, 1072, 1095, 1096, 1097, 1098, 1099, 1100, 1101, 1146, 1174, 1234, 1235, 1236, 1237, 1238, 1247, 1248, 1256, 1265, 1266, 1267, 1268, 1269, 1278, 1287, 1294, 1318, 1319, 1320, 1372, 1379, 1441, 1442, 1444, 1461, 1465, 1466, 1467, 1468, 1469, 1471, 1472, 1476, 1477, 1478, 1479, 1480, 1481, 1482, 1622, 1623, 1628, 1649, 1650, 1659, 1660, 1661, 1662, 1674, 1696, 1709, 1710, 1711, 1721, 1737.

Legal:

45, 425, 1094, 1265, 1266, 1372.

Sources/Inputs:

121, 122, 160, 254, 373, 448, 449, 678, 698, 699, 756, 1071, 1072, 1077, 1095, 1287, 1294, 1372, 1379, 1444, 1465, 1468, 1472, 1476, 1477, 1478, 1622, 1623, 1660, 1661, 1662, 1674, 1709, 1710, 1711, 1721.

Physical Processes:

664, 1372, 1471.

Chemical Processes:

121, 122, 310, 412, 413, 414, 764, 1021, 1072, 1248, 1256, 1269, 1318, 1372, 1379, 1444, 1461, 1468, 1472, 1623.

Biological Processes:

179, 225, 368, 369, 370, 371, 664, 699, 784, 785, 817, 819, 910, 911, 1002, 1018, 1019, 1020, 1072, 1097, 1101, 1248, 1256, 1257, 1258, 1259, 1260, 1261, 1267, 1268, 1269, 1372, 1442, 1466, 1650, 1796.

United Kingdom:

9, 10, 28, 31, 33, 129, 140, 285, 333, 334, 343, 346, 347, 348, 349, 351, 382, 386, 387, 417, 418, 419, 420, 421, 422, 423, 439, 440, 502, 519, 588, 671, 672, 673, 674, 675, 677, 711, 742, 755, 756, 757, 758, 759, 868, 924, 925, 964, 982, 983, 984, 1011, 1022, 1050, 1051, 1052, 1094, 1095, 1096, 1097, 1098, 1099, 1100, 1101, 1103, 1104, 1130, 1147, 1170, 1204, 1205, 1209, 1222, 1225, 1231, 1232, 1233, 1234, 1235, 1236, 1237, 1238, 1247, 1248, 1265, 1266, 1278, 1340, 1341, 1394, 1441, 1442, 1444, 1466, 1467, 1468, 1476, 1477, 1478, 1479, 1480, 1481, 1482, 1483, 1484, 1485, 1630, 1654, 1655, 1674, 1692, 1796, 1797, 1708, 1709, 1710, 1711, 1737.

United Kingdom Dumpsites:

St. Abb's and Bell Rock:

1022, 1024.

Bristol Channel:

1052.

Firth of Clyde:

140, 285, 386, 519, 588, 589, 755, 924, 925, 1170, 1394, 1467, 1630.

Firth of Forth:

28, 346, 347, 1023

Garroch Head:

1204, 1205.

Liverpool Bay:

419, 675, 868, 1147, 1476, 1480.

Outer Thames:

1101, 1340, 1341, 1469.

Plymouth:

422, 675.

River Humber:

1051.

F. R. of Germany:

367, 368, 369, 370, 371, 373, 385, 412, 413, 414, 448, 510, 530, 607, 664, 783, 784, 785, 817, 890, 910, 911, 1002, 1071, 1072, 1077, 1174, 1256, 1257, 1258, 1259, 1260, 1261, 1318, 1319, 1320, 1372, 1465, 1650.

The Netherlands:

449, 608, 779, 1471, 1622, 1623, 1649.

Finland:

74, 91, 861, 862, 1456.

Norway:

1, 187, 827, 828, 829, 830, 1350.

Sweden:

74, 91, 861, 862, 1456.

Denmark:

744.

BALTIC SEA

General:

55, 57, 58, 63, 66, 69, 74, 79, 91, 128, 628, 734, 768, 1018, 1019, 1379, 1487.

MEDITERRANEAN/BLACK SEA

General:

60, 70, 248, 393, 506, 767, 1058, 1145, 1270, 1741, 1742.

NORTH AMERICA

Canada

64, 167, 218, 243, 321, 363, 364, 365, 847, 922, 1436, 1632, 1636.

New York Bight/Apex:

General:

3, 4, 5, 14, 15, 16, 17, 18, 19, 20, 21, 22, 23, 24, 25, 26, 27, 29, 30, 62, 85, 86, 89, 96, 100, 101, 102, 103, 104, 105, 106, 107, 108, 109, 110, 111, 113, 114, 115, 116, 117, 118, 119, 120, 123, 132, 136, 139, 146, 148, 155, 156, 157, 158, 161, 162, 163, 164, 165, 166, 169, 177, 180, 184, 185, 188, 189, 190, 191, 192, 193, 194, 197, 199, 200, 201, 202, 203, 204, 205, 206, 207, 208, 209, 210, 211, 212, 213, 217, 219, 220, 221, 224, 232, 233, 234, 235, 236, 237, 239, 240, 245, 246, 247, 251, 253, 255, 259, 260, 262, 263, 264 265, 277, 285, 291, 292, 293, 294, 295, 296, 297, 298, 302, 303, 306, 308, 311, 312, 315, 316, 318, 319, 324, 329, 330, 331, 332, 341, 344, 353, 355, 362, 366, 367, 372, 377, 378, 388, 389, 390, 391, 392, 394, 395, 396, 397, 398, 399, 400, 401, 402, 403, 404, 405, 406, 407, 409, 411, 426, 428, 429, 430, 431, 434, 436, 437, 438, 444, 446, 450, 459, 479, 480, 481, 482, 483, 484, 485, 486, 487, 488, 489, 491, 496, 497, 498, 503, 504, 505, 515, 516, 520, 521, 526, 527, 528, 529, 533, 534, 535, 538, 539, 540, 542, 544, 547, 548, 552, 553, 554, 555, 556, 557, 558, 559, 560, 561, 562, 563, 564, 565, 566, 567, 568, 569, 571, 572, 573, 574, 575, 582, 583, 584, 585,

591, 592, 595, 601, 604, 605, 606, 609, 610, 611, 612, 613, 614, 619, 620, 621, 622, 623, 624, 625, 632, 633, 634, 635, 636, 637, 638, 639, 640, 641, 642, 643, 644, 645, 645, 647, 646, 649, 650, 651, 652, 660, 663, 670, 679, 686, 687, 688, 692, 693, 696, 697, 708, 718, 729, 730, 731, 735, 738, 739, 740, 741, 748, 749, 750, 751, 752, 753, 754, 763, 769, 774, 775, 781, 787, 798, 799, 801, 802, 805, 806, 807, 810, 811, 812, 814, 820, 826, 834, 835, 836, 841, 842, 845, 856, 858, 859, 860, 863, 864, 867, 880, 881, 887, 888, 891, 892, 893, 894, 895, 896, 897, 898, 902, 904, 905, 906, 907, 908, 912, 920, 928, 933, 934, 935, 936, 937, 938, 939, 940, 941, 942, 943, 944, 945, 946, 947, 949, 951, 952, 954, 959, 962, 967, 972, 975, 976, 977, 978, 979, 980, 981, 985, 986, 987, 994, 999,1000, 1006, 1012, 1013, 1014, 1025, 1028, 1029, 1030, 1031, 1032, 1037, 1038, 1039, 1040, 1041, 1042, 1043, 1044, 1045, 1057, 1061, 1062, 1063, 1064, 1065, 1066, 1068, 1069, 1076, 1080, 1081, 1082, 1083, 1084, 1085, 1086, 1090, 1102, 1108, 1110, 1111, 1112, 1113, 1114, 1115, 1116, 1117, 1118, 1119, 1120, 1121, 1122, 1128, 1133, 1134, 1135, 1136, 1138, 1139, 1142, 1161, 1167, 1169, 1171, 1172, 1175, 1176, 1177, 1179, 1180, 1181, 1182, 1183, 1184, 1185, 1186, 1188, 1189, 1190, 1191, 1211, 1214, 1254, 1255, 1264, 1271, 1279, 1281, 1284, 1285, 1288, 1295, 1296, 1297, 1299, 1300, 1301, 1304, 1307, 1308, 1310, 1311, 1312, 1315, 1321, 1325, 1327, 1328, 1329, 1330, 1331, 1336, 1337, 1338, 1347, 1353, 1354, 1355, 1358, 1361, 1362, 1363, 1365, 1375, 1378, 1380, 1384, 1385, 1388, 1389, 1390, 1391, 1392, 1395, 1396, 1397, 1398, 1399, 1400, 1404, 1405, 1408, 1409, 1410, 1414, 1415, 1416, 1417, 1419, 1420, 1421, 1422, 1423, 1424, 1425, 1426, 1427, 1428, 1429, 1430, 1431, 1432, 1433, 1434, 1435, 1440, 1451, 1452, 1453, 1454, 1457, 1459, 1560, 1462, 1463, 1464, 1465, 1469, 1474, 1475, 1488, 1494, 1507, 1508, 1509, 1513,

1515, 1516, 1518, 1523, 1524, 1525, 1526, 1527, 1528, 1535, 1538, 1539, 1540, 1544, 1545, 1546, 1547, 1548, 1549, 1550, 1551, 1554, 1555, 1557, 1558, 1559, 1560, 1561, 1562, 1563, 1564, 1565, 1567, 1596, 1598, 1599, 1611, 1613, 1625, 1626, 1631, 1637, 1640, 1641, 1652, 1667, 1669, 1670, 1672, 1673, 1677, 1678, 1679, 1682, 1683, 1686, 1687, 1689, 1690, 1691, 1693, 1707, 1713, 1717, 1718, 1719, 1720, 1722, 1723, 1726, 1727, 1728, 1729, 1730, 1731, 1732, 1733, 1738, 1739, 1740.

Contaminant Inputs/Sources:

17, 23, 24, 28, 29, 65, 78, 80, 83, 91, 92, 94, 125, 150, 178, 213, 214, 222, 227, 228, 230, 238, 241, 244, 249, 250, 252, 254, 276, 280, 285, 286, 288, 289, 301, 304, 307, 309, 320, 327, 329, 342, 343, 346, 360, 361, 363, 364, 365, 372, 378, 382, 395, 399, 407, 409, 415, 417, 424, 435, 448, 451, 460, 465, 474, 476, 494, 495, 501, 507, 513, 516, 517, 546, 552, 553, 554, 555, 556, 560, 561, 562, 564, 565, 566, 568, 583, 584, 620, 663, 718, 812, 820, 946, 1029, 1030, 1031, 1032, 1097, 1108, 1111, 1115, 1117, 1120, 1121, 1325, 1422, 1423, 1528, 1535, 1539, 1540, 1549, 1565, 1718, 1738, 1739, 1740.

Outfalls/Wastewater:

8, 24, 101, 161, 187, 211, 362, 372, 399, 407, 503, 553, 560, 622, 663, 820, 887, 946, 1171, 1538, 1539, 1564, 1591, 1667, 1668.

Acid Wastes:

20, 21, 22, 89, 154, 217, 378, 428, 429, 430, 431, 434, 548, 554, 555, 564, 798, 799, 805, 806, 810, 812, 897, 898, 1102, 1129, 1220, 1221, 1239, 1385, 1420, 1422, 1513, 1535, 1540, 1544, 1547, 1550, 1554, 1557, 1558, 1567, 1613, 1672, 1679.

Sewage Sludge:

62, 93, 139, 197, 253, 255, 356, 378, 388, 391, 397, 399, 400, 401, 402, 403, 404, 405, 407, 554, 555, 560, 561, 564, 573, 574, 575, 583, 610, 611, 612, 613, 620, 622, 625, 686, 749, 807, 812, 820, 904, 946, 951, 962, 1220, 1221, 1239, 1252, 1271, 1307, 1308, 1310, 1311, 1312, 1315, 1327, 1328, 1329, 1336, 1384, 1385, 1400, 1420, 1422, 1425, 1426, 1475, 1513, 1515, 1535, 1540, 1544, 1547, 1550, 1554, 1555, 1557, 1558, 1565, 1567, 1591, 1614, 1667, 1668.

Ocean Dumped Dredge Material - General:

184, 344, 353, 378, 388, 407, 482, 488, 491, 493, 554, 555, 561, 564, 583, 686, 812, 826, 1012, 1013, 1014, 1117, 1167, 1254, 1321, 1331, 1420, 1422, 1431, 1495, 1535, 1554, 1557, 1667, 1668, 1693.

Ocean Dumped Dredge Material - Bathymetry:

482, 483, 485, 564, 985, 1690, 1707.

Ocean Dumped Dredge Material - Contaminant Release:

182, 184, 493, 812, 1667, 1668.

Chemical Processes - General:

14, 17, 23, 96, 156, 296, 297, 302, 303, 374, 391, 392, 397, 398, 399, 400, 401, 402, 403, 404, 564, 573, 574, 575, 620, 621, 622, 623, 624, 625, 693, 812, 920, 1108, 1167, 1214, 1315, 1325, 1327, 1328, 1451, 1452, 1508, 1642.

Heavy Metals:

16, 18, 156, 247, 264, 265, 277, 302, 303, 312, 374, 392, 394, 395, 527, 528, 533, 534, 535, 538, 539, 540, 542, 544, 545, 548, 564, 592, 613, 614, 812, 1116, 1190, 1217, 1296, 1329, 1528, 1631, 1726.

Oxygen/Oxygen Depletion:

85, 96, 118, 120, 332, 504, 560, 564, 595, 812, 920, 936, 937, 939, 945, 1006, 1120, 1217, 1279, 1330, 1355, 1358, 1389, 1390, 1391, 1392, 1397, 1398, 1399, 1427, 1428, 1431, 1451, 1452, 1453, 1454, 1457, 1596.

Organochlorine Compounds:

168, 192, 193, 194, 306, 444, 696, 812, 1057, 1112, 1113, 1214, 1217, 1375, 1380, 1528, 1667, 1727.

Nutrients:

210, 324, 397, 398, 399, 400, 401, 402, 403, 404, 405, 407, 503, 504, 505, 812, 935, 937, 942, 943, 944, 945, 946, 987, 1028, 1111, 1171, 1217, 1315, 1338, 1384.

Transparency:

107, 366, 367, 396, 749, 1244.

Climate:

86, 224, 895.

Atmospheric - Waste Sources:

379, 394, 395, 859.

Modeling:

149, 180, 203, 233, 251, 438, 708, 812, 856, 986, 994, 1295, 1296, 1430, 1652.

Dispersion Processes:

102, 103, 104, 105, 184, 189, 253, 429, 430, 482, 564, 625, 751, 752, 812, 858, 1136, 1137, 1138, 1140, 1251, 1252, 1253, 1384, 1560, 1561, 1562.

Physical Processes - Currents/Circulation:

102, 104, 146, 148, 149, 158, 189, 204, 234, 235, 292, 294, 295, 296, 298, 388, 429, 529, 564, 601, 605, 660, 738, 739, 787, 812, 858, 860, 867, 891, 893, 894, 928, 949, 959, 967, 1169, 1172, 1404.

Physical Oceanography/Processes/Hydrography:

101, 102, 103, 104, 105, 108, 109, 111, 146, 147, 148, 149, 189, 205, 208, 209, 211, 237, 292, 294, 295, 296, 297, 381, 407, 564, 604, 606, 632, 633, 634, 635, 636, 637, 638, 639, 640, 641, 642, 643, 644, 645, 646, 647, 648, 649, 650, 651, 652, 679, 738, 739, 801, 812, 1025, 1167, 1169, 1217, 1251, 1328, 1392, 1396, 1417, 1451, 1508, 1682, 1719, 1734.

Physical Processes - Tides:

102, 104, 105, 146, 147, 204, 206, 207, 296, 398, 564, 858, 860, 892, 949, 959, 1421.

Salinity:

103, 521, 1139, 1408, 1409, 1410, 1717.

Sand Transport:

316, 318, 482, 489, 496, 564, 912, 1432, 1434, 1435, 1690, 1691, 1707, 1730, 1731, 1732, 1733.

Suspended Sediments:

157, 166, 177, 188, 189, 191, 192, 194, 316, 353, 381, 388, 389, 390, 406, 482, 556, 564, 749, 812, 891, 1076.

Bottom Sediments:

19, 157, 188, 189, 190, 191, 193, 194, 219, 240, 311, 316, 318, 331, 353, 388, 444, 481, 482, 483, 484, 487, 488, 527, 552, 557, 558, 559, 561, 564, 614, 763, 812, 845, 913, 985, 1380, 1414, 1415, 1416, 1417, 1433, 1451, 1452, 1453, 1475, 1687, 1689, 1690, 1691, 1707, 1727, 1730, 1731, 1732, 1733.

Field Surveys/Monitoring Cruise Reports:

3, 110, 111, 256, 257, 263, 312, 381, 431, 436, 437, 632, 633, 634, 635, 636, 637, 638, 639, 640, 641, 642, 643, 644, 645, 646, 647, 648, 649, 650, 651, 652, 841, 842, 880, 1061, 1109, 1110, 1271, 1296, 1337, 1351, 1365, 1388, 1389, 1390, 1391, 1424, 1429, 1451, 1453, 1454, 1475, 1497, 1558, 1598, 1677, 1683, 1734.

Socioeconomic:

245, 262, 459, 564, 836.

Public Health/Water Quality:

374, 527, 564, 565, 609, 1061, 1062, 1063, 1064, 1065, 1066, 1081, 1114, 1474.

Geography:

162, 262, 296, 450, 740, 1432, 1433, 1686, 1687, 1689, 1690, 1691, 1720.

Chemical Processes - PAH's:

1000, 1073.

Chemical Processes - H_2S.:

132, 936, 1358.

Biological/Ecological Effects (General):

89, 106, 247, 257, 391, 403, 404, 426, 688, 741, 750, 769, 781, 812, 904, 905, 906, 907, 908, 962, 972, 1068, 1069, 1073, 1118, 1119, 1122, 1141, 1167, 1297, 1299, 1300, 1301, 1304, 1305, 1306, 1307, 1310, 1311, 1331, 1347, 1353, 1354, 1357, 1385, 1388, 1430, 1507, 1508, 1514, 1537, 1583, 1584, 1585, 1586, 1587, 1588, 1591, 1600, 1613, 1640, 1641, 1706, 1723, 1728, 1739, 1740.

Bioassays/Toxicity Studies:

13, 25, 98, 247, 341, 374, 812, 1046, 1047, 1048, 1049, 1297.

Biological Processes - Microorganisms:

25, 119, 199, 200, 201, 232, 233, 245, 246, 287, 319, 374, 526, 527, 528, 812, 896, 902, 1090, 1122, 1250, 1304, 1306, 1307, 1308, 1310, 1311, 1312, 1405, 1462, 1463, 1474, 1475, 1545, 1546, 1598, 1670.

Phytoplankton/Plants:

27, 100, 155, 187, 212, 312, 315, 403, 404, 538, 775, 806, 881, 934, 935, 936, 937, 938, 940, 941, 942, 943, 944, 945, 946, 987, 1046, 1047, 1048, 1049, 1069, 1112, 1135, 1142, 1171, 1288, 1297, 1392, 1454, 1464, 1488, 1613, 1640, 1641, 1722, 1723.

Zooplankton:

341, 538, 548, 1214, 1613, 1640, 1641, 1679.

Benthic Invertebrates:

27, 185, 197, 212, 259, 260, 285, 306, 355, 520, 936, 972, 1000, 1122, 1175, 1176, 1177, 1179, 1180, 1181, 1182, 1183, 1184, 1185, 1186, 1188, 1190, 1191, 1217, 1284, 1285, 1295, 1304, 1307, 1311, 1358, 1395, 1399, 1400, 1459, 1460, 1613, 1637, 1729.

Shellfish:

212, 479, 480, 498, 544, 545, 592, 619, 936, 976, 979, 1122, 1279, 1281, 1355, 1357, 1358, 1380, 1398, 1457, 1525, 1564, 1626, 1727.

Finfish:

15, 86, 113, 114, 115, 116, 117, 118, 164, 165, 212, 219, 220, 221, 239, 312, 446, 497, 539, 544, 571, 572, 592, 788, 789, 896, 905, 906, 907, 908, 933, 939, 978, 979, 980, 981, 999, 1037, 1038, 1039, 1040, 1041, 1042, 1043, 1044, 1045, 1122, 1211, 1214, 1217, 1255, 1347, 1354, 1357, 1375, 1398, 1440, 1514, 1669, 1682, 1683, 1796, 1727, 1739, 1740.

Engineering Studies:

2, 8, 24, 32, 35, 38, 40, 54, 68, 84, 127, 285, 458, 463, 512, 603, 662, 724, 745, 838, 857, 919, 1162, 1193, 1199, 1200, 1201.

Remote Sensing:

123, 202, 217, 291, 293, 547, 591, 687, 692, 697, 748, 749, 750, 751, 752, 753, 754, 812, 856, 869, 897, 898, 928, 1085, 1128, 1129, 1142, 1161, 1220, 1221, 1497, 1527, 1611, 1673, 1678, 1734.

106-Mile Dumpsite:

99, 170, 171, 172, 173, 174, 175, 176, 226, 256, 257, 275, 278, 279, 285, 305, 313, 336, 338, 339, 375, 432, 433, 467, 468, 515, 516, 525, 535, 536, 541, 587, 602, 626, 627, 658, 659, 682, 690, 709, 761, 791, 792, 794, 795, 796, 812, 840, 849, 901, 909, 926, 927, 996, 1015, 1016, 1017, 1026, 1035, 1046, 1047, 1048, 1049, 1053, 1054, 1123, 1124, 1125, 1136, 1137, 1141,

1143, 1144, 1178, 1187, 1217, 1264, 1286, 1305, 1345, 1352, 1377, 1430, 1450, 1507, 1508, 1512, 1516, 1517, 1520, 1521, 1522, 1531, 1557, 1568, 1597, 1615, 1617, 1644, 1645, 1666, 1676, 1680, 1684, 1716.

Mid-Atlantic Municipal and Industrial Dumpsites:

53, 141, 142, 143, 147, 280, 281, 282, 284, 285, 287, 325, 327, 328, 452, 453, 469, 470, 471, 473, 494, 518, 537, 576, 577, 578, 579, 736, 746, 776, 777, 803, 812, 823, 824, 825, 870, 871, 872, 873, 874, 875, 876, 877, 878, 899, 954, 955, 956, 957, 958, 998, 1033, 1034, 1134, 1161, 1163, 1164, 1165, 1212, 1223, 1224, 1264, 1625, 1653.

Gulf of Mexico:

94, 95, 134, 268, 581, 600, 701, 726, 1309, 1310, 1322, 1323, 1470, 1552, 1553.

Caribbean Sea (Puerto Rico):

67, 700, 701, 1246, 1359, 1676, 1704, 1705.

West Coast U.S. (Outfalls):

80, 223, 249, 250, 299, 301, 516, 722, 843, 973, 974, 988, 989, 990, 992, 993, 1008, 1009, 1010, 1203, 1316, 1346, 1347, 1373, 1724, 1725, 1726, 1727.

Charts/Maps/Location of Dumpsites:

712, 714, 716, 719, 1226.

Marine Pollution/Ocean Dumping Bibliographies:

55, 57, 58, 63, 66, 69, 75, 79, 167, 248, 320, 321, 383, 692, 715, 717, 728, 884, 950, 961, 1014, 1348, 1386, 1510, 1541, 1651.

AUTHOR INDEX

Author/Reference No.

Author/Reference No.

Author/Reference No.

Author/Reference No.

Author/Reference No.

Author/Reference No.

Author/Reference No.

Author/Reference No.

Author/Reference No.

Author/Reference No.

Author/Reference No.

Author/Reference No.

Author/Reference No.

Author/Reference No.

Author/Reference No.

Author/Reference No.

Author/Reference No.

Author/Reference No.

Author/Reference No.

Author/Reference No.

Author/Reference No.

Author/Reference No.

Author/Reference No.

Author/Reference No.

Author/Reference No.

COMMON ABBREVIATIONS OF AGENCIES/ INSTITUTIONS WITH ADDRESSES

Aberdeen -

Aberdeen Marine Laboratory
P.O. Box 101
Victoria Road
Aberdeen, AB9 8DB
Scotland

CCIW -

Canada Centre for Inland Waters
P.O. Box 5050
Burlington, Ontario
Canada, L7R 4A6

CERC -

Coastal Engineering Research Center
U.S. Army Corps of Engineers
Kingman Building
Fort Belvoir, VA
USA

DHI -

Deutsches Hydrographisches Institut
Postfach 220
Bernhard - Nocht - Strasse 78
D-2000 Hamburg
Federal Republic of Germany

DMRP - WES

Dredged Material Research Project
U.S. Army Engineer Waterways Experiment Station
Environmental Laboratory
P.O. Box 631
Vicksburg, MS 39180
USA

EPA -

U.S. Environmental Protection Agency
Marine Protection Branch
Rm 2834 WH585
401 M St., S.W.
Washington, DC 20460
USA

EPA - Region II
U.S. Environmental Protection Agency
Marine & Wetland Protection
2SA - MWP
26 Federal Plaza
New York, N.Y. 10007
USA

EPA - Region III
U.S. Environmental Protection Agency
6th and Walnut Streets
Philadelphia, PA 19106
USA

EPA - Narragansett
U.S. Environmental Protection Agency
Narragansett Laboratory
Narragansett, RI 02882
USA

GESAMP - Joint Group of Experts on the Scientific Aspects of Marine Pollution
IMCO
101-104 Piccadilly
London, WIV OAE
UK

HELSINKI - Convention
Helsinki Commission
Baltic Marine Envir. Prot. Commission
Etelaesplanadi - 22C43
SF - 00130 Helsinki 13

IAEA -
International Atomic Energy Agency
Karntner Ring 11
P.O. Box 590
A-1011 Vienna, Austria

ICES - International Council for the Exploration of the Sea
Palaegade 2-4, DK-1261
Copenhagen K Denmark

IMCO - Inter-Governmental Maritime Consultation Organization
IMCO
101-104 Piccadilly
London, WIV OAE
UK

MAFF - Directorate of Fisheries Research
Fisheries Laboratory
Lowestoft
Suffolk NR 33 OHT
UK

MAFF -
MAFF Fisheries Laboratory
Remembrance Ave.
Burnham - on Crouch
Essex CNO 8HA
England

NACOA -
U.S. National Advisory Committee
on Oceans and Atmosphere
3300 Whitehaven Street, N.W.
Washington, DC 20235
USA

NAS - NRC
U.S. National Academy of Science
National Research Council
2101 Constitution Ave., N.W.
Washington, DC 20418
USA

NASA - Langley
U.S. National Aeronautics and
Space Administration
Langley Research Center
Hampton, VA 23665
USA

New York Sea Grant -
The Sea Grant Institute of the
State University of New York and
Cornell University
c/o The State University of New York
99 Washington Avenue
Albany, NY 12246
USA

NOAA/MESA - New York Bight Project
NOAA/RD/MPF26
Old Biology Bldg. 4,
State University of New York
Stony Brook, NY 11794
USA

NOAA/NMFS Middle Atlantic
Coastal Fisheries Center
NEFC/NMFS/NOAA
Sandy Hook Laboratory
Highlands, NJ 07732
USA

NOAA/ODP -Ocean Dumping Program
NOAA/RD/MP2
Rockwall Bldg., Room 320
11400 Rockville Pike
Rockville, MD 20852
USA

NOAA/OMPA - Office of Marine
Pollution Assessment (Hdgr.)
NOAA/RD/MP
Rockwall Bldg., Room 320
11400 Rockville Pike
Rockville, MD 20852
USA

NOAA/OMPA - Office of Marine
Pollution Assessment
NOAA/RD/MP3
325 Broadway
Boulder, CO 80303
USA

NTIS -
National Technical
Information Service
U.S. Department of Commerce
5285 Port Royal Road
Springfield, VA 22161
USA

Ocean Mining -
NOAA Office of Ocean Minerals
and Energy
2001 Wisconsin, Avenue, NW
Washington, DC 20235
USA

OSLO Convention -
OSLO & Paris Commissions
New Court
48 Carey Street
London, WC2A 2JE
UK

Oxford Lab -
    Pathology Section
    Oxford Lab
    NEFC/NMFS/NOAA
    Oxford, MD 21654
    USA

PHS - HEW
    Public Health Service, HEW
    Food & Drug Administration
    Shellfish Sanitation Branch
    CB Center, Build. S-26
    Davisville, RI 02854
    USA

RWS - North Sea Directorate
    Rijkswaterstaat
    North Sea Directorate
    Koopmansstraat 1,
    2288 BC Rijswijk (Z.H.)
    The Netherlands

SCCWRP -
    Southern California Coastal
      Water Research Project
    646 W. Pacific Coast Highway
    Long Beach, CA 90806
    USA

SUNY/MSRC -
    Marine Sciences Research Center
    State University of New York
    Stony Brook, NY 11794
    USA

TNO -
    TNO
    Schoemakerstraat 97
    NL 2600 DE Delft
    The Netherlands

UK - Dept. of the Environment
    Department of the Enviroment
    2 Marsham Street
    London SW1P 3EB
    UK

URI - University of Rhode Island
    Graduate School of Oceanography
    University of Rhode Island
    Kingston, RI 02881
    USA

WHOI -

Woods Hole Oceanographic Institution
Woods Hole, MA 02543
USA